Advances of Metal and Metal Oxide Nanocomposites: Synthesis, Characterization and Biomedical Applications

Advances of Metal and Metal Oxide Nanocomposites: Synthesis, Characterization and Biomedical Applications

Editors

Nagaraj Basavegowda
Kwang-Hyun Baek

MDPI • Basel • Beijing • Wuhan • Barcelona • Belgrade • Manchester • Tokyo • Cluj • Tianjin

Editors
Nagaraj Basavegowda
Department of Biotechnology
Yeungnam University
Gyeongsan, Gyeongbuk
Korea, South

Kwang-Hyun Baek
Department of Biotechnology
Yeungnam University
Gyeongsan, Gyeongbuk
Korea, South

Editorial Office
MDPI
St. Alban-Anlage 66
4052 Basel, Switzerland

This is a reprint of articles from the Special Issue published online in the open access journal *Molecules* (ISSN 1420-3049) (available at: www.mdpi.com/journal/molecules/special_issues/metal_nanocomposites_biomedical).

For citation purposes, cite each article independently as indicated on the article page online and as indicated below:

LastName, A.A.; LastName, B.B.; LastName, C.C. Article Title. *Journal Name* **Year**, *Volume Number*, Page Range.

ISBN 978-3-0365-7609-1 (Hbk)
ISBN 978-3-0365-7608-4 (PDF)

© 2023 by the authors. Articles in this book are Open Access and distributed under the Creative Commons Attribution (CC BY) license, which allows users to download, copy and build upon published articles, as long as the author and publisher are properly credited, which ensures maximum dissemination and a wider impact of our publications.

The book as a whole is distributed by MDPI under the terms and conditions of the Creative Commons license CC BY-NC-ND.

Contents

About the Editors vii

Preface to "Advances of Metal and Metal Oxide Nanocomposites: Synthesis, Characterization and Biomedical Applications" ix

Nagaraj Basavegowda and Kwang-Hyun Baek
Multimetallic Nanoparticles as Alternative Antimicrobial Agents: Challenges and Perspectives
Reprinted from: *Molecules* 2021, *26*, 912, doi:10.3390/molecules26040912 1

Soumya Menon, Santhoshkumar Jayakodi, Kanti Kusum Yadav, Prathap Somu, Mona Isaq and Venkat Kumar Shanmugam et al.
Preparation of Paclitaxel-Encapsulated Bio-Functionalized Selenium Nanoparticles and Evaluation of Their Efficacy against Cervical Cancer
Reprinted from: *Molecules* 2022, *27*, 7290, doi:10.3390/molecules27217290 23

Mohamed Abdel-Salam, Basma Omran, Kathryn Whitehead and Kwang-Hyun Baek
Superior Properties and Biomedical Applications of Microorganism-Derived Fluorescent Quantum Dots
Reprinted from: *Molecules* 2020, *25*, 4486, doi:10.3390/molecules25194486 41

VinayKumar Dachuri, Phil Hyun Song, Young Woo Kim, Sae-Kwang Ku and Chang-Hyun Song
Protective Effects of Traditional Polyherbs on Cisplatin-Induced Acute Kidney Injury Cell Model by Inhibiting Oxidative Stress and MAPK Signaling Pathway
Reprinted from: *Molecules* 2020, *25*, 5641, doi:10.3390/molecules25235641 69

Nadiyah Alahmadi and Mahmoud A. Hussein
Impact of Ag/ZnO Reinforcements on the Anticancer and Biological Performances of CA@Ag/ZnO Nanocomposite Materials
Reprinted from: *Molecules* 2023, *28*, 1290, doi:10.3390/molecules28031290 83

Sumitha Samuggam, Suresh V. Chinni, Prasanna Mutusamy, Subash C. B. Gopinath, Periasamy Anbu and Vijayan Venugopal et al.
Green Synthesis and Characterization of Silver Nanoparticles Using *Spondias mombin* Extract and Their Antimicrobial Activity against Biofilm-Producing Bacteria
Reprinted from: *Molecules* 2021, *26*, 2681, doi:10.3390/molecules26092681 97

Shwetha U R, Rajith Kumar C R, Kiran M S, Virupaxappa S. Betageri, Latha M S and Ravindra Veerapur et al.
Biogenic Synthesis of NiO Nanoparticles Using *Areca catechu* Leaf Extract and Their Antidiabetic and Cytotoxic Effects
Reprinted from: *Molecules* 2021, *26*, 2448, doi:10.3390/molecules26092448 111

Reham Samir Hamida, Mohamed Abdelaal Ali, Mariam Abdulaziz Alkhateeb, Haifa Essa Alfassam, Maha Abdullah Momenah and Mashael Mohammed Bin-Meferij
Algal-Derived Synthesis of Silver Nanoparticles Using the Unicellular *ulvophyte* sp. MBIC10591: Optimisation, Characterisation, and Biological Activities
Reprinted from: *Molecules* 2022, *28*, 279, doi:10.3390/molecules28010279 123

Russul M. Adnan, Malak Mezher, Alaa M. Abdallah, Ramadan Awad and Mahmoud I. Khalil
Synthesis, Characterization, and Antibacterial Activity of Mg-Doped CuO Nanoparticles
Reprinted from: *Molecules* 2022, *28*, 103, doi:10.3390/molecules28010103 149

Syed Ghazanfar Ali, Mohammad Jalal, Hilal Ahmad, Khalid Umar, Akil Ahmad and Mohammed B. Alshammari et al.
Biosynthesis of Gold Nanoparticles and Its Effect against *Pseudomonas aeruginosa*
Reprinted from: *Molecules* **2022**, *27*, 8685, doi:10.3390/molecules27248685 . . . . **167**

Sirajul Haq, Nadia Shahzad, Muhammad Imran Shahzad, Khaled Elmnasri, Manel Ben Ali and Alaa Baazeem et al.
Investigations into the Antifungal, Photocatalytic, and Physicochemical Properties of Sol-Gel-Produced Tin Dioxide Nanoparticles
Reprinted from: *Molecules* **2022**, *27*, 6750, doi:10.3390/molecules27196750 . . . . **183**

About the Editors

Nagaraj Basavegowda

Nagaraj Basavegowda works as an International Research Professor in the Department of Biotechnology, at Yeungnam University, South Korea. He received his Master's degree in Bioscience and Ph.D. in Chemistry at the University of Mysore. His research interests focus on nanoparticles production, especially multimetallic, such as Bi-, tri-metallic nanocomposites and characterization, applications like nanomedicine, animal models (in vivo biocompatibility tests), in vitro biocompatibility tests, and natural compounds for different therapeutic strategies. He is also a member of several academic societies, an international reviewer, editorial board member for top-ranked international journals in MDPI, Frontiers, Springer, and Elsevier. He has been awarded various research prizes, excellence awards, special honors and medals from international organizations. He has published over 110 scientific papers, reviews, book chapters, patents, and communications in international journals.

Kwang-Hyun Baek

Kwang-Hyun Baek serves as a Full Professor in the Department of Biotechnology, at Yeungnam University, Republic of Korea. He obtained a Ph.D. in Crop and Soil Sciences at Washington State University, Pullman, WA, U.S.A., and he has worked as an Assistant, Associate, and full professor at Yeungnam University. Professor Kwang-Hyun Baek's research focuses on the development of nanocomposites for antibiotics, sustainable production of nanoparticles, and elucidation of natural compounds for natural antibiotics. He has served as an editorial board member for top-ranked international journals in MDPI, Plant Pathology Journal, and Applied Biological Chemistry, and a committee member for Korea Research Funding Allocation.

Preface to "Advances of Metal and Metal Oxide Nanocomposites: Synthesis, Characterization and Biomedical Applications"

Over the past few decades, many pathogenic bacteria have become resistant to existing antibiotics, which has become a threat to infectious disease control worldwide. As a result, there has been an extensive search for new, efficient, and alternative sources of antimicrobial agents to combat multidrug-resistant pathogenic microorganisms. Nanotechnology deals with materials on a nanometer scale, which is a thousand times smaller than the micrometer scale. This is comparable in size to biomolecules and biomaterials, rather than to the plant and animal cells or viruses with which biotechnology works. The difference between a biomaterial and a nanomaterial is that even though biomolecules are engineered for various purposes, they are not considered nanomaterials because they are not manmade. It involves expertise in a variety of fields, from biological, biochemical, molecular biological, molecular engineering, and genetic engineering to agricultural knowledge. Nanobiotechnology deals with technology that incorporates nano-molecules into biological systems, or which miniaturizes biotechnology solutions to nanometer size to achieve greater reach and efficacy. This may result in more effective and inexpensive assays and therapies. Nano-based approaches are developed to improve traditional biotechnological methods and overcome their limitations, such as the side effects caused by conventional therapies.

Metal and metal oxide nanocomposites have received considerable attention as an alternative to conventional antimicrobial agents because of their diverse shape, size, high surface-to-volume ratio, chemical/physical stability, activity, and a greater degree of selectivity. The design and synthesis of metal and metal oxide nanocomposites (e.g. mono, bi-, tri-, and multi-metallic nanocomposites), as well as polymer-based nanocomposites and hydrogels with diversified nanostructures (e.g. nanoarrays, nanotubes, core-shell, nanosheets, and nanorods), has sparked considerable interest in terms of biomedical therapeutics, bioimaging, biosensors, drug delivery, and antimicrobial attributes. As opposed to focusing on individual components, combinations of two or three antimicrobial agents as synergy represent a potential area in need and are worthy of further investigation due to several substances with the ability to improve solubility. It is carrying the science of about incomprehensibly small devices closer to reality. These advancements will, at some stage, be so vast that they affect all the areas of science and technology. This reprint comprises all aspects of the synthesis, characterization, and biomedical application of different metal and metal oxide nanomaterials, and nanocomposites.

Nagaraj Basavegowda and Kwang-Hyun Baek
Editors

Review

Multimetallic Nanoparticles as Alternative Antimicrobial Agents: Challenges and Perspectives

Nagaraj Basavegowda and Kwang-Hyun Baek *

Department of Biotechnology, Yeungnam University, Gyeongsan, Gyeongbuk 38451, Korea; nagarajb2005@yahoo.co.in
* Correspondence: khbaek@ynu.ac.kr; Tel.: +82-53-810-3029

Abstract: Recently, infectious diseases caused by bacterial pathogens have become a major cause of morbidity and mortality globally due to their resistance to multiple antibiotics. This has triggered initiatives to develop novel, alternative antimicrobial materials, which solve the issue of infection with multidrug-resistant bacteria. Nanotechnology using nanoscale materials, especially multimetallic nanoparticles (NPs), has attracted interest because of the favorable physicochemical properties of these materials, including antibacterial properties and excellent biocompatibility. Multimetallic NPs, particularly those formed by more than two metals, exhibit rich electronic, optical, and magnetic properties. Multimetallic NP properties, including size and shape, zeta potential, and large surface area, facilitate their efficient interaction with bacterial cell membranes, thereby inducing disruption, reactive oxygen species production, protein dysfunction, DNA damage, and killing potentiated by the host's immune system. In this review, we summarize research progress on the synergistic effect of multimetallic NPs as alternative antimicrobial agents for treating severe bacterial infections. We highlight recent promising innovations of multimetallic NPs that help overcome antimicrobial resistance. These include insights into their properties, mode of action, the development of synthetic methods, and combinatorial therapies using bi- and trimetallic NPs with other existing antimicrobial agents.

Keywords: alternative antimicrobial materials; infectious diseases; multidrug resistance; multimetallic nanoparticles; synergistic effect

1. Introduction

Pathogenic bacteria are abundant in the environment. They spread quickly and can easily cause adverse reactions, long-lasting health effects, or even death. Infection originating from the invasion of pathogens into the body is an acute threat to humans, causing diseases, such as pneumonia, gastritis, and sepsis, which can lead to tissue damage, organ failure, and death [1]. Antibiotics have immense benefits in the fight against a diverse range of pathogens. However, mutations are one strategy that bacteria employ to enhance their resistance to antibiotics, leading to the advent of a large number of multidrug-resistant (MDR) strains, which markedly lowers the therapeutic ability of antibiotics. There is increasing concern regarding the generation of antibiotic resistance, as bacteria vigorously persist to emerge with flexible countersteps against conventional antibiotics [2]. This is one of the most notable health-related matters of the 21st century [3]. Infectious diseases caused by MDR bacteria and the abundance of these bacteria have increased at an alarming rate, especially penicillin-resistant *Streptococcus pneumoniae*, methicillin-resistant *Staphylococcus aureus*, vancomycin-resistant *Enterococcus faecium*, ceftazidime-resistant *Klebsiella pneumonia* and *Escherichia coli*, fluoroquinolone-resistant *Pseudomonas aeruginosa*, and multi-antibiotic resistant *Acinetobacter baumannii* [4,5]. In addition, various foodborne pathogens associated with Gram-positive bacteria, such as *Bacillus cereus*, *Campylobacter jejuni*, *Clostridium perfringens*, *Clostridium botulinum*, *Cronobacter sakazakii*, *E. coli*, *Listeria monocytogenes*, *Salmonella enteritidis*, *Shigella dysenteriae*, *S. aureus*, *Vibrio furnissii*, and *Yersinia enterocolitica*, are causing a large number of diseases, with major effects on public health and safety [6].

Citation: Basavegowda, N.; Baek, K.-H. Multimetallic Nanoparticles as Alternative Antimicrobial Agents: Challenges and Perspectives. *Molecules* **2021**, *26*, 912. https://doi.org/10.3390/molecules26040912

Academic Editor: Alexandru Mihai Grumezescu
Received: 15 January 2021
Accepted: 4 February 2021
Published: 9 February 2021

Publisher's Note: MDPI stays neutral with regard to jurisdictional claims in published maps and institutional affiliations.

Copyright: © 2021 by the authors. Licensee MDPI, Basel, Switzerland. This article is an open access article distributed under the terms and conditions of the Creative Commons Attribution (CC BY) license (https://creativecommons.org/licenses/by/4.0/).

The production of new antibiotics requires enormous economic and labor demands and is a time-consuming process. Hence, the development of alternative, unconventional strategies to treat infectious diseases has become highly advisable [7]. The combination of one antimicrobial agent with other antimicrobial agents has many advantages, such as increased biological activity, reduced adverse effects, and increased antimicrobial toxicity of the combined elements. In synergism, one antimicrobial agent influences the activity of the other, and they finally act more efficiently and effectively together due to their different mechanisms of individual action. This may be considered a new approach to solve the problem of bacterial resistance and reduced antimicrobial susceptibility. Thus far, many alternative strategies have been developed to combat MDR bacteria. Among them, several in vitro studies have confirmed the significant antimicrobial activities of combinations of essential oils/plant extracts, conventional antibiotics/plant extracts, and phytochemicals/antibiotics [8].

Nanotechnology has important commercial applications in the fields of biology and medicine, particularly in the areas of drug delivery, diagnosis, tissue engineering, imaging, and bacterial infections [9]. Nanomaterials have special properties owing to their small dimensions; electrical, mechanical, optical, and magnetic properties; thermal stability; and high surface-to-volume ratio [10]. Nanomaterials are considered favorable alternatives to antibiotics for controlling bacterial infections due to a diverse range of factors, such as their size, morphology, surface charge, stability, and concentration in the growth medium. The surface coatings of nanoparticles (NPs) play an important role in influencing the antimicrobial properties of nanomaterials [11,12]. Depending on the number of metals, metal and metal oxide NPs are divided into mono-, bi-, tri-, and quadrometallic types. Among these, bi-, tri-, and multimetallic NPs have attracted the greatest interest due to their enhanced catalytic properties and favorable characteristics compared with monometallic NPs [13]. Multimetallic NPs are novel materials that incorporate two or more metals to make alloys with different functionalities and tunable properties, such as catalytic and optical properties. Multimetallic NPs can be modified by controlling their structure, chemical composition, and morphology to achieve maximum synergistic performance [14]. The addition of one or more metals into the NPs is expected to bring combinatorial effects, such as alteration of the electron structure, deduction of the lattice distance, and improvements in the total electronic charge shift [15]. Therefore, the study of the multidimensional space is warranted.

Monometallic NPs possess only one type of metal with specific chemical and physical properties, such as Ag, Au, Zn, Pd, Cu, Ti, Si, Al, Se, and Mg, which have been used for their antimicrobial activity for centuries. Among these, Ag NPs are the most effective as they are able to kill both Gram-positive and Gram-negative bacteria, and they are even effective against drug-resistant species [16]. Moreover, metal oxide NPs, such as Ag_2O, ZnO, CuO, TiO_2, NiO, Fe_3O_4, α-Fe_2O_3, CaO, MgO, Al_2O_3, CeO_2, Mn_3O_4, and ZrO_2 NPs, have highly potent antibacterial effects against a wide spectrum of microorganisms [17]. Similarly, metal sulfide and metal–organic framework (MOF) nanomaterials, such as AgS-, FeS-, CdS-, and ZnS-MOFs and Mn-, Cu-, and Zn-based MOFs, have demonstrated antimicrobial activities [18]. Bimetallic NPs are formed via the integration of two different types of metal atoms to form a single nanometric material with varying structures, morphologies, and properties [19]. Bimetallic NPs can be tuned by selecting the appropriate metal precursors to achieve the desired shape, size, structure, and morphology according to the configuration of atoms, and this finally leads to the formation of alloy, core–shell, and aggregated nanoparticle types [13]. Bimetallic NPs, such as Ag/Au, Ag/Cu, Au/Pt, Au/Pd, Ag/Fe, Fe/Pt, Cu/Zn, Cu/Ni, Au/CuS, and Fe_3S_4/Ag NPs, have distinctive surface activities. In addition, bimetallic oxide NPs, such as MgO/ZnO, CuO/ZnO, and Fe_3O_4/ZnO NPs, due to tensile strain and synergism between the constituent metals, often exhibit unique antibacterial performance.

Similarly, trimetallic NPs are made from three different metals for lowering metal consumption, atomic ordering, and fine-tuning the size and morphology of these NPs.

Trimetallic NPs exhibit higher catalytic selectivity/activity and efficiency in various applications, such as biomedical, antimicrobial, catalytic, active food packaging, and sensing applications. Moreover, owing to the presence of three metals, there are some possibilities for different structures and morphologies, such as core–shell, mixed structure, subcluster segregated, and multishell [20]. To change their catalytic performance, trimetallic NPs were further designated as alloys and intermetallic NPs by altering the atomic distribution and surface compositions of different metals [21]. Trimetallic NPs exhibit innovative physicochemical properties owing to their synergistic or multifunctional effects for diverse potential applications when compared with monometallic and bimetallic nanomaterials. However, to date, only monometallic, bimetallic, and very few trimetallic NPs have been reported for their antimicrobial effects.

Hence, the distinctive properties of nanomaterials provide a favorable environment for antibacterial therapies when compared with their bulk forms. Many inorganic and organic nanocomposites exhibit potential antibacterial properties with fast and sensitive bacterial detection. Nanocomposites are designed with targeted and sustained release mechanisms, environmental responsiveness, and combinatorial delivery systems for antibacterial therapies [22]. In particular, metal and metal oxide NPs, with nanotoxic mechanisms and collective modes of action, cause membrane damage and produce reactive oxygen species (ROS) that act against bacterial cells [23], which is why pathogens can barely develop resistance against them. The release of metal and metal oxide ions is the main mechanism responsible for the antimicrobial properties of nanocomposites. The current review aimed to highlight the most recent literature on mono-, bi-, tri-, and multimetallic NPs and their antimicrobial abilities. We discuss the importance, properties, fabrication methods, antimicrobial activities, and mechanisms of multimetallic NPs. Additionally, the synergistic effects, toxicity, and future prospects of multimetallic NPs are discussed. This study is expected to enhance our understanding of the development of multimetallic NPs and the replacement of conventional antibiotics with alternative antimicrobial agents to combat MDR pathogens.

2. Multimetallic NPs

Multimetallic NPs, comprising two or more different metals to form alloy or core–shell nanocomposites, have attracted considerable attention as novel materials due to their unique functionalities. The combined action of different metals and metal oxides in a chemical transformation enhances the catalytic performance of multimetallic NPs [24]. Multimetallic NPs with binary, ternary, and quaternary combinations usually have special characteristics, with enhanced chemical, optical, and catalytic properties when compared with mono- and bimetallic NPs, because of the synergistic effects between different metals [14]. Metallic NPs are classified as monometallic, bimetallic, trimetallic, quadrometallic, and so on based on the number of metallic ingredients.

2.1. Monometallic NPs

As the name suggests, monometallic NPs compose a single metal species, which compels for the catalytic characteristics of the nanoparticle. Based on the type of metal atom and properties, monometallic NPs are in different forms, like metallic, magnetic, transition metal, and oxide. They can be synthesized by chemical reduction and green synthetic methods, and their structure can be stabilized by various functional groups. Many studies have reported on a wide range of monometallic and metal oxide NPs, and these are used in various catalytic, medical, agricultural, active food packaging, nano-biosensor construction, industrial, and environmental applications. The order of atoms at the nanoscale, which differs from the bulk materials, is due to not only the large surface-area-to-volume ratio but also the specific electronic structure, plasmon excitation, and quantum confinement. In addition, the increased number of kinks, short-range ordering, chemical properties, and ability to store excess electrons also enhance activity. Recent studies on the antimicrobial activity of monometallic and metal oxide NPs are summarized in Table 1, highlighting the size, bacterial strains tested, mode of action, and fabrication techniques used.

Table 1. Antimicrobial activity of monometallic and metal oxide nanoparticles (NPs).

NPs	Size (nm)	Bacteria	Mode of Action	Synthesis	Ref.
Ag	10	V. natriegens	DNA damage and cell membrane rupture by reactive oxygen species (ROS)	Green catalysis	[25]
Au	20	S. pneumoniae	Cell lysis	Chemical reduction	[26]
Pd	13–18	S. aureus, S. pyrogens, B. subtilis	Cell membrane destruction and apoptosis	Biosynthesis (plant)	[27]
Ga	305	M. tuberculosis	Reduction of the growth of mycobacterium	Homogenizer	[28]
Cu	15–25	S. aureus, B. subtilis	Synergistic effects of organic functional groups	Biosynthesis (plant)	[29]
Pt	2–5	E. coli, A. hydrophila	Decrease in the bacterial cell viability and ROS generation	Chemical reduction	[30]
Si	90–100	S. aureus, P. aeruginosa	Mechanical damage of the bacterial membrane	Laser ablation	[31]
Se	117	Klebsiella sp.	Production of ROS, disruption of the phospholipid bilayer	Biosynthesis (plant)	[32]
Se	55.9	B. subtilis, E. coli	Ionic interaction between NPs and bacteria-caused cell damage	Biosynthesis (plant)	[33]
Se	85	E. coli, S. aureus	Cell membrane damage due to action of ROS	Laser ablation	[34]
Ni	60	P. aeruginosa	Cell membrane destruction	Biosynthesis (plant)	[35]
Mn	50–100	S. aureus, E. coli	Inactivation of proteins and decrease in the membrane permeability	Biosynthesis (plant)	[36]
Fe	474	E. coli.	Attraction between negatively charged cell membrane and NPs	Biosynthesis (plant)	[37]
Bi	40	B. anthracis, C. jejuni, E. coli, M. arginini	Inhibition of protein synthesis	Chemical condensation	[38]
Ag_2O	10–60	S. mutans, L. acidophilus	Penetration of the cells and hindrance of the growth of the pathogen	Biosynthesis (plant)	[39]
CuO	60	B. cereus	Disturbance of various biochemical processes when copper ions invade inside the cells	Biosynthesis (plant)	[40]
ZnO	30	A. baumannii	Increase in the production of ROS	Sol–gel and biosynthesis	[41]
TiO_2	9.2	E. coli	Decomposition of outer cell membrane by ROS, primarily hydroxyl radicals (OH·)	Biosynthesis (plant)	[42]
NiO	40–100	B. subtilis, E. coli	Induction of membrane damage by oxidative stress created at the NiO NP interface	Hydrothermal	[43]
Fe_3O_4	25–40	S. aureus, E. coli, S. dysentery	Cellular enzyme deactivation and disruption in plasma membrane permeability	Coprecipitation	[44]
α-Fe_2O_3	16	B. subtilis, S. aureus, E. coli, K. pneumonia	Desorption of membrane by the generated free radicals, including $O_2\cdot$ and OH·	Biosynthesis (plant)	[45]

Table 1. Cont.

NPs	Size (nm)	Bacteria	Mode of Action	Synthesis	Ref.
CaO	58	E. coli, S. aureus, K. pneumonia	Cell membrane destruction	Biosynthesis (plant)	[46]
MgO	27	Bacillus sp., E. coli	Destruction of cell membrane integrity resulting in leakage of intracellular materials	Ultrasonication	[47]
Al_2O_3	30–50	F. oxysporum, S. typhi, A. flavus, C. violaceum	Decomposition of bacterial outer membranes by ROS	Biosynthesis (fungi)	[48]
CeO_2	5–20	L. monocytogenes, E. coli, B. cereus	ROS generation by CeO_2 as a pro-oxidant	Precipitation	[49]
Mn_3O_4	130	K. pneumonia, P. aeruginosa	Membrane damage of bacterial cells by the easy penetration of Mn_3O_4 NPs	Hydrothermal	[50]
ZrO_2	2.5	S. mutans, S. mitis, R. dentocariosa, R. mucilaginosa	Enhancement of the interactions between NPs and bacterial constituents	Solvothermal	[51]
Ag_2S	65	Phormidium spp.	Inhibition of cell membrane by Ag_2S NPs, resulting in harmful effects on other biological activities	Chemical reduction	[52]
ZnS	65	Streptococcus sp., S. aureus, Lactobacillus sp., C. albicans	Dischargement of ions, which react with the thiol groups in the proteins present on the cell membrane	Biosynthesis (bacteria)	[53]
CdS	25	Streptococcus sp., S. aureus, Lactobacillus sp., C. albicans	Impregnation and surrounding the bacterial cells by CdS NPs	Biosynthesis (bacteria)	[53]
FeS	35	S. aureus, E. coli, E. faecalis	NP internalization through the fine cell membrane	Hydrothermal	[54]
Mn-MOF	-	E. coli, E. faecalis, S. aureus, P. aeruginosa	Peptide–nalidixic acid conjugate formation	Mechanochemical	[55]
Mg-MOF	-	E. coli, E. faecalis, S. aureus, P. aeruginosa	Peptide–nalidixic acid conjugate formation	Mechanochemical	[55]
Ag-MOF	-	S. aureus	High stability in water and the existence of Ag ion	Solvothermal	[56]
Cu-MOF	-	S. aureus, E. coli, K. pneumonia, P. aeruginosa, S. aureus	Attachment to the bacterial surfaces by active surface metal sites in Cu-MOF	Hydrothermal	[57]
Zn-MOF	-	P. aeruginosa	Penetration inside the bacteria, causing cell damage by interaction with lipotropic acid	Solvothermal	[58]
Co-MOF	-	E. coli	Strong interaction with membranes containing glycerophosphoryl moieties	Hydro-solvothermal	[59]

2.2. Bimetallic NPs

Bimetallic NPs have attracted huge attention due to their modified properties, and they can be prepared in different sizes, shapes, and structure with a combination of different metals.

Extensive studies in the past decade investigated the use of bimetallic NPs as a new advancement in the field of research and as technological domains to increase efficiency. Owing to the distinct catalytic and synergistic properties between two different metals, bimetallic NPs have potential applications in more fields than their corresponding monometallic one. Depending on their physical and chemical interactions, the spatial overlapping and distribution of two atoms can lead to the formation of a core–shell or simply an alloy due to the impact of individual metals [60]. The addition of a second metal is a major technique for tuning the geometric and electronic structures of NPs to increase their catalytic activity and selectivity. The size, shape, and morphology of alloy or core–shell NPs are comparatively different from those of the individual metals, thereby creating novel opportunities for a range of biomedical applications [61]. The antimicrobial activity of bimetallic NPs has been assessed against numerous types of pathogenic bacteria, especially *E. coli*, *P. aeruginosa*, and *S. mutans*, which are mainly responsible for human epidemics. Bimetallic NPs exhibit remarkable performance compared with commonly used antibiotics and other antimicrobial treatments, as pathogens cannot develop resistance to them because they suppress the generation of biofilms and accelerate other correlated processes [62].

2.3. Trimetallic NPs

Trimetallic NPs have favorable properties, such as physical, chemical, and tunable properties, when compared with mono- and bimetallic NPs, which result in multiple applications for these NPs. These favorable properties are due to multifunctional or synergistic effects produced by the three metals present in the same system [20]. The addition of a third metal or metal oxide into the composite supposedly generates a combinatorial effect and introduces several possibilities for different morphologies, structures, and chemical compositions to improve catalytic activity, selectivity, and specific performance [63]. Trimetallic NPs exhibit improved reactivity because of the electronic and synergistic effects of the different elements and the geometric arrangements of the metal surrounding the absorbing atoms. At this coherence, the properties of the materials are altered due to electron transfer effects, lattice mismatching, and surface rearrangement [64]. Consequently, trimetallic NPs, such as Fe/Co/Ni supported on multiwalled carbon nanotubes (MWCNTs), show efficient and enhanced bifunctional performance for the oxygen reduction and oxygen evolution reactions [65]. Similarly, Cu/Au/Pt, with high catalytic activity and excellent killing performance for biosensing and cancer theranostics [66], and Pd/Cu/Au, as excellent temperature sensors and fluorescence detectors of H_2O_2 and glucose [67], have attracted attention as promising catalysts. Hence, trimetallic NPs exhibit potent antibacterial activity and have been found to be more effective agents than bimetallic and monometallic NPs, even at lower metal concentrations.

2.4. Quadrometallic NPs

Higher-order quadrometallic NPs are made from four different metals for various applications with different fabrication methods. Recently, solid-state dewetting of Ag/Pt/Au/Pd quadrometallic NPs on sapphire has been prepared successfully with tunable localized surface plasmon resonance [68]. Similarly, Ag/Cu/Pt/Pd quadrometallic NPs prepared by seed-mediated growth, where small Pt and Pd NPs were attached on the surface of AgCu Janus bimetallic NPs as seeds in an aqueous solution [69], and heterostructured Pt/Pd/Rh/Au tetrahexahedral multimetallic NPs were synthesized through alloying/dealloying with Bi in a tube furnace [70]. Moreover, to date, there has been no investigation into the antimicrobial properties of quadrometallic and multimetallic NPs. Table 2 summarizes bi- and trimetallic NPs used for antibacterial activity with an array of sizes, strains tested, mechanisms, and synthetic methods.

Table 2. Antimicrobial activity of bimetallic and trimetallic NPs.

NPs	Size (nm)	Bacteria	Mode of Action	Synthesis	Ref.
Ag/Au	9.7	E. coli, S. aureus	Increased production of ROS	Green	[71]
Ag/Cu	26	E. coli, B. subtilis	Permeability of copper and silver ions into the bacterial cell membrane	Biosynthesis (plant)	[72]
Au/Pt	2–10	S. aureus, P. aeruginosa, C. albicans	Release of Ag^+ ions, unbalance of cell metabolism, and ROS generation	Chemical reduction	[73]
Ag/Fe	110	S. aureus, P. aeruginosa	Release of Ag^+ ions and ROS generation	Electrical explosion	[74]
Ag/Pt	36	E. faecalis, E. coli	Increased production of ROS	Biosynthesis (plant)	[75]
Cu/Zn	100	A. faecalis, S. aureus, C. freundii	Synergistic properties of Zn^{2+} and Cu^{2+} ions together	Biosynthesis (plant)	[76]
Cu-Ni	25	S. mutans, S. aureus, E. coli	Strong adsorption of ions to the bacterial cells	Chemical reduction	[77]
Ag/ZnO	43	S. aureus, P. aeruginosa	Ag^+ leaching from metallic silver	Photoreduction	[78]
Ag/SnO$_2$	9	B. subtilis, P. aeruginosa, E. coli	Synergistic properties of Ag and SnO	Biosynthesis (plant)	[79]
Cu/FeO$_2$	32.4	B. subtilis, X. campestris	DNA damage induced by NPs	Hydrothermal	[80]
Au/CuS	2–5	B. anthracis	Disordered and damaged membranes	Seeded	[81]
Fe$_3$S$_4$/Ag	226	S. aureus, E. coli	Release of Ag^+ ions and ROS generation	Solvothermal	[82]
MgO/ZnO	10	P. mirabilis	Alteration of cell membrane activity, ion release, and ROS production	Precipitation	[83]
CuO/ZnO,	50 and 82	E. coli, S. aureus	Electrostatic interaction causing to change membrane permeability on account of depolarization	Electrical explosion	[84]
CuO/Ag	20–100	L. innocua, S. enteritidis	Binding of the ions released by μCuO/nAg to the thiol groups of many enzymes in cell membrane	Hydrothermal	[85]
Fe$_3$O$_4$/ZnO,	200–800	S. aureus, E. coli	Membrane stress, disrupting and damaging cell membrane	Coprecipitation	[86]
CeO$_2$/FeO$_2$	40 and 25	P. aeruginosa	Combination of NPs with antibiotic ciprofloxacin, causing inhibitory effect on bacterial growth and biofilm formation	Hydrothermal	[87]
Cu/Zn/Fe	42	E. faecalis, E. coli	Interruption of cellular processes by released ions, which can cross cell membranes	Chemical reduction	[88]
Au/Pt/Ag	20–40	E. coli, S. typhi, E. faecalis	Generation of ROS	Microwave	[89]
Cu/Cr/Ni	100–200	E. coli, S. aureus	Antibacterial activity of trimetallic NPs in comparison with pure metals	Biosynthesis (plant)	[63]
CuO/NiO/ZnO	7	S. aureus, E. coli	Ruptured and cracked bacterial cells by the release of intracellular components	Coprecipitation	[90]
Ag/ZnO/TiO$_2$	60–170	E. coli	Reduction in the bandgap energy by increasing the e^- & h^+ charge separation time	Sol–gel	[91]

3. Properties of Multimetallic NPs

The properties of multimetallic NPs depend on the atomic structure, thickness, composition, shape, surface morphology, and stability of the core and shell and the disorder of alloy NPs. Furthermore, the percentage of atoms on the surface of a substance or material becomes more important for these NPs. Their enhanced surface area in contrast to their bulk metal counterparts mainly influences the elemental properties and active antibacterial properties of multimetallic NPs. Many studies have shown that tri- or multimetallic NPs are more active than mono- and bimetallic NPs made with the same metal. The catalytic properties of tri- or multimetallic NPs depend on the structure of the core and shell and the composition of the alloy NPs. Our previous studies have shown strong correlations between the number of different metals in NPs used as nanocatalysts and their activity, where trimetallic NPs show higher catalytic activities along with excellent selectivity than mono- and bimetallic NPs [92,93]. However, since different metals and metal oxides have different physicochemical properties, there will be slight parameter changes, like phase separation, agglomeration, and detachment of multimetallic NPs, which leads to poor catalytic activity and stability. On the other hand, some active research aims to initiate effective protocols, for instance, nanomaterials prepared by carbothermal shock method can enhance the overall structural and chemical stability of the multimetallic NPs [94].

Similarly, alloy nanocomposites have the special advantage of photocatalytic properties, as they possess both catalytic and plasmonic metals that absorb visible or ultraviolet light, which are released as energy to promote catalysis. The optical properties of multimetallic NPs depend on their size and shape, and they are strongly affected by their component metals. Surface plasmon resonance (SPR) increases when combining two or more metals that are vibrant in the visible region; however, it decreases or compresses when one metal in the combination is vibrant in the ultraviolet region. In addition, SPR sensitivity greatly depends on the size and shape of the nanocomposites. For instance, by decreasing the size of any metal NPs, the emission light position changes from the near-infrared to the ultraviolet region, but the sensitivity is increased when the shape of the NPs is a sphere, cube, or rod. Ultimately, nanocomposites can lose their SPR and become photoluminescent because of the very small size of the NPs [95]. In bi-, tri-, or multimetallic NPs, the SPR can provide information on the internal distribution of the elements. For alloyed nanocomposites, the shift in the SPR absorption is maximum and linear with the composition. Similarly, for core–shell nanocomposites, only one mode (frequency) of SPR is observed in the metal shell.

The magnetic properties of multimetallic NPs are influenced by several factors, including particle size and shape, chemical composition, morphology, crystal lattice, coordination of the particle with the surrounding matrix, and the adjacent particles [96]. The addition of a second or third metal into a nanoparticle structure enhances the optical/plasmonic and magnetic properties of the nanocomposites. Metals, such as iron, copper, nickel, and cobalt, possess good catalytic, electronic, and magnetic properties. Furthermore, the union of 3d metals with 4d or 5d metals, which display strong spin–orbit coupling, creates multimetallic NPs with large atomic magnetic moments and high magnetic anisotropy. Iron-based multimetallic NPs have been used in multidisciplinary fields, such as magnetic resonance imaging, drug delivery, cancer treatment, tumor detection, separation processes, and other biological activities with high biocompatibility, easy surface modifications, and low toxicity in living cells [97].

4. Synthetic Methods of Multimetallic NPs

Different approaches have been developed for the fabrication of multimetallic NPs, based on "top–down" and "bottom–up" techniques, with diverse procedures for functional nanomaterial alterations. The synthetic methods used to prepare multimetallic NPs are depicted in Figure 1.

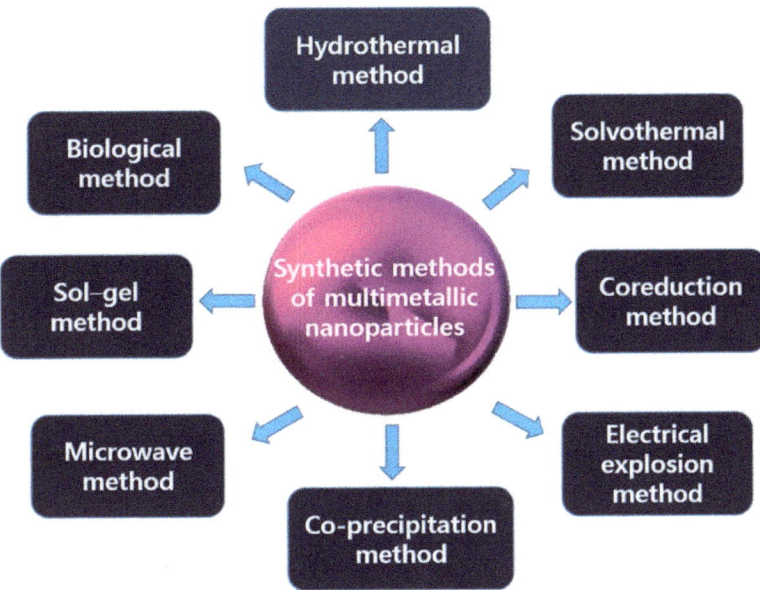

Figure 1. Synthetic methods for the preparation of multimetallic NPs.

4.1. Hydrothermal/Solvothermal Method

Hydrothermal and solvothermal techniques are the most important methods for the synthesis of various kinds of monodispersed and highly homogeneous nanomaterials. Using these methods, nanomaterials are fabricated using a typical wet-chemical approach, with high pressure and temperature, in aqueous solvents that dissolve and recover the materials. All the reactants are dissolved in an autoclave with a suitable solvent under low or high pressure and temperature conditions depending on the desired composition, crystal structure, size, and shape of the nanomaterials. The main advantage of this approach is that, with high vapor pressures and minimal loss of nanomaterials, the procedure is well controlled through liquid-phase or multiphase chemical reactions. Its disadvantages include the use of expensive equipment and high temperatures. Many bimetallic NPs, such as Ni–Fe, co-doped $Zn_{1-x}Co_xMn_2O$, and $NiFe_2O_4$ NPs, have been designed using the hydrothermal method [98–100]. Commonly, the hydrothermal process involves the use of solvent and surfactant; however, the surfactant-free synthesis of multimetallic NPs on an electrode surface was recently reported successfully [101].

4.2. Coreduction Method

Coreduction is a simple method to synthesize mono-, bi-, and trimetallic NPs and can be used to produce multimetallic NPs. The two metal precursors are first dissolved in a suitable solvent along with the stabilizing agent, and the transitional metals exist in their ionic states. The reducing agent is then added to convert them into zerovalent states; however, due to their lower reduction ability, light transitional metals undergo less reduction. These mild transition metals undergo oxidation very quickly when present in their zerovalent states and are, therefore, unstable. These metals play a prominent role in catalysis, and several methods have been developed to produce transition metal NPs. The coreduction method has moderate reaction conditions when compared with thermal decomposition. The reaction occurs under air atmosphere and low temperatures, and no toxic organic solvents are used [102]. Recently, mesoporous PtPdNi alloy NPs have been synthesized by this method using three metal precursors under constant sonication [103].

4.3. Electrical Explosion of Wires

The electrical explosion of wires (EEW) plays an important role in several applications, such as nanoparticle production, wire-array Z-pinch, exploding wire detonators, material property investigation under the most extreme conditions, high-temperature photography, and shockwave technology for promoting fossil energy [103]. EEW is one of the most promising technologies for the synthesis of metal NPs and is based on the increased activity generated when passing a pulse current with high density across a metal wire. During the EEW procedure, a capacitor bank releases a high current flowing through a desired metal wire over a short period. The amount of energy released is much larger than the sublimation energy of the metal wire, which ultimately leads to superheating, evaporation, dispersal of the wire material into the surrounding medium, and finally, formation of NPs by condensation of the vapor [104]. Recently, antimicrobial ZnxMe(100-x)/O nanocomposites have been designed by electrical explosion of a two-wire method in an oxygen-containing atmosphere [105].

4.4. Coprecipitation Method

Coprecipitation is one of the simplest and most widely used methods for the synthesis of NPs in various forms, such as hydroxides, oxides, sulfides, and carbonates, with controlled sizes and magnetic properties. In this method, aqueous salt solutions, such as nitrates or chlorides, precipitate the oxo-hydroxide under an inert atmosphere. Once the solution reaches a critical concentration, a short nucleation burst occurs, followed by a subsequent growth phase. Magnetite NPs are synthesized using different bases, such as KOH, NaOH, at $(C_2H_5)_4NOH$, at room temperature, depending on the desired crystal size, and the ratio of agglomeration promotes the formation of mesoporous structures. This chemical method is suitable for the fabrication of NPs because it does not require high temperature or pressure, and impurities are eliminated by washing and filtration. Recently, CuO-NiO-ZnO trimetallic oxide NPs have been synthesized using a coprecipitation method [90].

4.5. Microwave Method

The microwave heating technique is a simple, fast, reliable, versatile, and widely accepted method for the production of various nanocomposites with controlled sizes. It has a fast reaction rate and a short reaction time and high selectivity and yield when compared with conventional heating. The mechanism involved in this method affords uniform internal heating, facile nucleation, crystallization, dipolar polarization, and ionic conduction due to the force of dielectric heating. It has been found that the size and morphology of the nanocatalyst can be more easily controlled using the microwave method, when compared with conventionally prepared catalysts. Other advantages include fast ramping from temperatures of 150–250 °C in less than 1 min, compared with nearly 30 min with conventional heating methods. The microwave heating method has been employed for the preparation of trimetallic colloidal Au-Pt-Ag and Au-Pd-Pt nanocomposites using 160–800 W [106].

4.6. Sol–Gel Method

The sol–gel process is an effective method to produce metal oxide NPs at low temperatures compared with other physical and chemical methods. As the name suggests, a solution is transformed into a gel by a process based on the condensation reaction and hydrolysis of organometallic compounds in alcoholic solutions. The sol–gel method has excellent potential to control the reaction kinetics and the bulk and surface properties of the oxides. Moreover, it allows customizable microstructure, the addition of several functional groups, and ease of compositional modifications. The sol–gel method has been used to fabricate a variety of metal oxide NPs and bi- and trimetallic oxide nanocomposites. For instance, titanium and vanadium trimetallic nanocomposites have been synthesized with three other metals by the sol–gel technique using ammonia and hydrazine as reducing agents [107].

4.7. Biological Methods

Recently, green synthetic nanoparticle production methods have been developed based on biological sources, such as plant leaves, root tissues, algae, bacteria, yeast, fungi, and industrial and agricultural wastes. These methods have also been used to fabricate bi- and trimetallic alloy nanocomposites. In plants, secondary metabolites, such as flavonoids, alkaloids, terpenoids, heterocyclic compounds, polysaccharides, organic acids, proteins, and vitamins have been used as sources. In the case of microorganisms, sources, such as reductase enzymes, proteins, metal-resistant genes, and organic materials, have been used. Agricultural wastes, such as fruit peels, wild weeds, unwanted plants, burned herbs, and shrubs, are used as biological sources. In addition, industrial wastes, such as rice husks, eggshells, timber dust, and sugarcane bagasse, are used for the reduction of metal salts into NPs. Recently, Au/ZnO/Ag trimetallic nanocomposites have been synthesized using a green method that involves the use of an extract from *Melilotus officinalis* [108].

5. Antimicrobial Activity of Multimetallic NPs

The antimicrobial efficacies of metal and metal oxide NPs have been tested against various types of antimicrobial-susceptible and antimicrobial-resistant pathogenic strains. Pathogenic bacteria, including *E. coli*, *P. aeruginosa*, *S. mutans*, and *S. aureus*, are the main agents responsible for many infectious diseases. These pathogenic bacteria become a major challenge in the health-care system and pose a serious threat to human health. Metal and metal oxide NPs, especially multimetallics, such as bi-, tri-, and quadrometallic nanocomposites, show improved performance when compared with conventional antibiotics, as pathogens cannot develop resistance to these NPs because the nanocomposites affect biofilm formation and other associated processes. In comparison with monometallic nanomaterials, multicomponent metal nanocomposites have multiple functionalities and exhibit enhanced and cumulative properties due to the various synergistic effects of the individual components. Moreover, the catalytic and optical properties of multimetallic NPs can be successfully controlled by modifying their structure, morphology, and chemical composition. Hence, they have attracted considerable attention due to their wide range of applications in medical, sensing, and catalytic fields.

Numerous studies have demonstrated the antimicrobial activities of mono- and bimetallic NPs, as shown in Tables 1 and 2; however, very few have reported on trimetallic NPs, as shown in Table 2. Quadro- and multimetallic nanocomposites are at the initial stages of development and are still limited in types suitable for other applications. The effectiveness of the resultant antimicrobial activities is mainly influenced by two important parameters: the type of material (precursor) used to prepare the NPs and the particle size. Generally, nanomaterials have unique properties when compared with bulk materials, and they often change dramatically with nanoscale ingredients because the surface-to-volume ratio of the NPs increases significantly with a decrease in particle size. Additionally, at nanometer dimensions, the atomic fraction of the molecule increases significantly at the surface, consequently enhancing some properties of the particles, such as dissolution rate, catalytic activity, heat treatment, and mass transfer [109]. The nanomaterials used to fight against infectious pathogens comprise mainly transition metals and metal oxides because of their special characteristics, including decreased particle size, increased surface-to-volume ratio, hydrophobic interactions, nanoparticle stability, and electrostatic attraction. These properties allow bactericidal activity and microbiostatic activity against both Gram-positive and Gram-negative bacteria.

The bactericidal mechanism of action of nanocomposites is normally described as metal ion release and oxidative stress induction, but nonoxidative mechanisms may also occur simultaneously. Moreover, the production of ROS interrupts the antioxidant defense system and causes mechanical damage to the cell membrane. In addition, the majority of recent studies have described the mechanism of action of nanocomposites as follows: adhesion to microbial cells and cell wall destruction by physical contact, leading to the generation of ROS, particle penetration into the cell, damage to proteins and DNA, oxidative stress, and facilitation of internalization. The positively charged surfaces of metal and metal oxide nanocomposites promote their binding to negatively charged bacterial surfaces, which may result in the strengthening of their bactericidal effect. In addition to size, different formulations with different particle shapes and surface charges may also influence the intrinsic properties of the NPs and potentially influence their antibacterial activity. The different morphologies, crystal growth habits, and increased lattice constants of multimetallic NPs may enhance their antibacterial activity. Multimetal and metal oxide nanocomposites have been widely studied as alternative antimicrobial agents because of the favorable synergistic effects of their individual components.

6. Antibacterial Mechanisms of Multimetallic NPs

The mode of action of bacterial destruction depends on the type of NP. Metals, metal oxides, and their nanocomposites bind to the cell wall of bacteria and form membrane-penetrating pores due to the deposition of nanomaterials on the bacterial cell surface, which causes the formation of free radicals that are able to destroy the cell membrane. In addition, ions released from the nanomaterials inhibit the production of enzymes and increase the production of ROS, which in turn affects DNA transcription [110]. The mode of action of multimetallic NPs is summarized in Figure 2.

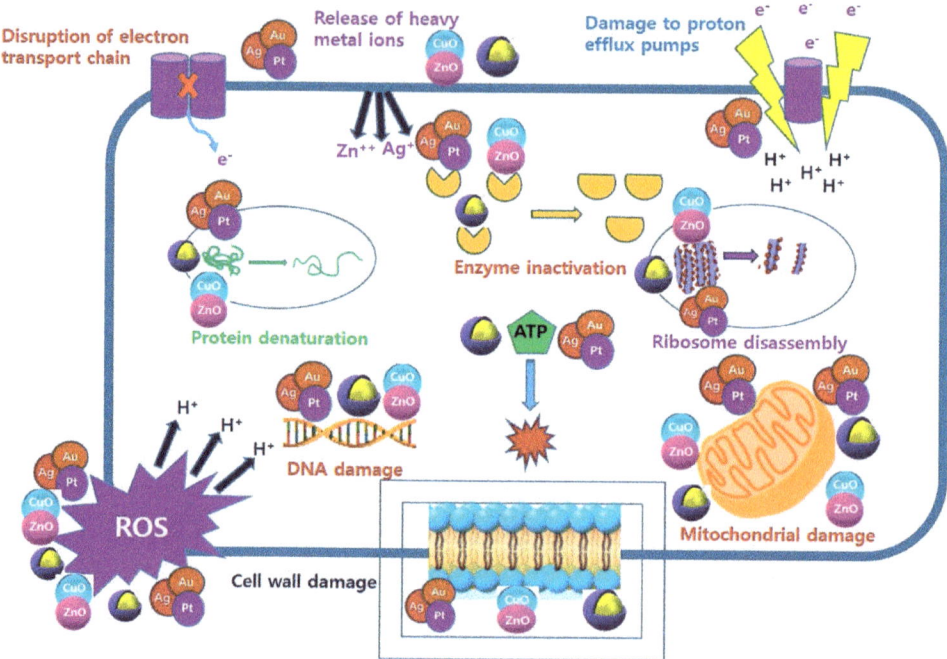

Figure 2. Antibacterial mechanism of multimetallic NPs.

6.1. Disruption of Cell Membrane

Metal- and metal oxide-based nanocomposites can damage the bacterial cell membrane by electrostatically binding to the cell wall and releasing metallic ions. The difference in charge between bacterial membranes (negative charge) and nanocomposites (positive charge) induces an electrostatic attraction and changes the permeability of the cell membrane. Hence, the disturbance of the bacterial membrane integrity is an effective mechanism of action. Based on the structure of the cell wall, bacteria are divided into two groups: Gram-positive bacteria, which have many layers of thick peptidoglycan in their cell wall, and Gram-negative bacteria, which have a cell wall consisting of a thin peptidoglycan layer with an additional outer membrane of lipopolysaccharide. The negatively charged properties of both Gram-positive and Gram-negative bacterial cell walls affect the interactions between the cell wall and nanoparticle ions.

Metal and metal oxide NPs exhibit higher antibacterial activity against Gram-negative bacteria than Gram-positive bacteria [111] because the negative charge of lipopolysaccharides allows the adherence of NPs to Gram-negative bacterial cell walls. Thus, a high NP-binding capability on these negative anionic areas may increase toxicity due to high NP concentrations. A layer of lipopolysaccharides covers the cell wall of the Gram-negative bacterium, *E. coli*, and peptidoglycans, whereas the cell wall of the Gram-positive bacterium, *S. aureus*, consists of a peptidoglycan layer and is much thicker than the Gram-negative bacterial cell wall. The difference in composition, structure, and thickness of the cell wall allows NPs to penetrate, resulting in less inhibition of *S. aureus* than *E. coli*, which shows significant inhibition even at low antibiotic concentrations. Using this approach, there is a clear correlation between the concentration of the NPs and the different classes of bacteria being treated due to differences in the chemical and structural organization of the cell wall.

6.2. Formation of ROS

ROS are highly reactive metabolic products with strong positive redox potential. They are produced in numerous cells by two cellular organelles, the endoplasmic reticulum and mitochondria. The production of ROS is an alternative mechanism by which nanomaterials kill bacteria. Different types of ROS, namely, the superoxide radical (O_2^-), singlet oxygen ($^1O_2$), hydrogen peroxide (H_2O_2), and the hydroxyl radical (OH), exhibit different levels of activity. Superoxide, an anion radical (O_2^-), is a potent oxidizing agent that is highly reactive with water and is produced mainly in the thylakoid-localized photosystem I (PSI) by one electron, as well as in other cellular compartments. Singlet oxygen is a strong reagent that facilitates undesirable oxidation inside the cell and causes severe damage to various molecules of biological importance. The production of singlet oxygen leads to peroxidation of cellular constituents, such as proteins and lipids. ROS react with H ions to produce hydrogen peroxide, which more easily penetrates cell membranes and destroys various cellular organelles, resulting in bacterial death. Hydroxyl radicals are fatal to pathogenic bacteria; however, some hydroxyl radicals that are negatively charged cannot easily penetrate the negatively charged cell membrane. The diffusion of metal ions from metal oxide NPs generates a large number of hydroxyl radicals due to the decomposition of bacterial cells. ROS are formed when oxygen enters reduction states and is converted into free radicals, such as superoxides and peroxides, instead of water. However, some pathogens can fight back in response to ROS by producing superoxide dismutase enzymes. ROS induce acute oxidative stress, inhibit enzymes, and cause damage to lipids, proteins, and DNA/RNA. Gold, magnesium oxide, and zinc oxide promote ROS formation through increased catalytic activity by producing H_2O_2 from glucose oxidase [112].

6.3. Dysfunction of Cytosolic Proteins

Protein dysfunction is another mode of action by which multimetallic NPs induce their antimicrobial response by binding to cytosolic proteins, such as DNA and enzymes. The metal-ion-catalyzed oxidation of the amino acid side-chain leads to the formation of protein-bound carbonyls. The carboxylation of protein molecules may serve as a better marker for oxidative damage to proteins. However, in the case of enzymes, carboxylation results in the loss of catalytic activity and stimulates protein degradation [113]. Silver and copper NPs inhibit DNA replication, cell division, and DNA degradation [114,115], whereas gold NPs interact with DNA within the cell by upregulating genes [111], leading to decreased membrane integrity and the accumulation of ROS within the cytosol. Many studies have reported that when bacterial cells are exposed to nanomaterials, the physical attachment of NPs leads to nuclear DNA fragmentation and damage due to the high affinity of metal ions for phosphate, which is highly abundant in DNA molecules [116].

7. Synergistic Effects of Multimetallic NPs

A synergistic effect is a process in which biological structures or chemical substances combine to create an effect that is greater than either one of them could have caused alone. Multimetallic NPs have attracted more attention than monometallic NPs from researchers in different fields, such as physics, biology, and medicine, because of their widespread applications in catalysis, sensing, and medical fields. Multimetallic NPs are obtained by incorporating three or more metal elements with various nanostructures and different properties into single nanomaterials called alloys. These multifunctional nanomaterials exhibit advanced properties with novel functions due to the synergistic effects of different elements. The continuous modulation of the electronic structure of multimetallic NPs enhances their catalytic performance because of the various binding forces acting on electrons between different metal atoms [117]. Hence, the synergistic effects of multimetallic nanocatalysts play a prominent role in the field of heterogeneous catalysis, especially in oxidation and reduction reactions [118]. Multimetallic NPs also show synergistic effects by the combination of two or more metals with an atomic ratio of various metals, ultimately achieving high catalytic efficiency, including high selectivity and catalytic activity [119].

The synergistic effect of nanocatalysts is influenced by geometric parameters, local strain, electronic states, and successful coordination at the surface [66]. In addition, multimetallic NPs modify the electronic structure of metals, which permits the tuning of the binding energy between nanocatalysts and reaction intermediates, thus initiating a synergistic effect that can enhance the durability and catalytic activity of the NPs [66]. In the case of tandem reactions, which require two or more catalysts, multimetallic NPs improve the catalytic properties by forming an interconnection area between two or more metals [120]. For instance, our previous studies showed that FeAgPt alloy trimetallic NPs performed better than FeAg and FePt bimetallic and Fe, Ag, and Pt monometallic NPs in reduction and decolorization reactions, with excellent selectivity and activity [92]. Similarly, AuFeAg hybrid NPs, a convenient green catalyst, showed greater catalytic activity than FeAu and FeAg bimetallic NPs and Au, Fe, and Ag monometallic NPs in the preparation of α,β- and β,β-dichloroenone from diazodicarbonyl and oxalyl chloride [93]. Recently, FeCoNi mixed oxide NPs supported on oxidized MWCNTs were used as catalysts for oxygen reduction and oxygen evolution reactions, and they demonstrated enhanced selectivity with respect to the reduction of O_2 to $OH-$ when compared with FeNi, FeCo, and CoNi bimetallic catalysts, thus suggesting synergistic effects among the metal oxide elements [65]. Similarly, the antimicrobial activity of trimetallic NPs shows synergistic effects when compared with bi- and trimetallic NPs, as shown in Table 2 [63,88–91]. Therefore, greater attention has been paid to synergistic interactions or combinatorial therapy as an alternative strategy to combat antibiotic-resistant infectious pathogens.

8. Toxicity of Multimetallic NPs

In recent decades, various efforts have been made to evaluate the toxicological impact and possible hazards and risks of different NPs on human health and the environment. Multimetal and metal oxide NPs have great potential to manage several diseases and infection by resistant bacterial strains because of their special physicochemical properties and small size, thus enabling the particles to be ingested or inhaled or to penetrate through the skin more readily. However, there is increasing concern about the safety of prolonged exposure to these NPs before their application in various fields and their large-scale production. The range of toxic effects depends on the nature of the metal and metal oxide NPs and their surface functional groups. The in vitro evaluation of different types of NP interactions with living cells of humans, animals, plants, and aquatic organisms has already been conducted. However, major issues, such as the behavior of NPs inside the cells, their tissue penetration, and the metabolic and immunological responses they induce are often ignored.

Predictions of toxicity are certainly useful for understanding the mechanisms of nanotoxicology, interactions of nanomaterials with biological systems, and careful assessment of nanomaterial properties. When evaluating toxicity, it is important to note that different organisms have different sensitivities to NPs, and differences in NP concentration and solubility, the presence of additives, and the synthetic methods used are clearly accounted for in some reported results. Metal NPs, including Au and Ag NPs, and metal oxide NPs, including MgO, CuO, Fe_3O_4, ZnO, and TiO_2 NPs, are the most promising antimicrobial agents, with low toxicity against human cells [121]; however, more research is needed to clarify the mechanisms of NP migration into the human body. The migration of NPs into the human body depends on the size, structure, solubility, and chemical composition of the NPs, and because of their small size, NPs can pass through different organs and settle in the central nervous system to stimulate an immune system response. The surface chemistry of NPs is another factor that determines their cytotoxicity and their effects on biological systems due to the resulting charge, roughness, and hydrophobic or hydrophilic nature [122]. In several studies, metal and metal oxide NPs have been shown to be toxic at higher concentrations in human fibroblasts, kidneys, liver cells, and macrophages [123].

9. Concluding Remarks and Future Perspectives

The use of alternative strategies to combat MDR pathogens is warranted because of their constantly increasing resistance to current antibiotics. Pathogens usually develop resistance to antibiotics as they are overused and misused in husbandry practices, with no effective management. Nanotechnology has been applied in almost every field of science, including materials science, physics, chemistry, computer science, biology, and engineering, over the last decade. In recent years, nanotechnology has also been applied to human health as a promising candidate for the treatment of infectious bacterial diseases. Nanomaterials possess superior properties due to their optical, electrical, mechanical, and magnetic properties, thermal stability, small dimensions, and high surface-to-volume ratio. Thus, nanomaterials are considered an alternative therapeutic option to control bacterial infections because of their size, shape, solubility, surface charge, stability, and surface coatings.

As discussed previously, various studies have been performed on a wide range of monometallic and metal oxide NPs, and these have demonstrated potential antimicrobial activities against MDR pathogens. Despite these potential advantages, very few studies have reported the synergistic effects of bi-, tri-, and quadrometallic and metal oxide nanocomposites. Multimetallic NPs, such as bi-, tri-, and quadrometallic NPs, have attracted great interest because of their favorable characteristics as advanced materials and their enhanced catalytic properties when compared with monometallic NPs. The combination of two or more nanomaterials to form a single system can enhance their special optical, catalytic, electronic, and magnetic properties. This is expected to bring combinatorial effects, such as alterations of the electronic structure and a reduction in the lattice distance. In addition, tuning or altering the size, shape, elemental composition, internal structure, and surface modification of nanomaterials can result in the formation of nano-alloys, core–shell structures, and heterodimers with enhanced catalytic performance. Furthermore, the use of multimetallic NPs as hybrid materials also provides specific plasmon excitation, an increased number of kinks, changes in the electronic and magnetic properties, short-range ordering, altered chemical properties, improvements in the overall electronic charge shift, and the ability to store more electrons.

The exact contribution of multimetallic NPs to the treatment of MDR pathogens has not yet been elucidated, and further research is needed to identify alternative antimicrobial agents by modulating the shape, size, and surface chemistry of multimetallic NPs. Thus, multimetallic NPs are a potential source of alternative antimicrobial agents and may play a significant and synergistic role in the near future. The present paper discussed findings on the in vitro antimicrobial activities of nanomaterials; however, further studies are essential to assess the synergistic effects of bi-, tri-, and multimetallic NPs in vivo. Furthermore, careful selection of the nanomaterial type, the appropriate dosage, and the choice of application is crucial for beneficial outcomes, because the majority of nanomaterials are metallic and may lead to potential uptake and accumulation. In addition, the management of environmental issues, health and safety aspects, risk assessment, potential toxicity, and hazards must be considered before introducing them as safe and effective antimicrobial agents. In conclusion, a broad understanding of these new and creative therapeutic strategies is essential as they may serve as alternatives to conventional antibiotics. The present study suggests that multimetallic NPs and their composites can be utilized to develop more effective antimicrobial agents in the near future to prevent the emergence and spread of bacterial resistance to conventional antibiotics, and they may play an important role in many medical applications.

Author Contributions: Data collection, N.B. and K.-H.B.; writing—original draft preparation, N.B.; writing—review and editing, K.-H.B. All authors have read and agreed to the published version of the manuscript.

Funding: This research was funded by PJ0157260, Rural Development Administration, Republic of Korea.

Institutional Review Board Statement: Not applicable.

Informed Consent Statement: The study did not involve humans or animals.

Data Availability Statement: Not applicable.

Acknowledgments: This work carried out with the support of by the Cooperative Research Program for Agriculture Science and Technology Development (Project No. PJ015726), RDA, Republic of Korea.

Conflicts of Interest: The authors declare no conflict of interest.

References

1. Guo, Z.; Chen, Y.; Wang, Y.; Jiang, H.; Wang, X. Advances and challenges in metallic nanomaterial synthesis and antibacterial applications. *J. Mater. Chem. B* **2020**, *8*, 4764–4777. [CrossRef] [PubMed]
2. Vimbela, G.V.; Ngo, S.M.; Fraze, C.; Yang, L.; Stout, D.A. Antibacterial properties and toxicity from metallic nanomaterials. *Int. J. Nanomed.* **2017**, *12*, 3941. [CrossRef]
3. Medina, E.; Pieper, D.H. *Tackling Threats and Future Problems of Multidrug-Resistant Bacteria*; Springer: Berlin/Heidelberg, Germany, 2016; Volume 398, ISBN 9783319492827.
4. Kon, K.V.; Rai, M.K. *Combining Essential Oils with Antibiotics and other Antimicrobial Agents to Overcome Multidrug-Resistant Bacteria*; Elsevier: Amsterdam, The Netherlands, 2013; ISBN 9780123985392.
5. Almasaudi, S.B. Acinetobacter spp. as nosocomial pathogens: Epidemiology and resistance features. *Saudi J. Biol. Sci.* **2018**, *25*, 586–596. [CrossRef] [PubMed]
6. Abad, M.J.; Bedoya, L.M.; Bermejo, P. Essential oils from the Asteraceae family active against multidrug-resistant bacteria. In *Fighting Multidrug Resistance with Herbal Extracts, Essential Oils and Their Components*; Elsevier: Amsterdam, The Netherlands, 2013; pp. 205–221.
7. Wei, T.; Yu, Q.; Chen, H. Responsive and synergistic antibacterial coatings: Fighting against bacteria in a smart and effective way. *Adv. Healthc. Mater.* **2019**, *8*, 1–24. [CrossRef]
8. Rakholiya, K.D.; Kaneria, M.J.; Chanda, S.V. *Medicinal Plants as Alternative Sources of Therapeutics against Multidrug-Resistant Pathogenic Microorganisms Based on Their Antimicrobial Potential and Synergistic Properties*; Elsevier: Amsterdam, The Netherlands, 2013; ISBN 9780123985392.
9. Huh, A.J.; Kwon, Y.J. "Nanoantibiotics": A new paradigm for treating infectious diseases using nanomaterials in the antibiotics resistant era. *J. Control. Release* **2011**, *156*, 128–145. [CrossRef]
10. Moghimi, S.M.; Hunter, A.C.; Murray, J.C. Nanomedicine: Current status and future prospects. *FASEB J.* **2005**, *19*, 311–330. [CrossRef]
11. Yaqoob, A.A.; Parveen, T.; Umar, K.; Mohamad Ibrahim, M.N. Role of nanomaterials in the treatment of wastewater: A review. *Water* **2020**, *12*, 495. [CrossRef]
12. Khan, S.T.; Musarrat, J.; Al-Khedhairy, A.A. Countering drug resistance, infectious diseases, and sepsis using metal and metal oxides nanoparticles: Current status. *Colloids Surf. B Biointerfaces* **2016**, *146*, 70–83. [CrossRef] [PubMed]

13. Sharma, G.; Kumar, A.; Sharma, S.; Naushad, M.; Prakash Dwivedi, R.; ALOthman, Z.A.; Mola, G.T. Novel development of nanoparticles to bimetallic nanoparticles and their composites: A review. *J. King Saud Univ. Sci.* **2019**, *31*, 257–269. [CrossRef]
14. Zhang, J.; Ma, J.; Fan, X.; Peng, W.; Zhang, G.; Zhang, F.; Li, Y. Graphene supported Au-Pd-Fe3O4 alloy trimetallic nanoparticles with peroxidase-like activities as mimic enzyme. *Catal. Commun.* **2017**, *89*, 148–151. [CrossRef]
15. Deepak, F.L.; Mayoral, A.; Arenal, R. Advanced transmission electron microscopy: Applications to nanomaterials. *Adv. Transm. Electron. Microsc. Appl. Nanomater.* **2015**, 1–272. [CrossRef]
16. Zhu, X.; Radovic-Moreno, A.F.; Wu, J.; Langer, R.; Shi, J. Nanomedicine in the management of microbial infection–overview and perspectives. *Nano Today* **2014**, *9*, 478–498. [CrossRef]
17. Karaman, D.Ş.; Manner, S.; Fallarero, A.; Rosenholm, J.M. Current approaches for exploration of nanoparticles as antibacterial agents. *Antibact. Agents* **2017**. [CrossRef]
18. Yaqoob, A.A.; Ahmad, H.; Parveen, T.; Ahmad, A.; Oves, M.; Ismail, I.M.I.; Qari, H.A.; Umar, K.; Mohamad Ibrahim, M.N. Recent advances in metal decorated nanomaterials and their various biological applications: A review. *Front. Chem.* **2020**, *8*, 341. [CrossRef]
19. Belenov, S.V.; Volochaev, V.A.; Pryadchenko, V.V.; Srabionyan, V.V.; Shemet, D.B.; Tabachkova, N.Y.; Guterman, V.E. Phase behavior of Pt–Cu nanoparticles with different architecture upon their thermal treatment. *Nanotechnol. Russ.* **2017**, *12*, 147–155. [CrossRef]
20. Ali, S.; Sharma, A.S.; Ahmad, W.; Zareef, M.; Hassan, M.M.; Viswadevarayalu, A.; Jiao, T.; Li, H.; Chen, Q. Noble metals based bimetallic and trimetallic nanoparticles: Controlled synthesis, antimicrobial and anticancer applications. *Crit. Rev. Anal. Chem.* **2020**, 1–28. [CrossRef] [PubMed]
21. Zaleska-Medynska, A.; Marchelek, M.; Diak, M.; Grabowska, E. Noble metal-based bimetallic nanoparticles: The effect of the structure on the optical, catalytic and photocatalytic properties. *Adv. Colloid Interface Sci.* **2016**, *229*, 80–107. [CrossRef]
22. Gao, W.; Thamphiwatana, S.; Angsantikul, P.; Zhang, L. Nanoparticle approaches against bacterial infections. *Wiley Interdiscip. Rev. Nanomed. Nanobiotechnol.* **2014**, *6*, 532–547. [CrossRef] [PubMed]
23. Lemire, J.A.; Harrison, J.J.; Turner, R.J. Antimicrobial activity of metals: Mechanisms, molecular targets and applications. *Nat. Rev. Microbiol.* **2013**, *11*, 371–384. [CrossRef] [PubMed]
24. Buchwalter, P.; Rosé, J.; Braunstein, P. Multimetallic catalysis based on heterometallic complexes and clusters. *Chem. Rev.* **2015**, *115*, 28–126. [CrossRef]
25. Dong, Y.; Zhu, H.; Shen, Y.; Zhang, W.; Zhang, L. Antibacterial activity of silver nanoparticles of different particle size against Vibrio Natriegens. *PLoS ONE* **2019**, *14*, 1–12. [CrossRef] [PubMed]
26. Ortiz-Benítez, E.A.; Velázquez-Guadarrama, N.; Durán Figueroa, N.V.; Quezada, H.; De Jesús Olivares-Trejo, J. Antibacterial mechanism of gold nanoparticles on: Streptococcus pneumoniae. *Metallomics* **2019**, *11*, 1265–1276. [CrossRef] [PubMed]
27. Mohana, S.; Sumathi, S. Multi-functional biological effects of palladium nanoparticles synthesized using *Agaricus bisporus*. *J. Clust. Sci.* **2020**, *31*, 391–400. [CrossRef]
28. Narayanasamy, P.; Switzer, B.L.; Britigan, B.E. Prolonged-acting, multi-targeting gallium nanoparticles potently inhibit growth of both HIV and mycobacteria in co-infected human macrophages. *Sci. Rep.* **2015**, *5*, 1–7. [CrossRef]
29. Keihan, A.H.; Veisi, H.; Veasi, H. Green synthesis and characterization of spherical copper nanoparticles as organometallic antibacterial agent. *Appl. Organomet. Chem.* **2017**, *31*, 1–7. [CrossRef]
30. Ayaz Ahmed, K.B.; Raman, T.; Anbazhagan, V. Platinum nanoparticles inhibit bacteria proliferation and rescue zebrafish from bacterial infection. *RSC Adv.* **2016**, *6*, 44415–44424. [CrossRef]
31. Smirnov, N.A.; Kudryashov, S.I.; Nastulyavichus, A.A.; Rudenko, A.A.; Saraeva, I.N.; Tolordava, E.R.; Gonchukov, S.A.; Romanova, Y.M.; Ionin, A.A.; Zayarny, D.A. Antibacterial properties of silicon nanoparticles. *Laser Phys. Lett.* **2018**, *15*, 105602. [CrossRef]
32. Menon, S.; Agarwal, H.; Rajeshkumar, S.; Jacquline Rosy, P.; Shanmugam, V.K. Investigating the antimicrobial activities of the biosynthesized selenium nanoparticles and its statistical analysis. *Bionanoscience* **2020**, *10*, 122–135. [CrossRef]
33. Cittrarasu, V.; Kaliannan, D.; Dharman, K.; Maluventhen, V.; Easwaran, M.; Liu, W.C.; Balasubramanian, B.; Arumugam, M. Green synthesis of selenium nanoparticles mediated from *Ceropegia bulbosa* Roxb extract and its cytotoxicity, antimicrobial, mosquitocidal and photocatalytic activities. *Sci. Rep.* **2021**, *11*, 1032. [CrossRef]
34. Geoffrion, L.D.; Hesabizadeh, T.; Medina-Cruz, D.; Kusper, M.; Taylor, P.; Vernet-Crua, A.; Chen, J.; Ajo, A.; Webster, T.J.; Guisbiers, G. Naked selenium nanoparticles for antibacterial and anticancer treatments. *ACS Omega* **2020**, *5*, 2660–2669. [CrossRef]
35. Din, M.I.; Nabi, A.G.; Rani, A.; Aihetasham, A.; Mukhtar, M. Single step green synthesis of stable nickel and nickel oxide nanoparticles from Calotropis gigantea: Catalytic and antimicrobial potentials. *Environ. Nanotechnol. Monit. Manag.* **2018**, *9*, 29–36. [CrossRef]
36. Kamran, U.; Bhatti, H.N.; Iqbal, M.; Jamil, S.; Zahid, M. Biogenic synthesis, characterization and investigation of photocatalytic and antimicrobial activity of manganese nanoparticles synthesized from Cinnamomum verum bark extract. *J. Mol. Struct.* **2019**, *1179*, 532–539. [CrossRef]
37. Katata-Seru, L.; Moremedi, T.; Aremu, O.S.; Bahadur, I. Green synthesis of iron nanoparticles using Moringa oleifera extracts and their applications: Removal of nitrate from water and antibacterial activity against Escherichia coli. *J. Mol. Liq.* **2018**, *256*, 296–304. [CrossRef]

38. Rieznichenko, L.S.; Gruzina, T.G.; Dybkova, S.M.; Ushkalov, V.O.; Ulberg, Z.R. Investigation of bismuth nanoparticles antimicrobial activity against high pathogen microorganisms. *Am. J. Bioterror. Biosecur. Biodef* **2015**, *2*, 1004.
39. Manikandan, V.; Velmurugan, P.; Park, J.H.; Chang, W.S.; Park, Y.J.; Jayanthi, P.; Cho, M.; Oh, B.T. Green synthesis of silver oxide nanoparticles and its antibacterial activity against dental pathogens. *3 Biotech* **2017**, *7*, 1–9. [CrossRef]
40. Qamar, H.; Rehman, S.; Chauhan, D.K.; Tiwari, A.K.; Upmanyu, V. Green synthesis, characterization and antimicrobial activity of copper oxide nanomaterial derived from Momordica charantia. *Int. J. Nanomed.* **2020**, *15*, 2541–2553. [CrossRef] [PubMed]
41. Tiwari, V.; Mishra, N.; Gadani, K.; Solanki, P.S.; Shah, N.A.; Tiwari, M. Mechanism of anti-bacterial activity of zinc oxide nanoparticle against Carbapenem-Resistant Acinetobacter baumannii. *Front. Microbiol.* **2018**, *9*, 1–10. [CrossRef]
42. Eisa, N.E.; Almansour, S.; Alnaim, I.A.; Ali, A.M.; Algrafy, E.; Ortashi, K.M.; Awad, M.A.; Virk, P.; Hendi, A.A.; Eissa, F.Z. Eco-synthesis and characterization of titanium nanoparticles: Testing its cytotoxicity and antibacterial effects. *Green Process. Synth.* **2020**, *9*, 462–468. [CrossRef]
43. Behera, N.; Arakha, M.; Priyadarshinee, M.; Pattanayak, B.S.; Soren, S.; Jha, S.; Mallick, B.C. Oxidative stress generated at nickel oxide nanoparticle interface results in bacterial membrane damage leading to cell death. *RSC Adv.* **2019**, *9*, 24888–24894. [CrossRef]
44. Saqib, S.; Munis, M.F.H.; Zaman, W.; Ullah, F.; Shah, S.N.; Ayaz, A.; Farooq, M.; Bahadur, S. Synthesis, characterization and use of iron oxide nano particles for antibacterial activity. *Microsc. Res. Tech.* **2019**, *82*, 415–420. [CrossRef]
45. Pallela, P.N.V.K.; Ummey, S.; Ruddaraju, L.K.; Gadi, S.; Cherukuri, C.S.L.; Barla, S.; Pammi, S.V.N. Antibacterial efficacy of green synthesized α-Fe2O3 nanoparticles using Sida cordifolia plant extract. *Heliyon* **2019**, *5*, e02765. [CrossRef] [PubMed]
46. Marquis, G.; Ramasamy, B.; Banwarilal, S.; Munusamy, A.P. Evaluation of antibacterial activity of plant mediated CaO nanoparticles using Cissus quadrangularis extract. *J. Photochem. Photobiol. B Biol.* **2016**, *155*, 28–33. [CrossRef] [PubMed]
47. Maji, J.; Pandey, S.; Basu, S. Synthesis and evaluation of antibacterial properties of magnesium oxide nanoparticles. *Bull. Mater. Sci.* **2020**, *43*, 1–10. [CrossRef]
48. Suryavanshi, P.; Pandit, R.; Gade, A.; Derita, M.; Zachino, S.; Rai, M. Colletotrichum sp.- mediated synthesis of sulphur and aluminium oxide nanoparticles and its in vitro activity against selected food-borne pathogens. *LWT* **2017**, *81*, 188–194. [CrossRef]
49. Pop, O.L.; Mesaros, A.; Vodnar, D.C.; Suharoschi, R.; Tăbăran, F.; Mageruşan, L.; Tódor, I.S.; Diaconeasa, Z.; Balint, A.; Ciontea, L.; et al. Cerium oxide nanoparticles and their efficient antibacterial application in vitro against gram-positive and gram-negative pathogens. *Nanomaterials* **2020**, *10*, 1614. [CrossRef] [PubMed]
50. Sreenivasa Kumar, G.; Venkataramana, B.; Reddy, S.A.; Maseed, H.; Nagireddy, R.R. Hydrothermal synthesis of Mn3O4 nanoparticles by evaluation of pH effect on particle size formation and its antibacterial activity. *Adv. Nat. Sci. Nanosci. Nanotechnol.* **2020**, *11*. [CrossRef]
51. Khan, M.; Shaik, M.R.; Khan, S.T.; Adil, S.F.; Kuniyil, M.; Khan, M.; Al-Warthan, A.A.; Siddiqui, M.R.H.; Nawaz Tahir, M. Enhanced antimicrobial activity of biofunctionalized zirconia nanoparticles. *ACS Omega* **2020**, *5*, 1987–1996. [CrossRef]
52. Liu, S.; Wang, C.; Hou, J.; Wang, P.; Miao, L.; Li, T. Effects of silver sulfide nanoparticles on the microbial community structure and biological activity of freshwater biofilms. *Environ. Sci. Nano* **2018**, *5*, 2899–2908. [CrossRef]
53. Malarkodi, C.; Rajeshkumar, S.; Paulkumar, K.; Vanaja, M.; Gnanajobitha, G.; Annadurai, G. Biosynthesis and antimicrobial activity of semiconductor nanoparticles against oral pathogens. *Bioinorg. Chem. Appl.* **2014**, *2014*. [CrossRef] [PubMed]
54. Argueta-Figueroa, L.; Torres-Gómez, N.; García-Contreras, R.; Vilchis-Nestor, A.R.; Martínez-Alvarez, O.; Acosta-Torres, L.S.; Arenas-Arrocena, M.C. Hydrothermal synthesis of pyrrhotite (Fex-1S) nanoplates and their antibacterial, cytotoxic activity study. *Prog. Nat. Sci. Mater. Int.* **2018**, *28*, 447–455. [CrossRef]
55. André, V.; Da Silva, A.R.F.; Fernandes, A.; Frade, R.; Garcia, C.; Rijo, P.; Antunes, A.M.M.; Rocha, J.; Duarte, M.T. Mg- and Mn-MOFs boost the antibiotic activity of nalidixic acid. *ACS Appl. Bio Mater.* **2019**, *2*, 2347–2354. [CrossRef]
56. Zhang, S.-S.; Wang, X.; Su, H.-F.; Feng, L.; Wang, Z.; Ding, W.-Q.; Blatov, V.A.; Kurmoo, M.; Tung, C.-H.; Sun, D. A water-stable Cl@Ag14 cluster based metal–organic open framework for dichromate trapping and bacterial inhibition. *Inorg. Chem.* **2017**, *56*, 11891–11899. [CrossRef]
57. Jo, J.H.; Kim, H.C.; Huh, S.; Kim, Y.; Lee, D.N. Antibacterial activities of Cu-MOFs containing glutarates and bipyridyl ligands. *Dalt. Trans.* **2019**, *48*, 8084–8093. [CrossRef] [PubMed]
58. Pezeshkpour, V.; Khosravani, S.A.; Ghaedi, M.; Dashtian, K.; Zare, F.; Sharifi, A.; Jannesar, R.; Zoladl, M. Ultrasound assisted extraction of phenolic acids from broccoli vegetable and using sonochemistry for preparation of MOF-5 nanocubes: Comparative study based on micro-dilution broth and plate count method for synergism antibacterial effect. *Ultrason. Sonochem.* **2018**, *40*, 1031–1038. [CrossRef]
59. Zhuang, W.; Yuan, D.; Li, J.R.; Luo, Z.; Zhou, H.C.; Bashir, S.; Liu, J. Highly potent bactericidal activity of porous metal-organic frameworks. *Adv. Healthc. Mater.* **2012**, *1*, 225–238. [CrossRef]
60. Godfrey, I.J.; Dent, A.J.; Parkin, I.P.; Maenosono, S.; Sankar, G. Structure of gold-silver nanoparticles. *J. Phys. Chem. C* **2017**, *121*, 1957–1963. [CrossRef]
61. Elemike, E.E.; Onuwdiwe, D.C.; Fayemi, O.E.; Botha, T.L. Green synthesis and electrochemistry of Ag, Au, and Ag–Au bimetallic nanoparticles using golden rod (Solidago canadensis) leaf extract. *Appl. Phys. A Mater. Sci. Process.* **2019**, *125*. [CrossRef]
62. Arora, N.; Thangavelu, K.; Karanikolos, G.N. Bimetallic nanoparticles for antimicrobial applications. *Front. Chem.* **2020**, *8*, 1–22. [CrossRef]

63. Vaseghi, Z.; Tavakoli, O.; Nematollahzadeh, A. Rapid biosynthesis of novel Cu/Cr/Ni trimetallic oxide nanoparticles with antimicrobial activity. *J. Environ. Chem. Eng.* **2018**, *6*, 1898–1911. [CrossRef]
64. Merrill, N.A.; Nitka, T.T.; McKee, E.M.; Merino, K.C.; Drummy, L.F.; Lee, S.; Reinhart, B.; Ren, Y.; Munro, C.J.; Pylypenko, S.; et al. Effects of metal composition and ratio on peptide-templated multimetallic PdPt nanomaterials. *ACS Appl. Mater. Interfaces* **2017**, *9*, 8030–8040. [CrossRef]
65. Kazakova, M.A.; Morales, D.M.; Andronescu, C.; Elumeeva, K.; Selyutin, A.G.; Ishchenko, A.V.; Golubtsov, G.V.; Dieckhöfer, S.; Schuhmann, W.; Masa, J. Fe/Co/Ni mixed oxide nanoparticles supported on oxidized multi-walled carbon nanotubes as electrocatalysts for the oxygen reduction and the oxygen evolution reactions in alkaline media. *Catal. Today* **2020**, *357*, 259–268. [CrossRef]
66. Ye, X.; He, X.; Lei, Y.; Tang, J.; Yu, Y.; Shi, H.; Wang, K. One-pot synthesized Cu/Au/Pt trimetallic nanoparticles with enhanced catalytic and plasmonic properties as a universal platform for biosensing and cancer theranostics. *Chem. Commun.* **2019**, *55*, 2321–2324. [CrossRef]
67. Nie, F.; Ga, L.; Ai, J.; Wang, Y. Trimetallic PdCuAu nanoparticles for temperature sensing and fluorescence detection of H2O2 and glucose. *Front. Chem.* **2020**, *8*, 1–10. [CrossRef]
68. Sui, M.; Kunwar, S.; Pandey, P.; Pandit, S.; Lee, J. Improved localized surface plasmon resonance responses of multi-metallic Ag/Pt/Au/Pd nanostructures: Systematic study on the fabrication mechanism and localized surface plasmon resonance properties by solid-state dewetting. *New J. Phys.* **2019**, *21*, 113049. [CrossRef]
69. Tang, Z.; Yeo, B.C.; Han, S.S.; Lee, T.-J.; Bhang, S.H.; Kim, W.-S.; Yu, T. Facile aqueous-phase synthesis of Ag–Cu–Pt–Pd quadrometallic nanoparticles. *Nano Converg.* **2019**, *6*, 1–7. [CrossRef]
70. Huang, L.; Lin, H.; Zheng, C.Y.; Kluender, E.J.; Golnabi, R.; Shen, B.; Mirkin, C.A. Multimetallic High-Index Faceted Heterostructured Nanoparticles. *J. Am. Chem. Soc.* **2020**, *142*, 4570–4575. [CrossRef] [PubMed]
71. Lomelí-Marroquín, D.; Cruz, D.M.; Nieto-Argüello, A.; Crua, A.V.; Chen, J.; Torres-Castro, A.; Webster, T.J.; Cholula-Díaz, J.L. Starch-mediated synthesis of mono- and bimetallic silver/gold nanoparticles as antimicrobial and anticancer agents. *Int. J. Nanomed.* **2019**, *14*, 2171–2190. [CrossRef] [PubMed]
72. Al-Haddad, J.; Alzaabi, F.; Pal, P.; Rambabu, K.; Banat, F. Green synthesis of bimetallic copper–silver nanoparticles and their application in catalytic and antibacterial activities. *Clean Technol. Environ. Policy* **2020**, *22*, 269–277. [CrossRef]
73. Formaggio, D.M.D.; de Oliveira Neto, X.A.; Rodrigues, L.D.A.; de Andrade, V.M.; Nunes, B.C.; Lopes-Ferreira, M.; Ferreira, F.G.; Wachesk, C.C.; Camargo, E.R.; Conceição, K.; et al. In vivo toxicity and antimicrobial activity of AuPt bimetallic nanoparticles. *J. Nanopart. Res.* **2019**, *21*. [CrossRef]
74. Lozhkomoev, A.S.; Lerner, M.I.; Pervikov, A.V.; Kazantsev, S.O.; Fomenko, A.N. Development of Fe/Cu and Fe/Ag bimetallic nanoparticles for promising biodegradable materials with antimicrobial effect. *Nanotechnol. Russ.* **2018**, *13*, 18–25. [CrossRef]
75. Yang, M.; Lu, F.; Zhou, T.; Zhao, J.; Ding, C.; Fakhri, A.; Gupta, V.K. Biosynthesis of nano bimetallic Ag/Pt alloy from Crocus sativus L. extract: Biological efficacy and catalytic activity. *J. Photochem. Photobiol. B Biol.* **2020**, *212*, 112025. [CrossRef] [PubMed]
76. Merugu, R.; Gothalwal, R.; Kaushik Deshpande, P.; De Mandal, S.; Padala, G.; Latha Chitturi, K. Synthesis of Ag/Cu and Cu/Zn bimetallic nanoparticles using toddy palm: Investigations of their antitumor, antioxidant and antibacterial activities. *Mater. Today Proc.* **2020**. [CrossRef]
77. Argueta-Figueroa, L.; Morales-Luckie, R.A.; Scougall-Vilchis, R.J.; Olea-Mejía, O.F. Synthesis, characterization and antibacterial activity of copper, nickel and bimetallic Cu-Ni nanoparticles for potential use in dental materials. *Prog. Nat. Sci. Mater. Int.* **2014**, *24*, 321–328. [CrossRef]
78. Andrade, G.R.S.; Nascimento, C.C.; Lima, Z.M.; Teixeira-Neto, E.; Costa, L.P.; Gimenez, I.F. Star-shaped ZnO/Ag hybrid nanostructures for enhanced photocatalysis and antibacterial activity. *Appl. Surf. Sci.* **2017**, *399*, 573–582. [CrossRef]
79. Sinha, T.; Ahmaruzzaman, M.; Adhikari, P.P.; Bora, R. Green and environmentally sustainable fabrication of Ag-SnO2 nanocomposite and its multifunctional efficacy as photocatalyst and antibacterial and antioxidant agent. *ACS Sustain. Chem. Eng.* **2017**, *5*, 4645–4655. [CrossRef]
80. Antonoglou, O.; Lafazanis, K.; Mourdikoudis, S.; Vourlias, G.; Lialiaris, T.; Pantazaki, A.; Dendrinou-Samara, C. Biological relevance of CuFeO 2 nanoparticles: Antibacterial and anti-inflammatory activity, genotoxicity, DNA and protein interactions. *Mater. Sci. Eng. C* **2019**, *99*, 264–274. [CrossRef] [PubMed]
81. Addae, E.; Dong, X.; McCoy, E.; Yang, C.; Chen, W.; Yang, L. Investigation of antimicrobial activity of photothermal therapeutic gold/copper sulfide core/shell nanoparticles to bacterial spores and cells. *J. Biol. Eng.* **2014**, *8*, 1–11. [CrossRef]
82. He, Q.; Huang, C.; Liu, J. Preparation, characterization and antibacterial activity of magnetic greigite and fe3s4/ag nanoparticles. *Nanosci. Nanotechnol. Lett.* **2014**, *6*, 10–17. [CrossRef]
83. Iribarnegaray, V.; Navarro, N.; Robino, L.; Zunino, P.; Morales, J.; Scavone, P. Magnesium-doped zinc oxide nanoparticles alter biofilm formation of Proteus mirabilis. *Nanomedicine* **2019**, *14*, 1551–1564. [CrossRef]
84. Lozhkomoev, A.S.; Bakina, O.V.; Pervikov, A.V.; Kazantsev, S.O.; Glazkova, E.A. Synthesis of CuO–ZnO composite nanoparticles by electrical explosion of wires and their antibacterial activities. *J. Mater. Sci. Mater. Electron.* **2019**, *30*, 13209–13216. [CrossRef]
85. Chen, X.; Ku, S.; Weibel, J.A.; Ximenes, E.; Liu, X.; Ladisch, M.; Garimella, S.V. Enhanced antimicrobial efficacy of bimetallic porous CuO microspheres decorated with Ag nanoparticles. *ACS Appl. Mater. Interfaces* **2017**, *9*, 39165–39173. [CrossRef] [PubMed]

86. Singh, S.; Barick, K.C.; Bahadur, D. Inactivation of bacterial pathogens under magnetic hyperthermia using Fe3O4-ZnO nanocomposite. *Powder Technol.* **2015**, *269*, 513–519. [CrossRef]
87. Masadeh, M.M.; Karasneh, G.A.; Al-Akhras, M.A.; Albiss, B.A.; Aljarah, K.M.; Al-azzam, S.I.; Alzoubi, K.H. Cerium oxide and iron oxide nanoparticles abolish the antibacterial activity of ciprofloxacin against gram positive and gram negative biofilm bacteria. *Cytotechnology* **2015**, *67*, 427–435. [CrossRef]
88. Alzahrani, K.E.; Aniazy, A.; Alswieleh, A.M.; Wahab, R.; El-Toni, A.M.; Alghamdi, H.S. Antibacterial activity of trimetal (CuZnFe) oxide nanoparticles. *Int. J. Nanomed.* **2018**, *13*, 77–87. [CrossRef] [PubMed]
89. Yadav, N.; Jaiswal, A.K.; Dey, K.K.; Yadav, V.B.; Nath, G.; Srivastava, A.K.; Yadav, R.R. Trimetallic Au/Pt/Ag based nanofluid for enhanced antibacterial response. *Mater. Chem. Phys.* **2018**, *218*, 10–17. [CrossRef]
90. Paul, D.; Mangla, S.; Neogi, S. Antibacterial study of CuO-NiO-ZnO trimetallic oxide nanoparticle. *Mater. Lett.* **2020**, *271*, 127740. [CrossRef]
91. Gupta, A.; Khosla, N.; Govindasamy, V.; Saini, A.; Annapurna, K.; Dhakate, S.R. Trimetallic composite nanofibers for antibacterial and photocatalytic dye degradation of mixed dye water. *Appl. Nanosci.* **2020**, *10*, 4191–4205. [CrossRef]
92. Basavegowda, N.; Mishra, K.; Lee, Y.R. Trimetallic FeAgPt alloy as a nanocatalyst for the reduction of 4-nitroaniline and decolorization of rhodamine B: A comparative study. *J. Alloys Compd.* **2017**, *701*, 456–464. [CrossRef]
93. Mishra, K.; Basavegowda, N.; Lee, Y.R. AuFeAg hybrid nanoparticles as an efficient recyclable catalyst for the synthesis of α,β- and β,β-dichloroenones. *Appl. Catal. A Gen.* **2015**, *506*, 180–187. [CrossRef]
94. Lacey, S.D.; Dong, Q.; Huang, Z.; Luo, J.; Xie, H.; Lin, Z.; Kirsch, D.J.; Vattipalli, V.; Povinelli, C.; Fan, W. Stable multimetallic nanoparticles for oxygen electrocatalysis. *Nano Lett.* **2019**, *19*, 5149–5158. [CrossRef] [PubMed]
95. Huynh, K.H.; Pham, X.H.; Kim, J.; Lee, S.H.; Chang, H.; Rho, W.Y.; Jun, B.H. Synthesis, properties, and biological applications of metallic alloy nanoparticles. *Int. J. Mol. Sci.* **2020**, *21*, 5174. [CrossRef]
96. Razmara, Z.; Razmara, F. Synthesis and magnetic properties of Fe-Ni-Zn, Fe-Co-Zn and Co-Ni-Zn nanoparticles by co-precipitation method. *Inorg. Nano Metal. Chem.* **2019**, *49*, 163–168. [CrossRef]
97. Chou, S.W.; Liu, C.L.; Liu, T.M.; Shen, Y.F.; Kuo, L.C.; Wu, C.H.; Hsieh, T.Y.; Wu, P.C.; Tsai, M.R.; Yang, C.C.; et al. Infrared-active quadruple contrast FePt nanoparticles for multiple scale molecular imaging. *Biomaterials* **2016**, *85*, 54–64. [CrossRef] [PubMed]
98. Liu, Y.; Shen, X. Preparation and characterization of NiFe bimetallic micro-particles and its composite with silica shell. *J. Saudi Chem. Soc.* **2019**, *23*, 1032–1040. [CrossRef]
99. Ma, L.; Wei, Z.; Zhu, X.; Liang, J.; Zhang, X. Synthesis and photocatalytic properties of co-doped Zn1-xCoxMn2O hollow nanospheres. *J. Nanomater.* **2019**, *2019*. [CrossRef]
100. Rák, Z.; Brenner, D.W. Negative surface energies of nickel ferrite nanoparticles under hydrothermal conditions. *J. Nanomater.* **2019**, *2019*. [CrossRef]
101. Park, J.H.; Ahn, H.S. Electrochemical synthesis of multimetallic nanoparticles and their application in alkaline oxygen reduction catalysis. *Appl. Surf. Sci.* **2020**, *504*, 144517. [CrossRef]
102. Srinoi, P.; Chen, Y.T.; Vittur, V.; Marquez, M.D.; Lee, T.R. Bimetallic nanoparticles: Enhanced magnetic and optical properties for emerging biological applications. *Appl. Sci.* **2018**, *8*, 1106. [CrossRef]
103. Li, C.; Xu, Y.; Li, Y.; Yu, H.; Yin, S.; Xue, H.; Li, X.; Wang, H.; Wang, L. Engineering porosity into trimetallic PtPdNi nanospheres for enhanced electrocatalytic oxygen reduction activity. *Green Energy Environ.* **2018**, *3*, 352–359. [CrossRef]
104. Kotov, Y.A. The electrical explosion of wire: A method for the synthesis of weakly aggregated nanopowders. *Nanotechnol. Russ.* **2009**, *4*, 415–424. [CrossRef]
105. Lozhkomoev, A.S.; Kazantsev, S.O.; Kondranova, A.M.; Fomenko, A.N.; Pervikov, A.V.; Rodkevich, N.G.; Bakina, O.V. Design of antimicrobial composite nanoparticles ZnxMe(100-x)/O by electrical explosion of two wires in the oxygen-containing atmosphere. *Mater. Des.* **2019**, *183*, 108099. [CrossRef]
106. Loganathan, B.; Karthikeyan, B. Tailored Au and Pt containing multi-metallic nanocomposites for a promising fuel cell reaction. *J. Clust. Sci.* **2017**, *28*, 1463–1487. [CrossRef]
107. Kumar, A.; Mishra, N.K.; Sachan, K.; Ali, M.A.; Gupta, S.S.; Singh, R. Trimetallic oxide nanocomposites of transition metals titanium and vanadium by sol-gel technique: Synthesis, characterization and electronic properties. *Mater. Res. Express* **2018**, *5*, 45037. [CrossRef]
108. Dobrucka, R. Biogenic synthesis of trimetallic nanoparticles Au/ZnO/Ag using Meliloti officinalis extract. *Int. J. Environ. Anal. Chem.* **2020**, *100*, 981–991. [CrossRef]
109. Rai, R.V.; Bai, J.A. Nanoparticles and their potential application as antimicrobials. In *Science against Microbial Pathogens: Communicating Current Research and Technological Advances*; Méndez-Vilas, A., Ed.; Formatex Research Center: Badajoz, Spain, 2011; Volume 1, pp. 197–209, ISBN 9788493984328.
110. Dakal, T.C.; Kumar, A.; Majumdar, R.S.; Yadav, V. Mechanistic basis of antimicrobial actions of silver nanoparticles. *Front. Microbiol.* **2016**, *7*, 1–17. [CrossRef]
111. Roy, A.; Bulut, O.; Some, S.; Mandal, A.K.; Yılmaz, M.D. Green synthesis of silver nanoparticles: Biomolecule-nanoparticle organizations targeting antimicrobial activity. *RSC Adv.* **2019**, *9*, 2673–2702. [CrossRef]
112. Zheng, K.; Setyawati, M.I.; Leong, D.T.; Xie, J. Antimicrobial gold nanoclusters. *ACS Nano* **2017**, *11*, 6904–6910. [CrossRef]
113. Vega-Jiménez, A.L.; Vázquez-Olmos, A.R.; Acosta-Gío, E.; Álvarez-Pérez, M.A. In vitro antimicrobial activity evaluation of metal oxide nanoparticles. *Nanoemulsions Prop. Fabr. Appl.* **2019**, 1–18. [CrossRef]

114. Durán, N.; Durán, M.; de Jesus, M.B.; Seabra, A.B.; Fávaro, W.J.; Nakazato, G. Silver nanoparticles: A new view on mechanistic aspects on antimicrobial activity. *Nanomed. Nanotechnol. Biol. Med.* **2016**, *12*, 789–799. [CrossRef]
115. Chatterjee, A.K.; Chakraborty, R.; Basu, T. Mechanism of antibacterial activity of copper nanoparticles. *Nanotechnology* **2014**, *25*. [CrossRef]
116. Jahnke, J.P.; Cornejo, J.A.; Sumner, J.J.; Schuler, A.J.; Atanassov, P.; Ista, L.K. Conjugated gold nanoparticles as a tool for probing the bacterial cell envelope: The case of Shewanella oneidensis MR-1. *Biointerphases* **2016**, *11*, 011003. [CrossRef]
117. Hu, S.; Che, F.; Khorasani, B.; Jeon, M.; Yoon, C.W.; McEwen, J.S.; Scudiero, L.; Ha, S. Improving the electrochemical oxidation of formic acid by tuning the electronic properties of Pd-based bimetallic nanoparticles. *Appl. Catal. B Environ.* **2019**, *254*, 685–692. [CrossRef]
118. Yun, R.; Hong, L.; Ma, W.; Wang, S.; Zheng, B. Nitrogen-rich porous carbon-stabilized Ni-Co nanoparticles for the hydrogenation of quinolines. *ACS Appl. Nano Mater.* **2019**, *2*, 6763–6768. [CrossRef]
119. Hansgen, D.A.; Vlachos, D.G.; Chen, J.G. Using first principles to predict bimetallic catalysts for the ammonia decomposition reaction. *Nat. Chem.* **2010**, *2*, 484–489. [CrossRef] [PubMed]
120. He, W.; Cai, J.; Zhang, H.; Zhang, L.; Zhang, X.; Li, J.; Yin, J.J. Formation of PtCuCo trimetallic nanostructures with enhanced catalytic and enzyme-like activities for biodetection. *ACS Appl. Nano Mater.* **2018**, *1*, 222–231. [CrossRef]
121. De Silva, G.O.; Abeysundara, A.T.; Aponso, M.M.W. A review on nanocomposite materials for food packaging which comprising antimicrobial activity. *Int. J. Chem. Stud.* **2017**, *5*, 228–230.
122. Amini, S.M.; Gilaki, M.; Karchani, M. Safety of nanotechnology in food industries. *Electron. Physician* **2014**, *6*, 962–968. [CrossRef] [PubMed]
123. Shavandi, Z.; Ghazanfari, T.; Moghaddam, K.N. In vitro toxicity of silver nanoparticles on murine peritoneal macrophages. *Immunopharmacol. Immunotoxicol.* **2011**, *33*, 135–140. [CrossRef] [PubMed]

Article

Preparation of Paclitaxel-Encapsulated Bio-Functionalized Selenium Nanoparticles and Evaluation of Their Efficacy against Cervical Cancer

Soumya Menon [1,†], Santhoshkumar Jayakodi [2,†], Kanti Kusum Yadav [3], Prathap Somu [2], Mona Isaq [4], Venkat Kumar Shanmugam [5], Amballa Chaitanyakumar [6,*] and Nagaraj Basavegowda [7,*]

1. Department of Chemistry, Indian Institute of Technology, Roorkee 247667, India
2. Department of Biotechnology, Saveetha School of Engineering, Saveetha Institute of Medical and Technical Science (SIMATS), Chennai 602105, India
3. Department of Biotechnology, Karunya Institute of Technology and Sciences (Deemed to be University), Karunya Nagar, Coimbatore 641114, India
4. Department of Biotechnology & Bioinformatics, Kuvempu University, Jnana Sahyadri, Shankaraghatta, Shivamogga 577451, India
5. School of Bio-Sciences and Technology, Vellore Institute of Technology, Vellore 632014, India
6. Department of Biotechnology, University Institute of Engineering and Technology, Guru Nanak University, Hyderabad 500085, India
7. Department of Biotechnology, Yeungnam University, Gyeongsan 712-749, Korea
* Correspondence: chaitu.05543@gmail.com (A.C.); nagarajb2005@yahoo.co.in (N.B.)
† These authors contributed equally to this work.

Abstract: The potentiality of nanomedicine in the cancer treatment being widely recognized in the recent years. In the present investigation, the synergistic effects of chitosan-modified selenium nanoparticles loaded with paclitaxel (PTX-chit-SeNPs) were studied. These selenium nanoparticles were tested for drug release analysis at a pH of 7.4 and 5.5, and further characterized using FTIR, DLS, zeta potential, and TEM to confirm their morphology, and the encapsulation of the drug was carried out using UPLC analysis. Quantitative evaluation of anti-cancer properties was performed via MTT analysis, apoptosis, gene expression analysis, cell cycle arrest, and over-production of ROS. The unique combination of phytochemicals from the seed extract, chitosan, paclitaxel, and selenium nanoparticles can be effectively utilized to combat cancerous cells. The production of the nanosystem has been demonstrated to be cost-effective and have unique characteristics, and can be utilized for improving future diagnostic approaches.

Keywords: *Mucuna pruriens* seed extract; chitosan; paclitaxel; selenium nanoparticles; flow cytometer; anticancer; cervical cancer

1. Introduction

Nanomedicines are emerging as a significant class of therapeutics. The most encouraging aspect of using nanoparticles in therapeutics is their capability to target or accumulate at the site of targeted tumour tissues with negligible toxicity. The nanomaterial, owing to its size, permeates through the kidney or liver due to surface decorations, and the cells diffuse through the blood vessels and vasculature of tumour tissues, thereby extending its presence in the blood circulation and this is explained through the phenomenon of EPR (enhanced permeation and retention). Moreover, nanoparticles have high surface area to volume proportions [1], yielding high adsorbing properties [2]. On this basis, nanoparticles are adsorbed with therapeutic drugs [3], phytochemicals with anticancer properties [4], imaging agents [5], or target-specific genes or peptides as targeting ligands to the cancer receptor cells [6].

Selenium (Se) is a trace element with broad pharmacological capabilities and physiological functions. It is an essential component of numerous anti-oxidant enzymes like

phospholipid hydroperoxide, glutathione peroxidase, and thioredoxin reductase. It is a micronutrient that plays various roles in human health, improving cardiovascular health or immunomodulatory functions, constraining cancer progression, preventing neurodegeneration, and regulating the thyroid hormone metabolism [7,8]. Selenium supplements traditionally used for cancer treatment have disadvantages of high toxicity or low absorption. Therefore, developing novel systems as transporters of selenium compounds that would promote bioavailability and permit its controlled release in the organism has become a significant requirement in this field. The characteristic features of Se can be explained through its antioxidant properties by protecting against oxidative stress generated by oxidative radicals or pro-oxidant properties by promoting a strong generation of ROS via the redox cycle and causing oxidative stress to cancer-infected cells [9–12]. The mechanism of chemotherapeutic action of SeNPs can be explained through the over-expression of antioxidant enzymes that induce the generation of ROS, eventually activating a sequence of events, such as induction of the apoptotic pathway and mitochondrial dysfunction that stimulates the release of cytochrome C, thereby inducing the activation of caspase cascade, cell cycle arrest, and DNA fragmentation [13]. The latest investigations report that the chemotherapeutic activity of SeNPs has been successful against various malignant cells, such as human cervical carcinoma (HeLa) [14], breast cancer (MCF-7) [15], human hepatocyte (HepG2) [16], human melanoma (A375), and lung cancer (A549) cells [17]. However, the major problem associated with the application of SeNPs as anticancer activity is reported to be undesired toxicity against healthy cells [10]. Thus, we need to overcome this undesired toxicity without compromising its anticancer activity. The functionalization of the NP surface with biopolymers is one of the best techniques reported, to date, for improving biocompatibility with reduced aggregation.

Thus, chitosan is used for the functionalization. Chitosan is a cationic hydrophilic linear polysaccharide composed of an acetylated unit (N-acetyl-d-glucosamine) and a deacetylated unit (β-(1–4)-linked D-glucosamine). These structural characteristics of chitosan confer membrane permeability, biocompatibility with tissues, and mucoadhesive properties. Additionally, they possess a pH-dependent profile, where the primary amines become positively charged by protonation in a low pH environment. The material becomes easily soluble in the medium, thereby assisting in the encapsulation of drugs. At high pH, chitosan becomes insoluble, as the primary amines are in deprotonated conditions. Therefore, pH plays a major role in this delivery system as the body may have various regions with significantly different levels of pH. In the stomach, chitosan can easily permeate the acidic surroundings without degradation by digestive enzymes and remain stable at physiological conditions, due to its mucoadhesive properties [18–20].

Moreover, SeNPs might also be used for drug delivery applications, thereby providing a synergistic therapeutic effect against cancer. Paclitaxel (PTX) is a compound derived from the bark of *Taxus brevifolia* [21] and is a well-established anticancer drug that can prohibit the mechanism of mitosis or polymerize the β-microtubules, thereby arresting the cell cycle and causing cell death [22]. Regardless of its exclusive mechanism and effectiveness, it lacks properties such as high therapeutic index and adequate solubility [23]. To improve its efficiency of solubility, PTX was conjugated with biosynthesized selenium nanoparticles modified with chitosan. In the present study, the Box–Behnken design was selected to optimise PTX-chit-SeNPs and was compared with the biosynthesized SeNPs for its anticancer progression activities and its cytotoxicity analysis against HeLa cells. Alternative formulations, such as nanoparticles, liposomes, self-emulsions, and micelles can aid in improvising the effective delivery of PTX to tumour cells with low toxicity to healthy cells. Among these, nanoparticles that are functionalized with phytochemicals from seed extract as novel agents can provide a promising delivery system with the advantage of the EPR effect for passive targeting of PTX. The results of this study suggest that this nanocombinational system may be a promising low toxicity alternative for cancer treatment; however, further analysis of the in vivo or ex vivo studies can still be reviewed in future studies.

2. Material and Methods

2.1. Materials

The seeds of *Mucuna pruriens* were bought from a local drug store. Sodium selenite, paclitaxel (PTX), chitosan with 90% deacetylation (MWCO 55 kDa), methanol, and acetonitrile (HPLC grade) were purchased from Hi-media. All of the chemicals were analytical grade and no further purification was required.

2.2. Cell Culture

HeLa cells were purchased from NCCS, Pune, India. The cells were maintained using DMEM, 10% (fetal bovine serum) and (1%) penicillin–streptomycin antibiotic and were incubated in a cell culture incubator with 5% CO_2 at 37 °C in a humidified atmosphere. Chemicals such as PI (propidium iodide) reagent, RNase, Invitrogen Annexin V/FITC dead cell kit by Thermo Fisher Scientific (Waltham, MA, USA), 3-(4,5-dimethylthiazolyl-2)-2,5-diphenyltetrazolium bromide MTT reagent, DCFDA-cellular ROS Assay kit, PBS (10X at pH 7.4) were purchased from Hi-Media, and a Primescript RT reagent kit was purchased from Takara Bio (Kusatsu, Japan).

2.3. Preparation of Biosynthesized Selenium Nanoparticles

The biosynthesis of SeNPs using seed extract of *Mucuna pruriens* and its standardization was published in our previous study [24]. In brief, the seeds were double-washed with double-distilled water (Mili-Q), shade-dried for 5–6 days, and the dried seeds were crushed, weighed, and then added to 100 mL of Mili-Q. The liquid was kept in ta water bath at 80 °C for 45 min, and filtered using Whatman 42 filter paper. The filtrate was stored at 4 °C for further use. A ratio of 9:1 was maintained for precursor and extract concentration. The reaction continued till the colorless solution changed to the dark brown reaction mixture. The synthesized nanoparticles were further purified with double-distilled water and ethanol solution; this helps in the removal of excess phytochemicals or untreated sodium selenite from the SeNPs solution. The pellet was then dialyzed against distilled water for a period of 12 h. The purified nanoparticles were lyophilized for further characterization and storage.

2.4. Preparation of Chitosan-Modified and Paclitaxel-Loaded Selenium Nanoparticles

The chitosan solution of 1% concentration (w/v) was prepared in glacial acetic acid where the solution was kept in a magnetic stirrer for 2 h until it was completely dissolved, followed by the addition of 1% of the biosynthesized SeNPs with continuous stirring at RT for 10–12 h. After the 12 h of stirring, paclitaxel in DMSO (at the required concentration) was added drop-wise, at 1 mL/min, to the chitosan-coated selenium nanoparticles and kept overnight for continuous stirring. The PTX-chit-SeNPs were then centrifuged at 10,000 rpm for 15 min, the pellet obtained was washed with Mili-Q water and the purified solution was then lyophilized. The moisture-free samples were then stored at room temperature for further analysis and characterization.

2.5. Entrapment Efficiency (EE%) through UPLC Analysis

The paclitaxel-coated and chitosan-modified selenium nanoparticles were evaluated using UPLC (ultra-performance liquid chromatography) method and the samples were passed through the C-18 column, with 10 µL injection volume with PDA detector, at RT, equipped with Empower 3 software. The flow rate of the sample was 1.0/min and the PTX drug detection peak was at 240 nm. The mobile phase selected was acetonitrile–methanol in the ratio of 60:40. The encapsulation efficiency of the drug PTX in PTX-chit-SeNPs was calculated according to Equation (1), as described in [25].

$$\text{Encapsulation efficiency \%} = \frac{\text{Total PTX used} - \text{amount of free PTX present in the supernatant}}{\text{Total PTX used}} \times 100 \quad (1)$$

2.6. Characterization of the Chitosan-Modified Selenium Nanoparticles Loaded with Paclitaxel

The confirmation of the paclitaxel-loaded chitosan-modified selenium nanoparticles was obtained using Fourier transform infrared spectroscopy (FTIR) by Shimadzu Spectroscopy (Model IR Affinity-1, which helped in the identification of the functional groups involved in the interactions of the paclitaxel with the modified selenium nanoparticles and was compared with paclitaxel, chitosan, and biosynthesized SeNPs that have a wavelength range from 400 to 4000 cm^{-1}. The morphological analysis was conducted using transmission electron microscopy (TEM), where the samples were sputtered on the copper grids.

2.7. In Vitro Drug Release Analysis

The in vitro release of PTX from the nanosystem was studied using the dialysis method. In a dialysis bag (MWCO 60 kDa), 10 mL of nanosuspension containing PTX-chit-SeNPs was immersed in 100 mL of pH 7.4 and 5.5 phosphate buffer solution (PBS) and was dialyzed at 37 °C under mild agitation, thereby mimicking the physiological conditions. At a regular interval of time, 2 mL of the external solution was extracted and again replenished with 2 mL of fresh medium. The solution was centrifuged at 10,000 rpm for 15 min and the supernatant containing the free drug was evaluated for the concentration of PTX using a spectrophotometer. A standard graph was generated for PTX, where the R^2 value was 98.85, and the drug concentration was calculated using slope and intercept values [26].

2.8. Cytotoxicity Studies of Paclitaxel-Loaded Chitosan-Modified Selenium Nanoparticles

The nanosystems-induced cytotoxicity was evaluated with the most commonly used MTT assay, where the viability percentage of the cells is measured when they transformed MTT to purple formazan precipitate. In a 96-well cell culture plate, 5000 cells/well were seeded, then incubated until 80% confluence was reached for 24 h at 37 °C. The next day, the spent media was removed and replaced with fresh media containing the treatment drugs at various concentrations (5 µg/mL to 100 µg/mL) for the biosynthesized SeNPs, SeNPs coated with chitosan, and PTX-chit-SeNPs. After incubation of 24 h, the media was again removed and MTT reagent (5 mg/mL) was added to each well and was kept for an incubation period of 4 h. The medium was then removed and replaced with 100 µL DMSO in each well, which was added to solubilize the crystalline formazan precipitate. The absorbance was recorded at 590 nm using an ELISA microplate reader. All the experiments were performed in triplicates and were expressed as the mean ± standard error [27].

2.9. Stimulation of Apoptosis by PTX-chit-SeNPs Compared with Biosynthesized SeNPs

The cell apoptosis was detected with FITC-Annexin V/PI (propidium iodide) double staining assay, according to the protocol (Invitrogen). The cells (1×10^5) were seeded in 6-well culture plates and were incubated for 24 h at 37 °C. The cells were then treated with selected IC$_{50}$ dose values for PTX-chit-SeNPs and biosynthesized SeNPs. The apoptotic and live cells were analysed with flow cytometry and were analysed with CytExpert 2.3 software.

2.10. Stimulation of Cell Cycle Arrest by PTX-chit-SeNPs Compared with Biosynthesized SeNPs

The cells (1×10^5) were seeded in 6-well culture plates and were incubated for 24 h at 37 °C. The cells were supplemented with IC$_{50}$ dose values for PTX-chit-SeNPs and biosynthesized SeNPs for 24 h. They were trypsinized and washed with 1x PBS, the obtained pellet was ethanol-fixed with chilled 70% ethanol solution for 2 h at low temperature. The pellet was washed to remove any traces of ethanol and was then incubated with the PI solution (1 mL) containing RNAase and PBS solution for 15 min at RT prior to analysis with a flow cytometer.

2.11. RNA Isolation and RT-PCR Assay of the Apoptotic Genes

The cells with a concentration of 1×10^5 were seeded for 24 h at 37 °C in 6-well plates, then the cells that were treated with IC_{50} dose values for PTX-chit-SeNPs and biosynthesized SeNPs for 24 h were selected for the analysis. The cells were trypsinized, washed with 1x PBS, and then homogenized with triazole reagent (0.5 mL). Then, 0.2 mL of chloroform was added, and after 15 min incubation, the solution was centrifuged at 10,000 rpm for 10 min at 4 °C. The upper phase of transparent solution was transferred to fresh tubes, which were incubated with 0.5 mL isopropanol for 10 min at RT, followed by centrifugation. The RNA pellets obtained were dispersed in RNase-free water (0.1 mL) and washed with 75% ethanol. The dried RNA pellet was quantitively evaluated with a NanoDrop spectrophotometer. Then, 1 µg RNA samples were reverse-transcribed to synthesize cDNA samples and the cycling program was followed according to the manufacturer's protocol given in the Primescript RT reagent kit (Takara). The apoptotic genes were selected, as shown in Table 1, and standard GAPDH-normalized with other genes.

Table 1. Forward and Reverse primers of apoptotic genes used for RT-PCR.

Genes	Forward Primer (5'-3')	Reverse Primer (5'-3')
Bcl_2	5'-CTTTTGCTGTGGGGTTTTGT-3'	5'-GTCATTCTGGCCTCTCTTGC-3'
Bax	5'-GGAGCTGCAGAGGATGATTG-3'	5'-CCTCCCAGAAAAATGCCATA-3'
GAPDH	5'-GAAGGTGAAGGTCGGAGT-3'	5'-GAAGATGGTGATGGGATTTC-3'

The relative gene expression was calculated according to the given equation:

$$\text{Relative expression values} = 2^{(Ct\ GAPDH\ -\ Ct\ target\ gene)} \qquad (2)$$

where Ct = Threshold level cycle [28].

2.12. Intracellular ROS Generation by PTX-chit-SeNPs Compared with Biosynthesized SeNPs

For the flow cytometer analysis, HeLa cells (1×10^5) were cultured in 6-well plates for 24 h at 37 °C, after the incubation period, they were exposed to IC_{50} dose values for PTX-chit-SeNPs and biosynthesized SeNPs under similar conditions. The cells were trypsinized, washed with 1x PBS, and the resulting pellet was resuspended in PBS containing 10 µM DCFDA solution, followed by an incubation period of 30 min. The treated cells were analyzed using a flow cytometer and evaluated using CytExpert software.

2.13. Statistical Analysis

The experimental runs were made in triplicate, and the values were analyzed using the ANOVA statistical method. The results were statistically significant when the p-value was <0.05 and was expressed as the mean ± standard error.

3. Result and Discussion

3.1. Encapsulation Efficiency

The encapsulation efficiency was performed using UPLC analysis; it was observed that at 0.5 min, the absorbance peak for the standard PTX was at 0.393, which was almost similar to PTX-SeNPs-chit at 0.403, as shown in Figure 1. Therefore, the result signifies that there was a visible coating of the drug on the nanosystem. It can be assumed that the mechanism behind the encapsulation is that a hydrophobic drug, such as paclitaxel, interacts with a polymer, such as a chitosan, via hydrophobic–hydrophilic interactions, thereby increasing its solubility. The PTX molecule is composed of a hydrophobic region and a moderately hydrophilic region containing secondary amine and hydroxyl groups that can produce hydrogen bonds with chitosan molecules [20,29]. The encapsulation efficiency was calculated to be approximately 80%, according to the given equation.

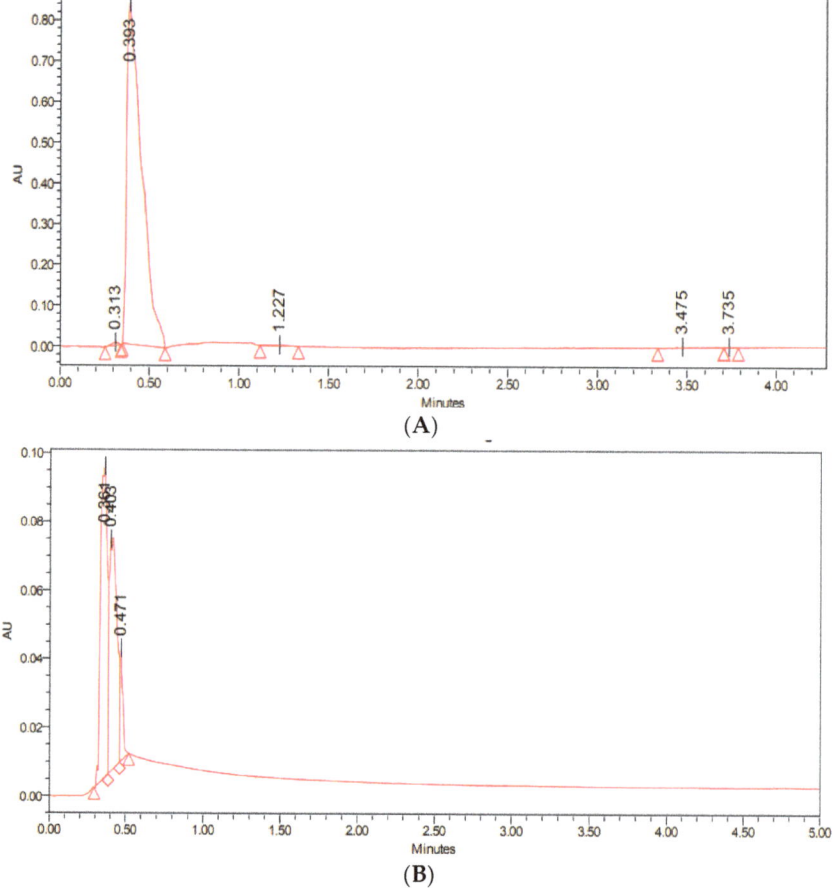

Figure 1. UPLC analysis of (**A**) standard drug paclitaxel, and (**B**) PTX-SeNPs-chit.

3.2. Characterization of PTX-chit-SeNPs

The morphology and surface topography of the nanoparticles show a significant role in transportation across the cell membranes. These particles deliver the active therapeutic agent in a controlled and target-specific manner. According to the HR-TEM analysis in Figure 2A, the biosynthesized SeNPs are dispersed and spherical with an average size range of 98 nm. In the case of chitosan-modified SeNPs in Figure 2B; the chitosan moiety surrounded the spherical-shaped nanoparticles and the average size range was calculated to be 142.8 nm. While the size range for PTX-chit-SeNPs was 167 nm with spherical shape SeNPs coated with drug molecule and modified with chitosan (indicated through arrows) as shown in Figure 2C. The dark spherical form is the SeNPs, while the smaller round particles are PTX, which is covered with chitosan on the exterior. Similar results were observed for Siqi Zeng et al. [30] and Kaikai Bai et al. [31]. The spherical morphology of NPs can enable cellular internalisation. The smaller size range is possible due to either ionic or hydrophobic interactions between the polymer and the drug.

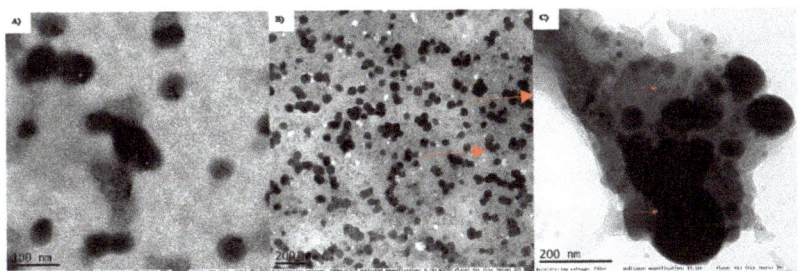

Figure 2. TEM analysis for (**A**) biosynthesized SeNPs using seed extract, (**B**) SeNPs modified with chitosan, (**C**) PTX (small rounds)-chit (covered as background)-SeNPs (larger spherical).

The major functional groups involved in the stabilization or reduction of SeNPs were —OH or carboxylic acids stretching, at 3294.78 cm^{-1}; C=C with medium intensity, at 1628.73 cm^{-1}; secondary alcohol, at 118.93 cm^{-1}; and halogen compounds, at 717.8 cm^{-1}, as shown in Figure 3D. The groups assigned for chitosan include amine, at 3356.92 cm^{-1}; alkane, at 2886.43 cm^{-1}; C=O stretching, at 1660.85 cm^{-1}; aromatic, at 1591.10 cm^{-1}; and strong C-O, at 1035.58 cm^{-1}, as shown in Figure 3C. For PTX, the functional groups involved at 3474.15 cm^{-1} were —OH and —NH; at 3001.96 cm^{-1}, alkenes or aromatic groups; at 2908.61 cm^{-1}, alkanes —CH or —CO groups; at 1440.56 cm^{-1}, aromatic groups; at 1309.91 cm^{-1}, —CN; at 1034.62 cm^{-1}, —CO; at 948.75 cm^{-1}, —C-C; at 696.17 cm^{-1}, alkene or aromatic; and at 511.86 cm^{-1}, halogen compounds, as shown in Figure 3B. For PTX-chit-SeNPs, the functional groups assigned include 3319.27 cm^{-1} for —OH (alcohol or carboxylic acid) with strong and broad intensity, 2928.62 cm^{-1} and 2869.78 cm^{-1} with medium intensity for the alkane group, 1557.71 cm^{-1} for diketones, 1404.06 cm^{-1} for CH$_2$ bending and CH$_3$ deformation, 1016.32 cm^{-1} for —C-O stretching for alcohol or phenol groups, and 739.6 cm^{-1} for halogen compounds, as shown in Figure 3A. The FTIR analysis reports suggest that peaks distinctly available in PTX-chit-SeNPs are mainly due to the coating of individual components.

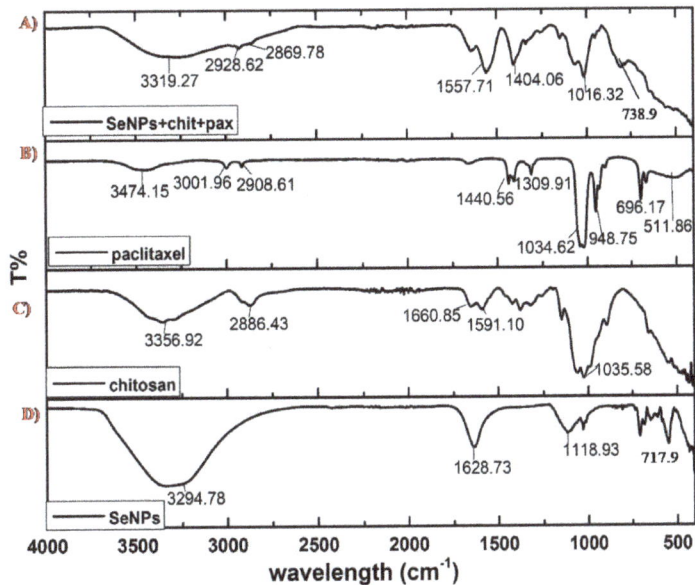

Figure 3. FTIR analysis of (**A**) SeNPs+chit+PTX, (**B**) PTX, (**C**) chitosan, and (**D**) biosynthesized.

SeNPs

The EDAX analysis showed elements such as Se, Na, Cl, C, and O. The presence of Cl helped maintain the 3D integrity of the spherical structure, which was present at the end of Se chains and allowed the interaction between the chains. The presence of C and O was due to the polymer structure of the chitosan with maximum atomic weight. Se at peak 1.3 keV, Na at 1.04 keV, Cl at 2.621 keV, C at 0.277 keV, and O at 0.525 keV were in accordance with the energy table for EDS analysis (JEOL certificated), as shown in Figure 4.

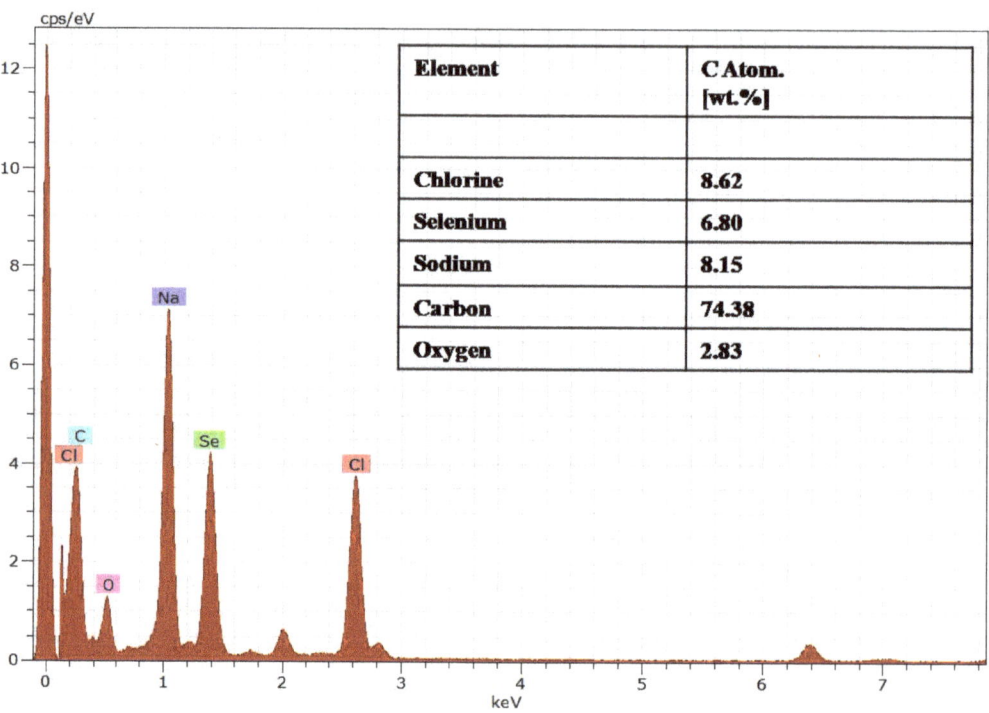

Figure 4. Energy diffraction spectroscopy (EDAX) of the PTX-chit-SeNPs.

Zeta potential offers information on the colloidal stability (> +30 mV or < −30 mV) and the electrostatic potential of particles in the reaction solution. The surface charge potential of the biosynthesized SeNPs was −23.4 mV. Still, the intensity changed with chitosan modification with +53.8 mV and the intensity further changed to +65.2 mV for PTX-loaded chitosan-modified SeNPs, as shown in Figure 5A–C. The interactions were mainly due to the ionic interactions between the positively charged PTX and negatively charged SeNPs. The hydrophobic–hydrophilic interaction between the chitosan and drug are mainly responsible for effective coating [32]. The positive charge is observed due to the protonated R-NH$_3$+ form of the chitosan, which helps in the electrostatic interface of the R-NH3+ and the negative charge present on the cellular membrane or mucosal surfaces. Thereby, this interaction helps in the reversible structural rearrangement of the proteins of tight junctions, which benefits in the opening of these junctions and consequently assists in endocytosis [33].

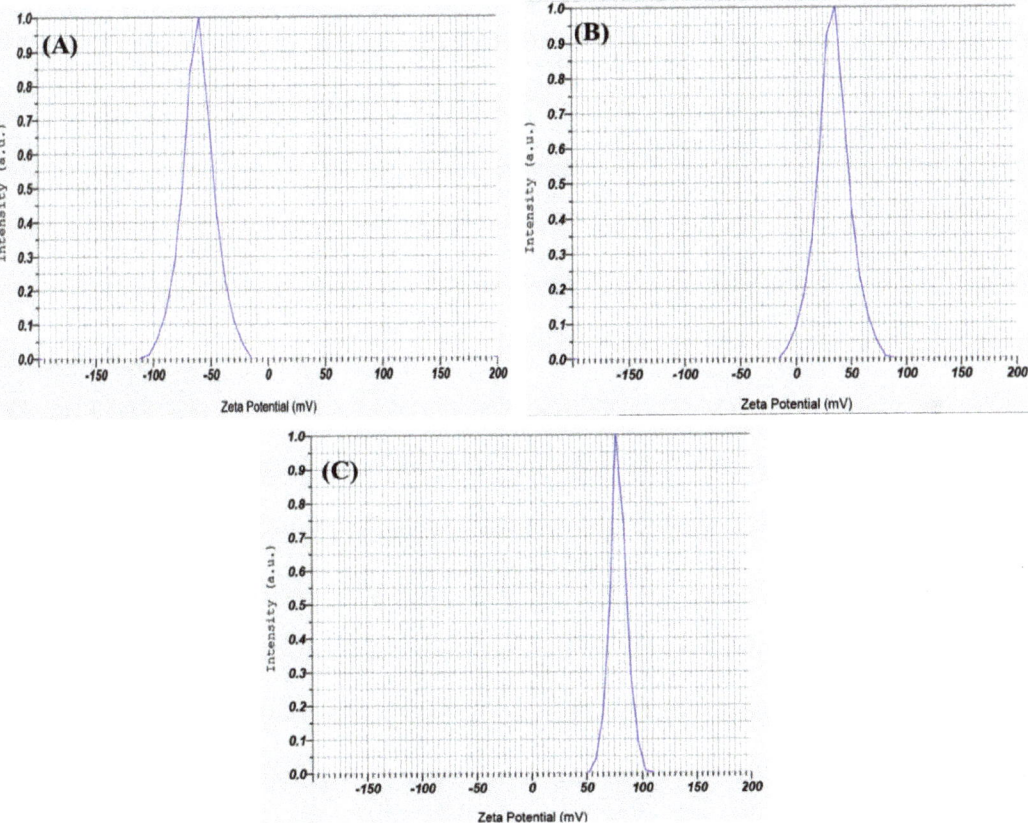

Figure 5. Zeta potential for (**A**) biosynthesized SeNPs using seed extract, (**B**) SeNPs modified with chitosan, and (**C**) PTX-loaded chitosan-modified SeNPs.

The DLS analysis of the biosynthesized SeNPs confirmed that they are polydispersed with a hydrodynamic diameter of 101 nm. The hydrodynamic diameter of chitosan-modified SeNPs and PTX-loaded chitosan-modified SeNPs were found to be 156 nm and 172 nm, respectively which is almost according to TEM data (Figure 6). The average PI (polydispersion index) value of 0.33 was less than the standard 0.5, as shown in Figure 6A–C. The increase in particle size can be assumed to be due to the presence of electrostatic repulsion of the protonated amino groups and interchain hydrogen bonding attractions [34]. They remained in equilibrium at certain conditions, but with increasing chitosan concentration, there was a partial growth in the intermolecular cross-linking, thereby producing larger particles. The intermolecular cross-linking was reduced at decreased chitosan concentration, producing smaller particles [18].

3.3. In Vitro Drug Release Analysis

The release pattern of PTX was examined under pH conditions of 5.5 and 7.4 at 37 °C for 72 h in PBS solution. According to Figure 7, the release graph was dependent upon the pH of the medium. In the 5 h there was an increase in the release at pH 5.5 which continued to reach a maximum level of 92% for 72 h, while the maximum level remained 20% at pH 7. These results prove that the release of PTX can be at a maximum level when it is exposed to an acidic environment when compared to neutral pH. Through the process of endocytosis, the nanoparticles can easily be exposed to the acidic microenvironment and initiate a rapid release of the drug from the nanosystem, which can ultimately enhance the cytotoxic effect on cancer cells [35,36]. Therefore, the toxicity against normal cells can be minimized with negligible releases at physiological pH of 7.4. Similar results were observed when PTX was encapsulated in chitin nanoparticles, in which slow release of the drug was observed at neutral pH [37]. The effective release of PTX from the chitosan matrix at acidic conditions can also be explained as being due to the protonation of the amine groups, which causes a hydrophilization of the hydrophobic core or swelling of chitosan [38], and this eventually leads to the release of the drug [39].

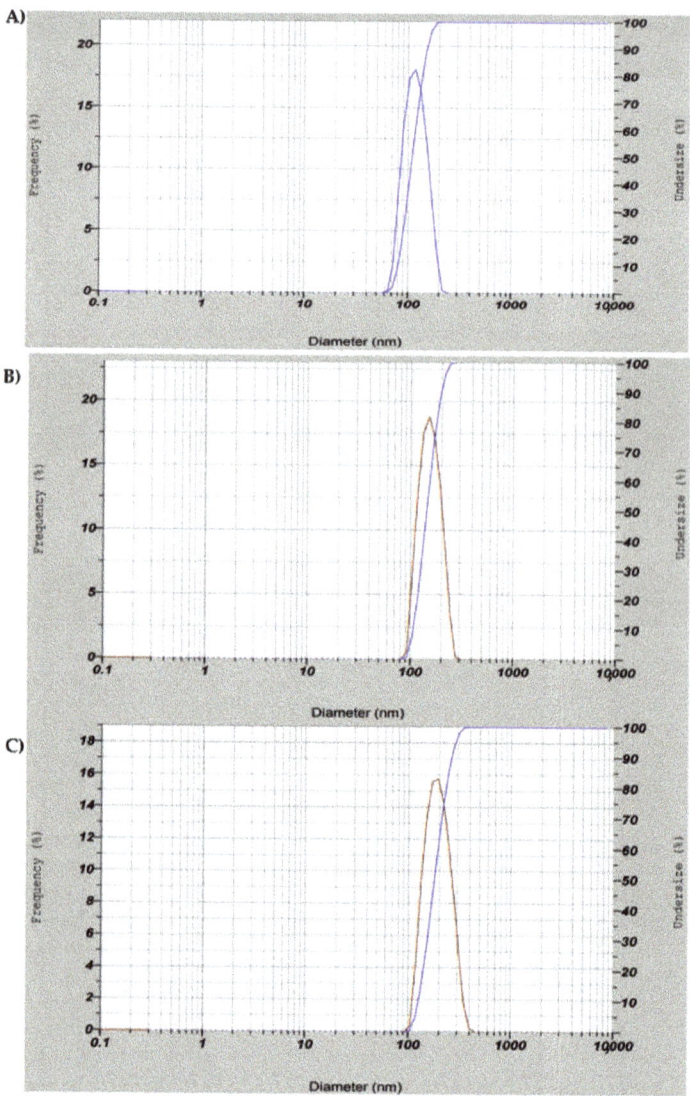

Figure 6. DLS analysis for (**A**) biosynthesized SeNPs using seed extract, (**B**) SeNPs modified with chitosan (**C**) PTX-loaded chitosan-modified SeNPs.

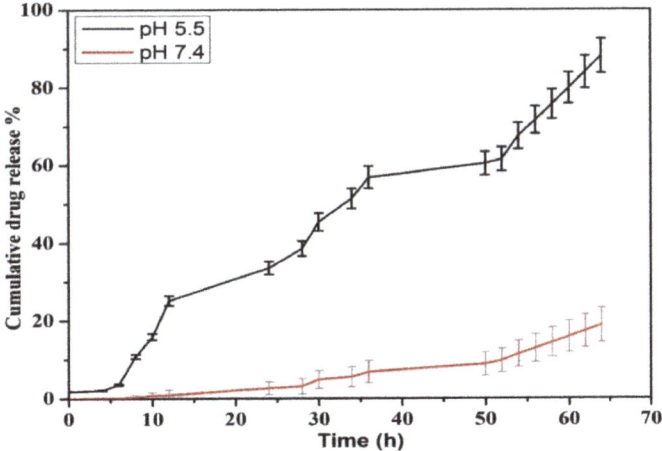

Figure 7. Drug release kinetics for PTX released from PTX-chit-SeNPs.

3.4. In Vitro Cytotoxicity Analysis against HeLa Cell Line

The viability analysis was conducted for biosynthesized SeNPs, chitosan modified SeNPs, and PTX-chit-SeNPs against HeLa cell lines at concentrations within 5–100 µg/mL, as shown in Figures 8 and 9A–D. The arrows indicate the fragmented apoptotic cells when exposed to the treatment nanosystems compared with control cells. There was a dose-dependent decrease in the inhibition rate at 24 h of incubation of treatment. The IC_{50} value considered for biosynthesized SeNPs, chitosan-modified SeNPs, and PTX-chit-SeNPs are 60 µg/mL, 45 µg/mL, and 30 µg/mL, and the viability rate decreased to approximately 20% for PTX-chit-SeNPs. The inhibition rate can be attributed to PTX assisting in the polymerization of microtubules [22], while the polymeric coating by chitosan enhances the cellular uptake of the hydrophobic drug via endocytosis [33], and the presence of SeNPs helps in the pro-oxidant activity, which helps in the generation of ROS, causing cell death [40]. The chitosan-modified SeNPs showed a similar effect on cytotoxicity, but it improved with the conjugation of PTX. Similar results were reported by T. Mary et al. [10,41] and J. Zang et al. [25].

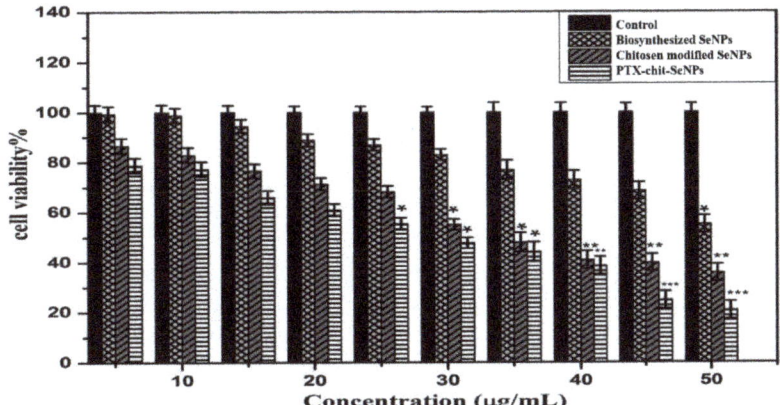

Figure 8. MTT analysis for biosynthesized SeNPs, chitosan-modified SeNPs, and PTX-chit-SeNPs against HeLa cell lines. Data are represented as mean ± SD for experiment in triplicate * indicates $p < 0.05$ versus control, ** indicates $p < 0.01$, and *** indicates $p < 0.001$.

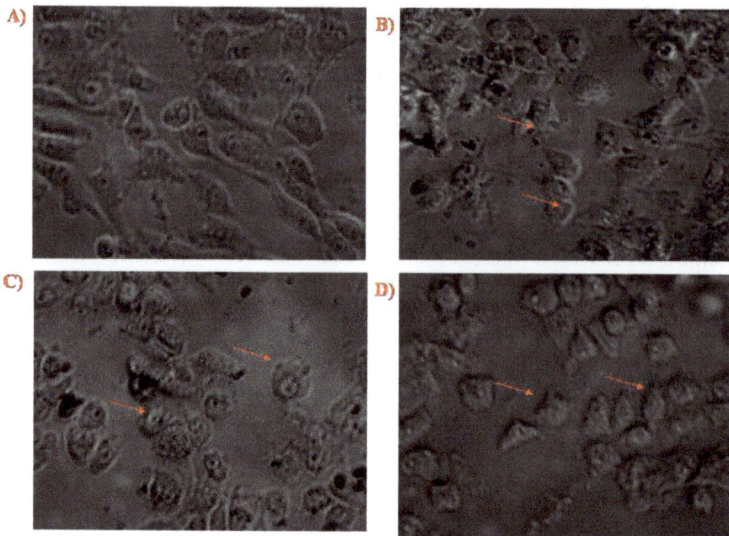

Figure 9. (**A**) Control cells, (**B**) cells treated with biosynthesized SeNPs, (**C**) cells treated with chitosan-modified SeNPs, and (**D**) cells treated with PTX-chit-SeNPs.

3.5. Apoptosis Analysis for PTX-chit-SeNPs against HeLa Cell Lines

PTX stimulates the polymerization of the microtubule by binding to the β-subunit of the α/β-tubulin dimerin, which eventually restrained the growth of progressively dividing cells that result in programmed cell death or apoptosis [22]. Annexin V is a Ca^{2+}- dependent protein with an affinity for the phospholipid phosphatidyl-serine, which forms cellular membranes. Dead cells undergo a translocation of this lining outward, which helps bind Annexin V protein labelled with FITC. PI stain distinguishes between viable and non-viable cells, and the stain can easily permeate through the plasma membrane of dead cells. The cell populations were differentiated as live, early, late, and necrotic or dead to quantify the rate of apoptosis [30,42]. HeLa cells were quantified for apoptosis analysis using PI and Annexin V-FITC dual staining to confirm this phenomenon. The apoptotic rate depended on the treatment dosage, as shown in Figure 10A–C. The early-stage and late-stage control cells with dual staining were observed to have cell populations of 8.77% and 0.05%, respectively. For the biosynthesized SeNPs at concentration 60 µg/mL, the population of cells was observed to be 29.35% at the early stage, and 10.90% at the later stage. The early and late stages for HeLa cells, when exposed to PTX-chit-SeNPs, were 47.61% and 6.50%, respectively. Therefore, the anticancer drug paclitaxel modified with chitosan on the biosynthesized SeNPs exhibited better ability to stimulate apoptosis in HeLa cells compared to the biosynthesized SeNPs without any surface modifications. The apoptotic characteristics such as nuclear condensation, chromatin condensation, or development of apoptotic bodies [41].

3.6. Cell Cycle Analysis of PTX-Chit-SeNPs Compared with Biosynthesized SeNPs

The cell cycle distribution was evaluated using CytExpert 2.3 software. DNA histograms were selected for analysing the population of cells in each phase, such SubG1, G0/G1, S, and G2/M. The apoptotic cell population containing only DNA content was measured by quantifying the sub-G1 peak, and 100,000 events were selected for each experiment, per sample. As observed in Figure 11, the data suggests that when the permeabilized cells were exposed to PTX-chit-SeNPs for 24 h, they showed a significant increase in the cell population in sub-G1 phase from 6.89% for control, 29.40% for biosynthesized SeNPs at 60 µg/mL concentration, and 49.35% for PTX-chit-SeNPs at 30 µg/mL concentration.

The results suggest that PTX-chit-SeNPs induced cell death due to activation of apoptosis. Similar results were observed in [19].

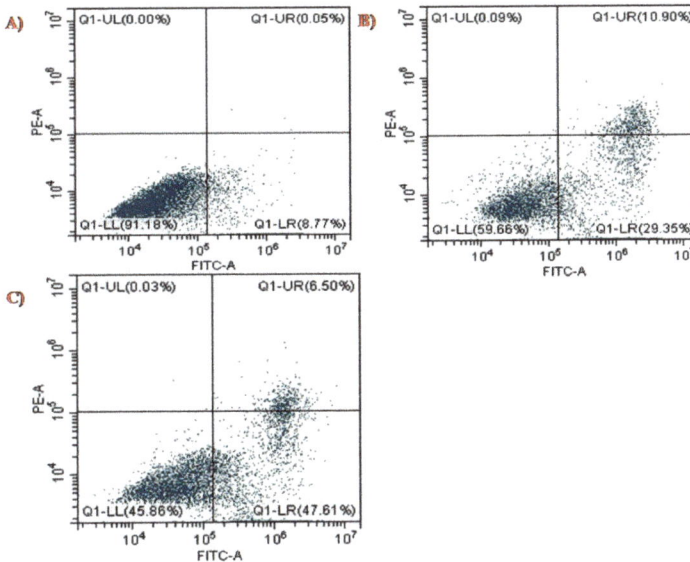

Figure 10. Apoptosis analysis of (**A**) control with dual staining of FITC+PI, (**B**) biosynthesized SeNPs at 60 μg/mL concentration, and (**C**) PTX-chit-SeNPs at 30 μg/mL concentration against HeLa cell lines.

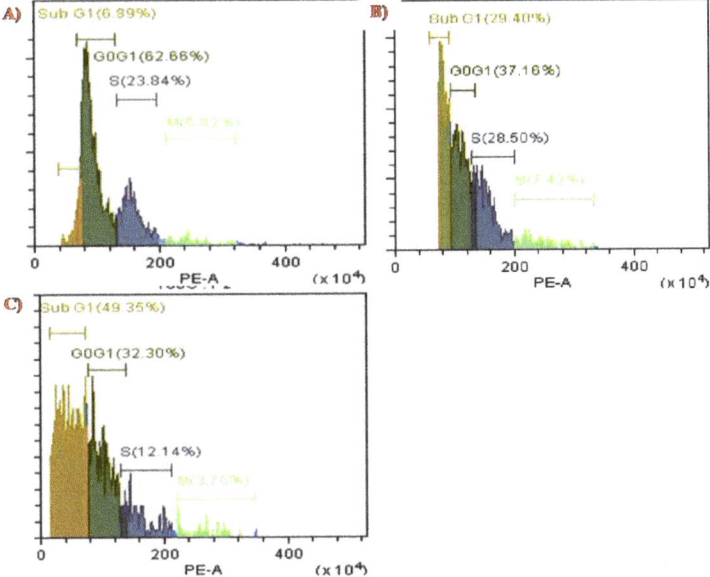

Figure 11. Cell cycle analysis of (**A**) control with PI stain, (**B**) biosynthesized SeNPs at 60 μg/mL concentration, and (**C**) PTX-chit-SeNPs at 30 μg/mL concentration.

3.7. Intracellular Investigation of ROS Analysis

The pro-oxidant effect and toxicity of the nanosystem against HeLa cells were examined when the cells were exposed to the nanosystem. The principle behind the nanosystem's mechanism is that, when the cells are exposed to it antioxidant enzymes, such as phospholipid hydroperoxide, glutathione peroxidase, thioredoxin, etc. are induced to produce superoxide radicals which cause oxidative stress to the cells and may lead to cell death as a consequence of events such as DNA damage, cell cycle arrest, apoptosis, activation of caspase proteins, and so on [43,44]. According to Figure 12A, the control cells were treated with DCDFA and demonstrated a higher number of live cells (up to 99.81%), the biosynthesized SeNPs in Figure 12B were evaluated and demonstrated 41.63% live cells and 55.80% dead cells at concentration 60 µg/mL. The PTX-chit-SeNPs in Figure 12C, when exposed, demonstrated 27.25% live cells and 71.73% dead cells. The surface-decorated SeNPs with anticancer drug and polymer showed better results than green-synthesized SeNPs mediated through aqueous seed extract. Therefore, this combination can be utilized in the treatment of various diseases caused by oxidative stress, such as Leigh syndrome, cancer, diabetes, skin disease, and others [45].

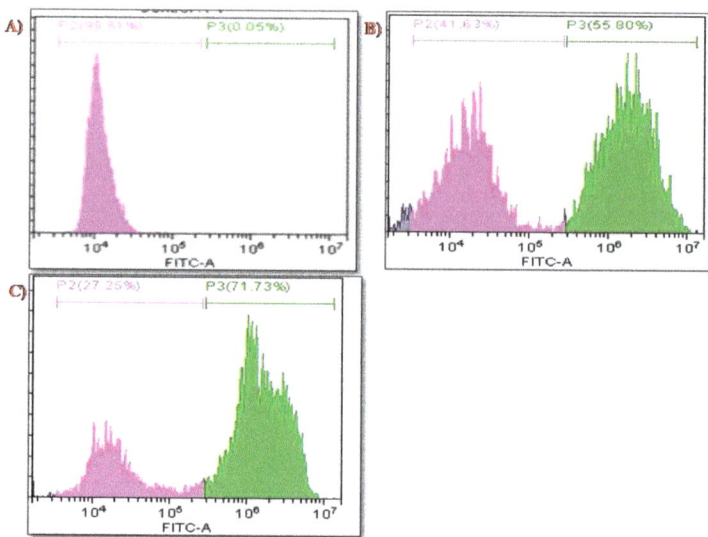

Figure 12. ROS analysis for (**A**) control cell population, (**B**) biosynthesized SeNPs, (**C**) PTX-chit-SeNPs-treated population.

3.8. Comparative Gene Expression Analysis of Biosynthesized SeNPs and PTX-chit SeNPs against Apoptotic Genes

The RT-PCR technique was used for analysing the gene expression of BAX and BCL-2 genes at a dosage of 60 µg/mL for biosynthesized SeNPs, and then at 30 µg/mL for PTX-chit-SeNPs, with a treatment period of 24 h. According to Figure 13, the PTX-chit-SeNPs significantly down-regulated the expression of BCL-2 and up-regulated the expression of BAX when compared to biosynthesized SeNPs. The treatment of PTX-chit-SeNPs is a predisposed mechanism in the apoptotic induction of chemoresistant cervical cells. The Bcl-2 protein family consists of both pro-apoptotic (Bax and Bak) and anti-apoptotic proteins (Bcl-2, Bcl-XL). The anti-apoptotic proteins inhibit the release of cytochrome c from mitochondria into the cytoplasm. In contrast, the pro-apoptotic proteins help release cytochrome c, activating the apoptotic cascade, and thereby inducing cell death. The pro-apoptotic protein BAX is one of the members of the BCL-2 family of apoptotic signalling proteins and an increase in the level of BAX has led to the enhancement of the rate of apoptosis [10,28].

In our investigation, the treatment with PTX-chit-SeNPs at a concentration of 30 μg/mL instigated apoptosis by increasing the expression in the Bax gene in HeLa cells after the incubation period of 24 h. Similar results were reported in [46].

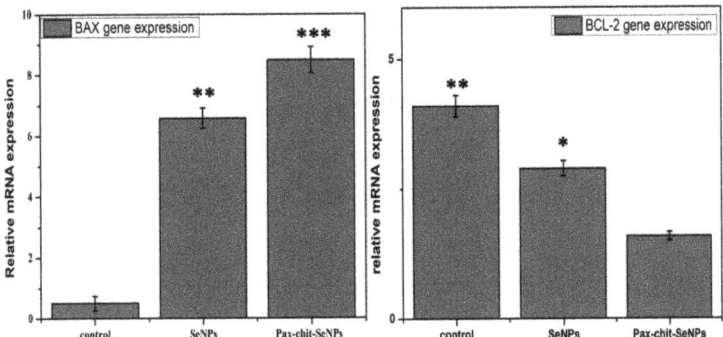

Figure 13. BAX and Bcl-2 gene expression normalized with GAPDH when treated with biosynthesized SeNPs at 60 μg/mL, and PTX-chit-SeNPs at 30 μg/mL. Data are represented as mean ± SD for experiment in triplicate; * indicates $p < 0.05$ versus control, ** indicates $p < 0.01$, *** indicates $p < 0.001$.

4. Conclusions

The encapsulation of anticancer drugs such as paclitaxel into green-synthesized selenium nanoparticles modified with a hydrophilic biocompatible polymer, such as chitosan, and its synergistic effects against cervical cancer was demonstrated. The nanoformulation was characterized using various microscopic and spectroscopic approaches, with an average size of 170 nm. The successful loading of a hydrophobic anticancer drug, such as paclitaxel, along with targeted drug release analysis at pH 5.5, mimicking an acidic microenvironment, was also demonstrated. The mechanistic in vitro study analysis of PTX-chit-SeNPs was performed at a concentration of 30 μg/mL against HeLa cell lines. Based on these results, the synthesized nanosystem can be considered as a biocompatible, environment-friendly, and cost-effective anticancer therapeutic agent developed for cancer diagnostics. It can be used to improve human health.

Author Contributions: Conceptualization, S.M., S.J., P.S., V.K.S., A.C. and N.B.; methodology, S.M., S.J., K.K.Y., P.S. and M.I.; validation, S.M. and P.S.; formal analysis, S.M., K.K.Y. and M.I.; investigation, S.J., K.K.Y., P.S. and M.I.; resources, V.K.S., A.C. and N.B.; data curation, K.K.Y., P.S. and M.I.; writing—original draft preparation, S.M., S.J. and P.S.; writing—review and editing, V.K.S., A.C. and N.B.; visualization, V.K.S., A.C. and N.B.; supervision, V.K.S., A.C. and N.B.; project administration, V.K.S. and N.B. All authors have read and agreed to the published version of the manuscript.

Funding: This research received no external funding.

Institutional Review Board Statement: Not applicable. No human or animal studies were conducted.

Informed Consent Statement: Not applicable.

Data Availability Statement: Not applicable.

Acknowledgments: The authors did not face any disagreement while doing the work and would like to thank the Vellore Institute of Technology for the encouragement, seed fund, and support bestowed upon us.

Conflicts of Interest: The authors declare no conflict of interest.

References

1. Chowdhury, S.; Yusof, F.; Salim, W.W.A.W.; Sulaiman, N.; Faruck, M.O. An overview of drug delivery vehicles for cancer treatment: Nanocarriers and nanoparticles including photovoltaic nanoparticles. *J. Photochem. Photobiol. B Biol.* **2016**, *164*, 151–159. [CrossRef] [PubMed]
2. Liu, P.; Zhang, R.; Pei, M. Design of pH/reduction dual-responsive nanoparticles as drug delivery system for DOX: Modulating controlled release behavior with bimodal drug-loading. *Colloids Surf. B Biointerfaces* **2017**, *160*, 455–461. [CrossRef] [PubMed]
3. Kumari, S.; Ram, B.; Kumar, D.; Ranote, S.; Chauhan, G.S. Nanoparticles of oxidized-cellulose synthesized by green method. *Mater. Sci. Energy Technol.* **2018**, *1*, 22–28. [CrossRef]
4. Mičová, J.; Buryi, M.; Šimek, D.; Drahokoupil, J.; Neykova, N.; Chang, Y.-Y.; Remeš, Z.; Pop-Georgievski, O.; Svoboda, J.; Im, C. Synthesis of zinc oxide nanostructures and comparison of their crystal quality. *Appl. Surf. Sci.* **2018**, *461*, 190–195. [CrossRef]
5. Raj, R.K. β-Sitosterol-assisted silver nanoparticles activates Nrf2 and triggers mitochondrial apoptosis via oxidative stress in human hepatocellular cancer cell line. *J. Biomed. Mater. Res. Part A* **2020**, *108*, 1899–1908. [CrossRef]
6. Vairavel, M.; Devaraj, E.; Shanmugam, R. An eco-friendly synthesis of Enterococcus sp.–mediated gold nanoparticle induces cytotoxicity in human colorectal cancer cells. *Environ. Sci. Pollut. Res.* **2020**, *27*, 8166–8175. [CrossRef]
7. Chellapa, L.R.; Shanmugam, R.; Indiran, M.A.; Samuel, S.R. Biogenic nanoselenium synthesis, its antimicrobial, antioxidant activity and toxicity. *Bioinspired Biomim. Nanobiomater.* **2020**, *9*, 184–189. [CrossRef]
8. Singh, K.R.; Nayak, V.; Singh, J.; Singh, A.K.; Singh, R.P. Potentialities of bioinspired metal and metal oxide nanoparticles in biomedical sciences. *RSC Adv.* **2021**, *11*, 24722–24746. [CrossRef]
9. Chaudhary, S.; Umar, A.; Mehta, S. Surface functionalized selenium nanoparticles for biomedical applications. *J. Biomed. Nanotechnol.* **2014**, *10*, 3004–3042. [CrossRef]
10. Mary, T.A.; Shanthi, K.; Vimala, K.; Soundarapandian, K. PEG functionalized selenium nanoparticles as a carrier of crocin to achieve anticancer synergism. *RSC Adv.* **2016**, *6*, 22936–22949. [CrossRef]
11. Kirupagaran, R.; Saritha, A.; Bhuvaneswari, S. Green synthesis of selenium nanoparticles from leaf and stem extract of leucas lavandulifolia sm. and their application. *J. Nanosci. Technol.* **2016**, *2*, 224–226.
12. Yanhua, W.; Hao, H.; Li, Y.; Zhang, S. Selenium-substituted hydroxyapatite nanoparticles and their in vivo antitumor effect on hepatocellular carcinoma. *Colloids Surf. B Biointerfaces* **2016**, *140*, 297–306. [CrossRef] [PubMed]
13. Kumari, M.; Purohit, M.P.; Patnaik, S.; Shukla, Y.; Kumar, P.; Gupta, K.C. Curcumin loaded selenium nanoparticles synergize the anticancer potential of doxorubicin contained in self-assembled, cell receptor targeted nanoparticles. *Eur. J. Pharm. Biopharm.* **2018**, *130*, 185–199. [CrossRef] [PubMed]
14. Srivastava, P.; Braganca, J.M.; Kowshik, M. In vivo synthesis of selenium nanoparticles by Halococcus salifodinae BK18 and their anti-proliferative properties against HeLa cell line. *Biotechnol. Prog.* **2014**, *30*, 1480–1487. [CrossRef] [PubMed]
15. Zhao, S.; Yu, Q.; Pan, J.; Zhou, Y.; Cao, C.; Ouyang, J.-M.; Liu, J. Redox-responsive mesoporous selenium delivery of doxorubicin targets MCF-7 cells and synergistically enhances its anti-tumor activity. *Acta Biomater.* **2017**, *54*, 294–306. [CrossRef] [PubMed]
16. Estevez, H.; Garcia-Lidon, J.C.; Luque-Garcia, J.L.; Camara, C. Effects of chitosan-stabilized selenium nanoparticles on cell proliferation, apoptosis and cell cycle pattern in HepG2 cells: Comparison with other selenospecies. *Colloids Surf. B Biointerfaces* **2014**, *122*, 184–193. [CrossRef] [PubMed]
17. Rajendran, P.; Maheshwari, U.; Muthukrishnan, A.; Muthuswamy, R.; Anand, K.; Ravindran, B.; Dhanaraj, P.; Balamuralikrishnan, B.; Chang, S.W.; Chung, W.J. Myricetin: Versatile plant based flavonoid for cancer treatment by inducing cell cycle arrest and ROS–reliant mitochondria-facilitated apoptosis in A549 lung cancer cells and in silico prediction. *Mol. Cell. Biochem.* **2021**, *476*, 57–68. [CrossRef]
18. Gaur, P.K.; Puri, D.; Singh, A.P.; Kumar, N.; Rastogi, S. Optimization and Pharmacokinetic Study of Boswellic Acid–Loaded Chitosan-Guggul Gum Nanoparticles Using Box-Behnken Experimental Design. *J. Pharm. Innov.* **2021**, *17*, 485–500. [CrossRef]
19. Fang, X.; Wu, X.; Zhou, B.; Chen, X.; Chen, T.; Yang, F. Targeting selenium nanoparticles combined with baicalin to treat HBV-infected liver cancer. *RSC Adv.* **2017**, *7*, 8178–8185. [CrossRef]
20. Sathyamoorthy, N.; Magharla, D.; Chintamaneni, P.; Vankayalu, S. Optimization of paclitaxel loaded poly (ε-caprolactone) nanoparticles using Box Behnken design. *Beni-Suef Univ. J. Basic Appl. Sci.* **2017**, *6*, 362–373. [CrossRef]
21. Chowdhury, P.; Nagesh, P.K.; Hatami, E.; Wagh, S.; Dan, N.; Tripathi, M.K.; Khan, S.; Hafeez, B.B.; Meibohm, B.; Chauhan, S.C. Tannic acid-inspired paclitaxel nanoparticles for enhanced anticancer effects in breast cancer cells. *J. Colloid Interface Sci.* **2019**, *535*, 133–148. [CrossRef] [PubMed]
22. Gupta, U.; Sharma, S.; Khan, I.; Gothwal, A.; Sharma, A.K.; Singh, Y.; Chourasia, M.K.; Kumar, V. Enhanced apoptotic and anticancer potential of paclitaxel loaded biodegradable nanoparticles based on chitosan. *Int. J. Biol. Macromol.* **2017**, *98*, 810–819. [CrossRef] [PubMed]
23. He, X.; Yang, F.; Huang, X.A. Proceedings of Chemistry, Pharmacology, Pharmacokinetics and Synthesis of Biflavonoids. *Molecules* **2021**, *26*, 6088. [CrossRef] [PubMed]
24. Menon, S.; Shanmugam, V. Cytotoxicity analysis of biosynthesized selenium nanoparticles towards A549 lung cancer cell line. *J. Inorg. Organomet. Polym. Mater.* **2020**, *30*, 1852–1864. [CrossRef]
25. Zhang, Z.; Wang, X.; Li, B.; Hou, Y.; Yang, J.; Yi, L. Development of a novel morphological paclitaxel-loaded PLGA microspheres for effective cancer therapy: In vitro and in vivo evaluations. *Drug Deliv.* **2018**, *25*, 166–177. [CrossRef]

26. Luesakul, U.; Puthong, S.; Neamati, N.; Muangsin, N. pH-responsive selenium nanoparticles stabilized by folate-chitosan delivering doxorubicin for overcoming drug-resistant cancer cells. *Carbohydr. Polym.* **2018**, *181*, 841–850. [CrossRef]
27. Manivasagan, P.; Bharathiraja, S.; Bui, N.Q.; Jang, B.; Oh, Y.-O.; Lim, I.G.; Oh, J. Doxorubicin-loaded fucoidan capped gold nanoparticles for drug delivery and photoacoustic imaging. *Int. J. Biol. Macromol.* **2016**, *91*, 578–588. [CrossRef]
28. Khalilia, W.M.; Özcan, G.; Karaçam, S. Gene Expression and Pathway Analysis of Radiation-Induced Apoptosis in C-4 I Cervical Cancer Cells. *J. Autoimmun. Res.* **2017**, *4*, 1014.
29. Ye, Y.-J.; Wang, Y.; Lou, K.-Y.; Chen, Y.-Z.; Chen, R.; Gao, F. The preparation, characterization, and pharmacokinetic studies of chitosan nanoparticles loaded with paclitaxel/dimethyl-β-cyclodextrin inclusion complexes. *Int. J. Nanomed.* **2015**, *10*, 4309.
30. Zeng, S.; Ke, Y.; Liu, Y.; Shen, Y.; Zhang, L.; Li, C.; Liu, A.; Shen, L.; Hu, X.; Wu, H. Synthesis and antidiabetic properties of chitosan-stabilized selenium nanoparticles. *Colloids Surf. B Biointerfaces* **2018**, *170*, 115–121. [CrossRef]
31. Bai, K.; Hong, B.; He, J.; Hong, Z.; Tan, R. Preparation and antioxidant properties of selenium nanoparticles-loaded chitosan microspheres. *Int. J. Nanomed.* **2017**, *12*, 4527. [CrossRef] [PubMed]
32. Majedi, F.S.; Hasani-Sadrabadi, M.M.; VanDersarl, J.J.; Mokarram, N.; Hojjati-Emami, S.; Dashtimoghadam, E.; Bonakdar, S.; Shokrgozar, M.A.; Bertsch, A.; Renaud, P. On-chip fabrication of paclitaxel-loaded chitosan nanoparticles for cancer therapeutics. *Adv. Funct. Mater.* **2014**, *24*, 432–441. [CrossRef]
33. Ahmed, T.A.; Aljaeid, B.M. Preparation, characterization, and potential application of chitosan, chitosan derivatives, and chitosan metal nanoparticles in pharmaceutical drug delivery. *Drug Des. Dev. Ther.* **2016**, *10*, 483. [CrossRef] [PubMed]
34. Harish, T.; Nirmala, G.; Vimala, R. Biosorption of di-butyl phthalate from aqueous solutions using Pleurotus ostreatus: Isotherm and kinetic study. *J. Chem. Pharm. Res.* **2015**, *7*, 697–706.
35. Fu, X.; Yang, Y.; Li, X.; Lai, H.; Huang, Y.; He, L.; Zheng, W.; Chen, T. RGD peptide-conjugated selenium nanoparticles: Antiangiogenesis by suppressing VEGF-VEGFR2-ERK/AKT pathway. *Nanomed. Nanotechnol. Biol. Med.* **2016**, *12*, 1627–1639. [CrossRef] [PubMed]
36. Khurana, A.; Tekula, S.; Saifi, M.A.; Venkatesh, P.; Godugu, C. Therapeutic applications of selenium nanoparticles. *Biomed. Pharmacother.* **2019**, *111*, 802–812. [CrossRef]
37. Smitha, K.; Anitha, A.; Furuike, T.; Tamura, H.; Nair, S.V.; Jayakumar, R. In vitro evaluation of paclitaxel loaded amorphous chitin nanoparticles for colon cancer drug delivery. *Colloids Surf. B Biointerfaces* **2013**, *104*, 245–253. [CrossRef]
38. Xu, J.; Ma, L.; Liu, Y.; Xu, F.; Nie, J.; Ma, G. Design and characterization of antitumor drug paclitaxel-loaded chitosan nanoparticles by W/O emulsions. *Int. J. Biol. Macromol.* **2012**, *50*, 438–443. [CrossRef]
39. Xie, F.; Ding, R.-L.; He, W.-F.; Liu, Z.-J.-L.; Fu, S.-Z.; Wu, J.-B.; Yang, L.-L.; Lin, S.; Wen, Q.-L. In vivo antitumor effect of endostatin-loaded chitosan nanoparticles combined with paclitaxel on Lewis lung carcinoma. *Drug Deliv.* **2017**, *24*, 1410–1418. [CrossRef]
40. Drake, E. Cancer chemoprevention: Selenium as a prooxidant, not an antioxidant. *Med. Hypotheses* **2006**, *67*, 318–322. [CrossRef]
41. Liao, W.; Yu, Z.; Lin, Z.; Lei, Z.; Ning, Z.; Regenstein, J.M.; Yang, J.; Ren, J. Biofunctionalization of selenium nanoparticle with dictyophora indusiata polysaccharide and its antiproliferative activity through death-receptor and mitochondria-mediated apoptotic pathways. *Sci. Rep.* **2015**, *5*, 1–13. [CrossRef] [PubMed]
42. Liao, X.-Z.; Gao, Y.; Zhao, H.-W.; Zhou, M.; Chen, D.-L.; Tao, L.-T.; Guo, W.; Sun, L.-L.; Gu, C.-Y.; Chen, H.-R. Cordycepin reverses cisplatin resistance in non-small cell lung cancer by activating AMPK and inhibiting AKT signaling pathway. *Front. Cell Dev. Biol.* **2021**, *8*, 1640. [CrossRef] [PubMed]
43. Zonaro, E.; Lampis, S.; Turner, R.J.; Qazi, S.J.S.; Vallini, G. Biogenic selenium and tellurium nanoparticles synthesized by environmental microbial isolates efficaciously inhibit bacterial planktonic cultures and biofilms. *Front. Microbiol.* **2015**, *6*, 584. [CrossRef] [PubMed]
44. Sonkusre, P.; Cameotra, S.S. Biogenic selenium nanoparticles induce ROS-mediated necroptosis in PC-3 cancer cells through TNF activation. *J. Nanobiotechnol.* **2017**, *15*, 43. [CrossRef] [PubMed]
45. Schieber, M.; Chandel, N.S. ROS function in redox signaling and oxidative stress. *Curr. Biol.* **2014**, *24*, R453–R462. [CrossRef] [PubMed]
46. Swarnalatha, Y. Quantification of Bcl-2/Bax genes in A549 Lung Cancer Cell Lines Treated with Heptamethoxy Flavones. *Asian J. Pharm. (AJP)* **2018**, *12*. [CrossRef]

Review

Superior Properties and Biomedical Applications of Microorganism-Derived Fluorescent Quantum Dots

Mohamed Abdel-Salam [1,†], Basma Omran [2,3,†], Kathryn Whitehead [4] and Kwang-Hyun Baek [2,*]

1. Analysis and Evaluation Department, Nanotechnology Research Center, Egyptian Petroleum Research Institute (EPRI), Nasr City, Cairo PO 11727, Egypt; dabdelsalam2008@epri.sci.eg
2. Department of Biotechnology, Yeungnam University, Gyeongbuk, Gyeongsan 38541, Korea; obasma@ynu.ac.kr
3. Department of Processes Design & Development, Egyptian Petroleum Research Institute (EPRI), Nasr City, Cairo PO 11727, Egypt
4. Microbiology at Interfaces, Manchester Metropolitan University, Chester Street, Manchester M1 5GD, UK; k.a.whitehead@mmu.ac.uk
* Correspondence: khbaek@ynu.ac.kr; Tel.: +82-53-810-3029; Fax: +82-53-810-4769
† These authors contributed equally to this work.

Academic Editor: Sheshanath Bhosale
Received: 2 September 2020; Accepted: 29 September 2020; Published: 30 September 2020

Abstract: Quantum dots (QDs) are fluorescent nanocrystals with superb photo-physical properties. Applications of QDs have been exponentially increased during the past decade. They can be employed in several disciplines, including biological, optical, biomedical, engineering, and energy applications. This review highlights the structural composition and distinctive features of QDs, such as resistance to photo-bleaching, wide range of excitations, and size-dependent light emission features. Physical and chemical preparation of QDs have prominent downsides, including high costs, regeneration of hazardous byproducts, and use of external noxious chemicals for capping and stabilization purposes. To eliminate the demerits of these methods, an emphasis on the latest progress of microbial synthesis of QDs by bacteria, yeast, and fungi is introduced. Some of the biomedical applications of QDs are overviewed as well, such as tumor and microRNA detection, drug delivery, photodynamic therapy, and microbial labeling. Challenges facing the microbial fabrication of QDs are discussed with the future prospects to fully maximize the yield of QDs by elucidating the key enzymes intermediating the nucleation and growth of QDs. Exploration of the distribution and mode of action of QDs is required to promote their biomedical applications.

Keywords: nanobiotechnology; fluorescent quantum dots; microbiological synthesis; biomedical applications

1. Introduction

Semiconductor nanocrystals are innovative nano-sized materials with several beneficial properties. They are also designated as quantum rods (QRs), quantum dots (QDs), and quantum particles. QDs are defined as "almost rounded shaped nano-sized materials confined in a three-dimensional structure with a size ranging between 1–10 nm and they possess quantum properties due to Bohr radii" [1]. They are often called synthetic atoms, since they possess traits identical to that of normal atoms, such as being spatially localized in a 3D structure and possessing discrete energy levels [2]. QDs are named as such because they demonstrate a quantum confinement regime, which means that their electron wave function is related to particle size [3]. QDs were first discovered by the Russian physicist Alexey I. Ekimov in the early 1980s. Ekimov and Onushchenko investigated the quantum configuration of CuCl-immersed silicate in glass ceramics [4]. Since then, several investigations have been performed to discover a wide variety of QD-immersed glass ceramics [5–7]. Glass ceramics containing QDs have gained considerable notice because they can be successfully applied in optoelectronic devices, sensors, and photocatalysts [8]. The systematic development of QD science was motivated by Luis Brus, who derived a relationship between the size and band energy gap (BEG) of semiconductor nanoparticles [9]. The term "quantum dot" was used for the first time to characterize a three-dimensionally confined semiconductor quantum well by Brus et al. [10]. In 1993, Horset Weller was the first to identify the quantum confinement features of semiconductor QD particles [11]. QD absorption and emission spectra are usually size tunable. The difference in energy levels between both the activated and resting states of QDs is designated as the BEG of QDs. When QDs absorb fluorescent light, they become excited. The frequency or magnitude of absorbed light is related to both the BEG and the size of the QDs. To return to their original resting state, the QDs emit the same frequency of absorbed light. However, QD size is indirectly correlated to their BEG level (i.e., small-sized QDs release high-energy light, which is blue in color, whereas the large-sized QDs emit low-energy light, which is red in color) [12].

A new and reproducible method for synthesizing high-quality as well as monodispersed QDs was further developed in 1993 by Murray et al. [12]. In 1998, QDs were introduced for the first time as fluorophores owning distinctive features that make them more advantageous over conventional organic dye molecules [13]. Since that time, considerable progress has been observed to discover their potency in nanomedicine and nanobiotechnology as promising tools for use in diagnosis and therapies. It took nearly a decade since 1998 for new advances in QD research to allow the production of a successful preparation of colloidal cadmium-based chalcogenides (CdX), where X refers to sulfur (S), selenium (Se), or tellurium (Te) [14].

QDs are characterized by powerful light absorbance, symmetric and narrow emission bands, bright fluorescence with a broad range of excitation, high photo-stability, and slow excited-state decay rates [15]. QDs are usually synthesized via several top-down and bottom-up approaches. In a top-down approach, QDs are synthesized by thinning the bulk semiconductor material. Some top-down approaches employed to synthesize small-sized QDs involve reactive-ion etching, electron beam lithography, and wet chemical etching [1]. These techniques, however, retain impurities in the produced QDs, which lead to structural imperfections. Bottom-up production of QDs includes wet-chemical methods such as competitive reaction chemistry, micro-emulsion, sonic waves, microwaves, sol-gel, hot-solution decomposition, and electrochemistry [16].

Although the chemical synthesis of QDs opened up new avenues for preparing semiconductor nanocrystals with tunable shapes and sizes with desirable optical features, there are some disadvantages of the chemical preparation of QDs [17], including high energy requirements, release of hazardous organic and inorganic byproducts to the surrounding environment, and the high costs. Several organophosphorus solvents involved in the chemical synthesis of QDs can reach up to 90% of the total production costs [18]. Accordingly, the selection of solvents is an important issue affecting the exponential impacts of the chemical synthesis routes. Furthermore, instability has limited their potential applications, particularly in biomedical fields; therefore, more research is required to overcome such problems.

The implementation of eco-friendly, feasible, and biocompatible synthesis procedures is immensely desired. Synthesis of QDs by biological routes proposes cost-effective, low-toxic, and environmentally friendly alternatives to the chemical synthesis methodologies. Microorganisms are superior biological nano-factories which can be beneficial in synthesizing QDs. Biosynthesis of QDs has diverse of potent benefits which can be exploited particularly in biomedical applications. Microorganisms contain huge pools of diversity which confer them the inherent potential to mediate the synthesis of QDs. Microorganisms such as bacteria, yeast, and fungi are preferred for the synthesis of QDs due to several factors, including easy cultivation and their ability to grow at ambient conditions of pressure, temperature, and pH [19]. Additionally, due to their high adaptation and tolerating capacity to toxic metal containing environments, microorganisms possess unique intrinsic potency to mediate the synthesis of QDs by applying reduction mechanisms either extracellularly or intracellularly [20]. Microbial enzymes play the key role for transforming the precursor metal ions to their nanoparticle states.

The key advantage of bacterial-based synthesis of nanomaterials is their defense or resistance mechanism. Bacteria transform toxic metal ions to their nanoscale forms as a result of the stresses caused by the metal ions toward the bacterial cells [21]. Interaction takes place between the negatively charged bacterial cell wall and the positively charged metal ions. Nonetheless, certain restrictions occur such as the poor control over the sizes, geometries, and distributions of the prepared particles.

Fungi are more advantageous compared to bacteria in many ways involving: (i) fungal mycelia have the capacity to resist agitation, flow pressure, and many other harsh conditions in bioreactors; (ii) they are easy to handle; (iii) they can be easily monitored in downstream processing; (iv) out of microbes, they are the most potential candidates for large scale production because they secrete large amounts of enzymes leading to high formation of nanoparticles; and (v) they possess a very high cell wall binding capacity [22–24]. However, the use of fungi to mediate the mycosynthesis of nanoparticles is time-consuming and must be challenged in order to establish an economical method for up-scale production [25]. Moreover, problems such as slow reaction time, control over particle size and identification of the exact biochemical and molecular mechanisms for the microbial synthesis of QDs are among the research challenges that need extensive investigations.

An ideal fluorescent agent for biomedical applications has to possess the following criteria: (i) emissible of high fluorescence quantum yield; (ii) cyto-compatible; (iii) highly stable; (iv) reproducible; and (v) easily functionalized [13]. Traditional fluorescent probes or fluorophores used in biomedical applications involve organometallic and organic dyes such as rhodamines, fluoresceins, and cyanins. However, they have prominent defects such as broad emission spectra and narrow excitation spectra with discrete absorption bands [26]. In addition, they are susceptible to photo-bleaching, and the photochemical stability is usually poor with short fluorescence lifetime. Thermal stability and dispersibility are usually dye class-dependent. They also have limited usages particularly for multi-color detection because of spectral overlapping [27]. Fluorescence intensity and lifetime of organic dyes are influenced by viscosity, ionic strength, polarity, and pH [28]. The toxicity ranged from low to high is dye-dependent.

Contrary to traditional fluorescent probes, superb optical and physico-chemical properties are endowed to QDs because of their unique atomic configuration and particle size [15]. QDs have narrow emission spectra and broad excitation spectra with steady increasing bands [26]. They are resistant to photo-bleaching with high photo-chemical stability and also have a long-time fluorescence intensity. Thermal stability and dispersibility are dependent on the core–shell structure and ligand chemistry; QDs are ideal for multiplexing experiments because they generate high fluorescence quantum yield [29]. Toxicity can be due to the leaching of the heavy metal and inorganic element contents, therefore can be reduced by using specific biocompatible materials during the synthesis processes. Adopting suitable biocompatible materials helps to modify the QD surface, thereby minimizing the leaching of metal ions [30]. Additionally, reproducibility can be achieved by controlling the process parameters during the synthesis of QDs.

This review introduces a comprehensive outline of the basic structure of fluorescent QDs and their unique properties. The fabrication of QDs mediated by microbial machineries, e.g., bacteria, fungi, and yeast, is reviewed in depth. This review depicts the versatile biomedical applications of QDs as extremely promising tools, such as in disease detection, drug delivery, single-protein tracking, biosensors, and cellular labeling.

2. Structural Composition of QDs

The quantum confinement regime of QDs usually occurs within 1–10 nm sized nanoparticles [31]. QDs structurally consist of a semiconductor core overlapped by an external shell, which is coated with ligands. The inorganic core structure controls the optical (e.g., light absorption and radiation) and semiconducting (e.g., electrical conductivity) properties. Ligands are key factors behind the development of QD properties, such as their colloidal solubility, stability, particle morphology, and particle size distribution (Figure 1) [32]. An ideal ligand should meet the following criteria: (i) provide the QDs with high stability and solubility in biological buffers; (ii) retain a high resistance toward photo-bleaching and other photophysical characteristics in aqueous media; (iii) contain functional groups to facilitate conjugation with biomolecules; and (iv) reduce the overall hydrodynamic size [33]. The most common ligands used include carboxylic acids (-COOH), alcohols (-OH), primary amines (-NH_2), long-chain organophosphates, and thiols (-SH) [34]. The shell is mainly comprised of Type II–VI and IV–VI elements, which include configurations such as ZnO, ZnS, MgO, HgS, CdSe, and CdS [35]. The outer coatings play a vital role as physical barriers to protect the core from the surrounding medium. Representative examples of Type I core-shell materials involve (CdSe) InAs [36] as well as (CdSe) CdS and (ZnS) CdSe [37]. Inverse Type I core-shell materials are usually comprised of (CdS) CdSe [38], (CdS) HgS [39], and (ZnSe) CdSe [40]. QDs of Type II core-shell materials are primarily composed of (CdTe) CdSe [41] as well as (CdSe) ZnTe and (CdSe) ZnTe [42]. Single shell, multi-shell, and graded alloyed structure are the three major kinds of shell coatings [43]. The three forms of coatings vary mutually, according to the BEG and possible location of electrons' energy state of both QD core and shell.

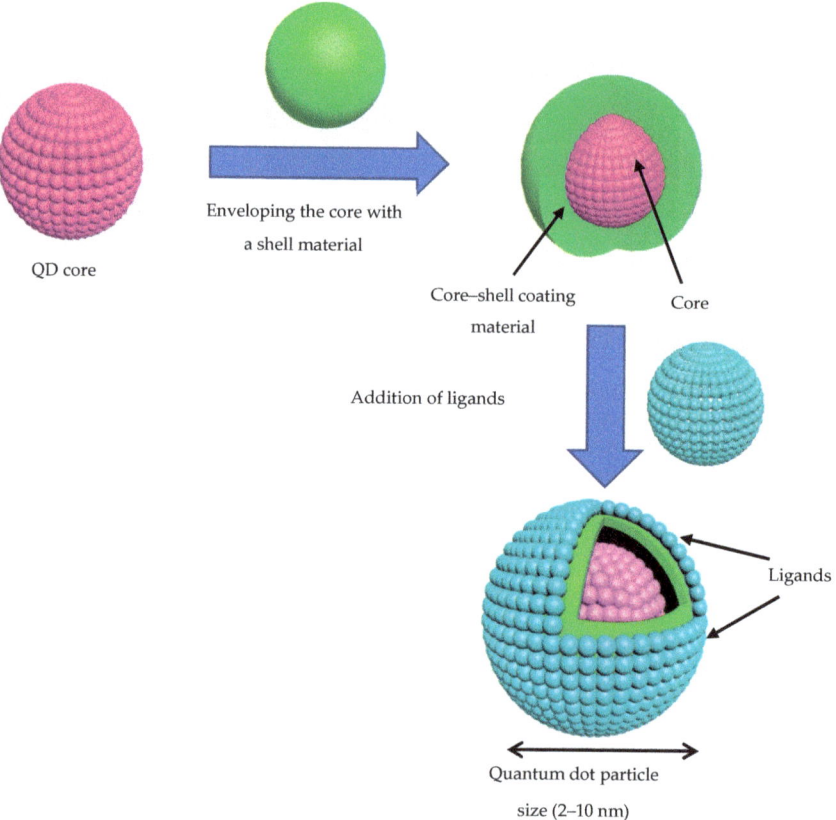

Figure 1. Quantum dot basic structure (core, shell, and ligands).

Electrons occur in a series of different energy levels, which are constant in bulk semiconductor material. At the nanoscale size, these rates become distinct because of the quantum confinement features.

After a trigger, the electron hops from the valence to conduction level via the BEG, which leads to the formation of a hole behind. This hole has a positive charge. An exciton refers to the bound electron-hole pair in semiconductor materials and the effect of quantum confinement arises from the electrons that are physically confined in the 3D structure [44]. The exciton Bohr radius describes the mean physical space between electron-hole pairs, and this space is variable across different semiconductors. When the size of a semiconductor material resembles the exciton Bohr radius, its characteristics start to resemble QDs rather than the bulk semiconductor material. When electrons become excited to the conduction level, they return to their valence position, releasing electromagnetic radiation that differs from the initial stimulus. This frequency of emission is viewed as fluorescence and is size dependent on the BEG that can be modified by adjusting the QD size and surface chemistry [45].

3. Physicochemical Properties of QDs

QDs possess outstanding physio-chemical features as a result of the quantum confinement effect [46]. When material's size is reduced, the quantum size features become prominent [47]. When the radius of nanoparticles is under the Bohr limit of excitation, the charge carrier's Bohr radius becomes greater than that of the sphere. Thus, the charging carrier's energy increases owing to containment within the sphere [48]. QD emission colors vary according to their size, chemical structure, and surface coating. The fluorescence of QDs can be adjusted over a broad wavelength (i.e., from 400 to 4000 nm). This facilitates measurements under the UV, visible, as well as near-infrared regions (NIRs) (Figure 2) [45]. Relatively large-sized QDs close to 10 nm in diameter emit red or orange emission at longer wavelengths with low-intensity radiation. In contrast, small-sized QDs close to 1 nm emit light at shorter wavelengths and possess green or blue emission with appropriate high levels of radiation intensity [35]. The emission wavelength of the core/shell CdSe/ZnS particles can be adjusted according to the particles' size [49]. The emission wavelength changes from 480 (i.e., particle size: 2 nm) to 660 nm (i.e., particle size: 8 nm). Variations of wavelength emission of CdTe/CdSe QDs have also been shown to vary from 650 to 850 nm for 4 and 8 nm sized-QD particles, respectively [50].

3.1. Blinking

QD optical features are size tunable and are regulated by the quantum confinement effects since QDs are tolerant to photo-bleaching. Photo-bleaching refers to the procedure at which the luminescent materials decompose irretrievably because of optical excitation (or light-prompted response), which leads to a reduction in fluorescence strength [51]. Since QDs are resistant to photo and chemical degradation, they serve as potential candidates for use as imaging probes over long time periods [32]. Furthermore, QDs possess blinking behavior, which takes place throughout continuous molecular excitation. Blinking is thought to occur primarily due to the photo-induced carriers trapping and de-trapping, and therefore QDs oscillate between emitting and non-emitting levels. It is worth noting that such intensity differences on short periods are referred to as "quantum jumps". Blinking takes place via two mechanisms: A-type and B-type blinking [52]. In A-type blinking, the presence of extra charges in the QDs leads to non-radiative decay rates. These charges appear due to the release of hole pairs photo-activated electrons. A temporal quenching of photoluminescence of a charged QDs occurs due to the recombination process between the excited hole pair electrons and the spare charges. However, no such correlation takes place in B-type blinking. To inhibit the blinking phenomenon in QDs, the charge carrier trapping process and trapping sites must be eliminated. Charge carrier trapping hinders charge transfer and recombination in QDs, hence limiting their efficiency [53]. A huge diversity of charge trapping sites has been speculated, experimentally measured, and analyzed via different theories. These states might be related to the surrounding ligands and medium. From the optical point of view, these states could be recognized in QDs via the incidence of broad and red-shifted emission peaks because of the rearrangement of the trapped charges. Accordingly, trapping minimizes the charge transfer in QD arrays. Further, withdrawing charges to a trap state reduces the overlapping of electron and hole wave functions inside the QDs and minimizes the recombination efficacy [53].

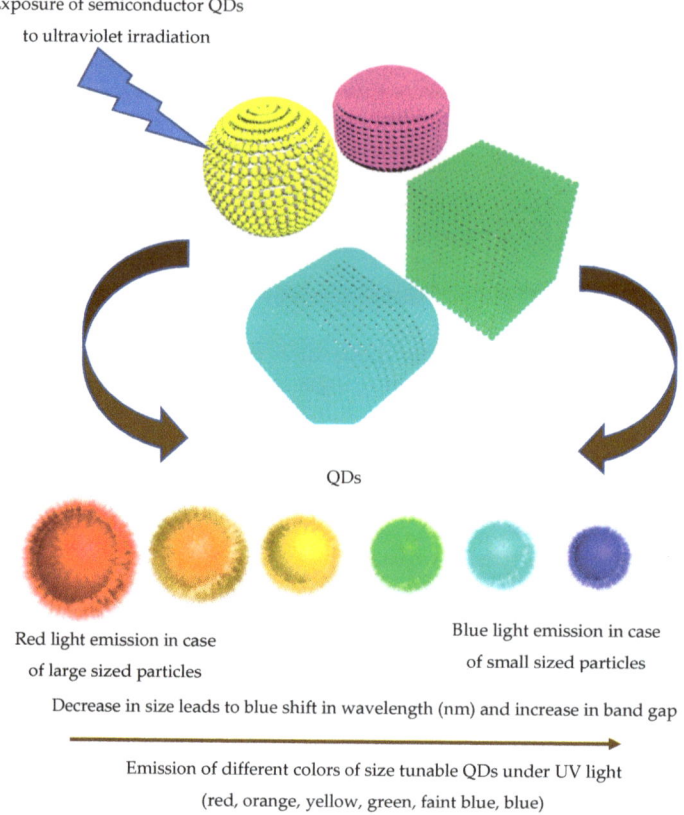

Figure 2. Quantum size confinement effect; irradiation of colloidal quantum dot (QD) particles under a UV light. Emission of different colors is dependent on the shape and size of the prepared particles.

Several studies have demonstrated that, by increasing shell thickness, electronic insulation will take place, leading to the prevention of ionization [52]. Highly crystallographic structure of CdSe/CdS QDs with dense crystalline shells (i.e., made from CdS) have displayed a complete absence of blinking phenomenon, which was ascribed to the shell thickness [54]. Therefore, it was proposed that QDs that have a thick crystalline shell (e.g., ZnSe, CdS, and ZnS) do not blink [55]. Accordingly, the shell materials should be carefully chosen depending on the core of the prepared QD particles. It has been noted that tuning thicker shells made up of 4–6 monolayers offers more protection for the core material against degradation and photo-oxidation. This, in turn, reduces the lattice tightness between both the core and shell and increases the photophysical features of the QDs.

3.2. Stokes Shift

In the middle of the nineteenth century, and particularly in 1852, G.G. Stokes thoroughly investigated the fluorescence phenomenon and inspected how this phenomenon differs from that of incident light. In 1852, Stokes proposed his conclusions to London Royal Society [56]. His findings illustrated that there is one theory related to fluorescence namely, refrangibility of light (i.e., the degree of refraction is indirectly proportional to wavelength), which is shifted by dispersion to longer wavelengths [57]. This became known as Stokes' law, which showed that fluorescence irradiation occurred at a higher wavelength than that of incident light. This law was accordingly named the

Stokes shift in his honor. The Stokes shift is a significant feature of fluorophores that demonstrates the energy difference between both the absorption and emission states [58]. The Stokes shift is defined as "the variance between absorption and emission maximum bands" [59]. A large Stokes shift is commonly observed in semiconductor QDs and is an important feature used to identify their optical properties. Determining the Stokes shift is advantageous since, by using such criteria, fluorescence detection becomes easier, even at very low signal intensities [60]. The fluorophore is propelled to the excited levels (S_n) after absorption, then quick relaxation takes place to the previous excited level (S_1, internal conversion), and then returns to the original ground state (S_0, fluorescence). The loss of excited energy during the internal conversion leads to red-shifted emissions (as opposed to absorption) and the incidence of Stoke shift. Small Stokes shifts (e.g., less than 50 nm) are observed in fluorescent proteins and organic dyes by photoluminescence spectroscopy, which cause an overlay between the absorption and emission spectra. Accordingly, the fluorophore absorbs the fluorescence emission in a process referred to as "self-absorption". Self-absorption is defined as "a fluorescence quenching mechanism of which dramatically affects the optical features of fluorophores" [61]. Self-absorption causes problems, including aggregation-caused quenching and concentration quenching of fluorophores. This, in turn, can impede their applications in optoelectronic devices as well as their biological applications [62]. Therefore, several studies aimed to increase the Stokes shift of QDs. Among the applied strategies, doping of QDs [63], the use of alloy QDs [64], and the use of noble metal nanoclusters [65] have been utilized.

4. Microbial Synthesis of QDs

With the advent of nanobiotechnology, biological synthesis studies of nanomaterials managed to use "green chemistry" approaches, which are globally sustainable and economically viable. Microbial nanotechnology is a bio-driven scientific discipline that interconnects microbial biotechnology and nanotechnology. Numerous biological resources have been exploited for the biosynthesis of nanoparticles, including bacteria [66], fungi [67], algae [68], viruses [69], plant extracts [70], and agro-industrial wastes [71]. Fungi, yeast, and bacteria are currently receiving particular attention for the biological production of nanoparticles owing to their capacity to biotransform and bioaccumulate metals (Table 1). Among the investigated microorganisms, fungi are favored for the synthesis of QDs as they are efficient secretors of several biological molecules. Further advantages include the ease in biomass handling and economic viability. Nevertheless, bacterial-mediated synthesis of nanoparticles confers advantages since bacteria can be genetically manipulated for the expression of particular enzymes involved during the synthesis of metallic nanoparticles [72].

Table 1. List of microorganisms (bacteria, yeast, and fungi) mediating quantum dots (QD) production and different types of microbially-produced QDs.

Microorganisms	QDs	Factors Optimization	References
Bacteria			
Desulfovibrio desulfuricans NCIMB 8307	ZnS	-	[12]
Genetically engineered *Escherichia coli*	CdS	Reactant concentrations, reaction time	[73]
Stenotrophomonas maltophilia	CdS	Reaction time	[74]
Clostridiaceae sp.	ZnS	-	[75]
Acidithiobacillus ferrooxidans, *A. thiooxidans* and *A. caldus*	CdS	pH	[76]
Pseudomonas putida KT2440	CdS	$CdSO_4$ concentration and exposure time	[77]
E. coli BW25113	CdS	-	[78]
E. coli	CdS	Reaction time	[79]
E. coli	CdTe	-	[80]
P. chlororaphis CHR05	CdS	$CdSO_4$ concentration, temperature, time and pH	[81]

Table 1. Cont.

Microorganisms	QDs	Factors Optimization	References
Yeast			
Saccharomyces cerevisiae	CdS	-	[82]
S. cerevisiae MTCC 2918	ZnS	Reaction time and different concentrations of yeast biomass and $ZnSO_4$	[83]
S. cerevisiae	CdSe	Effect of S. cerevisiae growth phase, selenite concentration, cadmium concentration, effects of selenite and cadmium incubating time	[84]
Schizosaccharomyces pombe	CdS	-	[85]
Rhodotorula mucilaginosa	CdSe	Different concentrations of Na_2SeO_3 and $CdCl_2$ and pH	[86]
Fungi			
Fusarium oxysporum	CdTe	-	[87]
Phanerochaete chrysosporium	CdS	-	[88]
F. oxysporum f. sp. lycopersici	CdS	Reaction time	[89]
Rhizopus stolonifera	CdTe and CdS	-	[90]
Pleurotus ostreatus	CdS	-	[91]
Aspergillus terreus	PbSe	-	[92]
Aspergillus sp.	ZnS	Reaction time, temperature, pH	[93]
Penicillium sp.	ZnS	-	[94]
Trametes versicolor	CdS	-	[95]

4.1. Mechanisms of Microbial Synthesis of QDs

Developing facile, low cost, and measurable methodologies for preparing QDs with controlled structures and distinctive properties represents a major challenge. Generally, two different mechanisms are applied: intracellular and extracellular syntheses mechanisms (Figure 3).

4.1.1. Intracellular Microbial Synthesis of QDs

The intracellular microbial synthesis of QDs involves the penetration of dissolved ions to microbial cell cytoplasm via the manganese or magnesium transferring systems. When transported into the cell cytoplasm, the metal ions are processed into nanoparticles using intracellular enzymes [96]. When metallic nanoparticles are synthesized intracellularly, it gets difficult to handle downstream processes and the cost of recovering the nanoparticles may increase. To recover the QDs, the genetically engineered cells may be lysed, centrifuged, freeze-thawed, and purified through columns of anion exchange [97]. Therefore, alternate pathways for QD biosynthesis need to be investigated, which can overcome subsequent procedures required for QD extraction and purification. This is extremely crucial for designing reliable routes for the development of QDs on a large-scale utilizing microbial community.

4.1.2. Extracellular Microbial Synthesis of QDs

To monitor the downstream procedures needed for QD intracellular microbial synthesis, one-pot extracellular manufacturing of QDs in microorganisms was developed. The microbial extracellular synthesis is rapid, scalable, and can be carried out under ambient conditions. The extracellular microbial synthesis of QDs is carried out with the help of the enzymes that reside in the cellular membrane or deposited in the growth medium [98]. The developed QDs are therefore either adsorbed on the cellular membrane or deposited in the growth medium. For instance, the extracellular biosynthesis of CdS QDs via the fungus *Fusarium oxysporum* was reported by Ahmad et al. [99]. Microorganisms produce specific enzymes such as reductases as a defense strategy. Such reductase enzymes could be involved during the QD extracellular biosynthesis [100]. The sections below introduce in depth overview of the synthesis of different QD particles via microbial machineries.

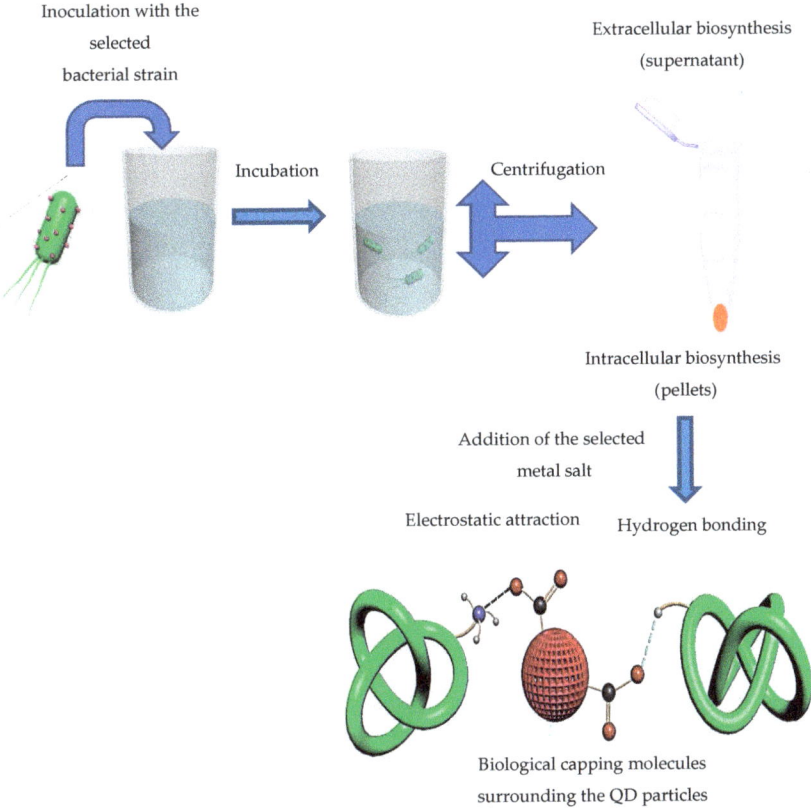

Figure 3. Extracellular and intracellular synthesis mechanisms of QD particles using bacteria.

4.2. Different Types of Microbially Fabricated QDs

4.2.1. Bacterial-Mediated Synthesis of QDs

Escherichia coli ATCC 29,181 is an ideal bio-factory to microbially fabricate CdTe QDs extracellularly [100]. CdTe QDs are synthesized from Cd and Te precursor metal salts via one-pot reaction using *E. coli*. Uniform spherical-shaped QDs with a size of approximately 2–3 nm were successfully observed by transmission electron microscope (TEM). X-ray diffraction (XRD) pattern demonstrated a sharp peak at 27.5 corresponding to the (111) planes, which indicated the cubic structure of CdTe. Since *E. coli* releases a large amount of proteins, especially metal-binding proteins, such proteins aid in the formation of the capping layer surrounding CdTe QDs. The hydrodynamic size of the *E. coli*-derived CdTe QDs was found to be larger compared to that of the hydrothermally-prepared CdTe QDs. This suggested the presence of capping/coating layers surrounding the microbially-produced CdTe QDs. Zeta potential measurements recorded −36.4 and −19.1 mV for *E. coli* and the hydrothermally fabricated CdTe QDs, respectively. The difference in zeta potentials significantly indicated the presence of several coating materials surrounding the prepared CdTe QDs. Further, the highly negative zeta potential of *E. coli*-mediated CdTe QDs revealed the high stability of the biosynthesized crystals. Fourier transform infrared (FTIR) spectroscopy was used to detect the possible chemical configuration of the surface of the microbially synthesized CdTe QDs. Two absorption peaks were centered at 1668 and 1545 cm^{-1} that could be attributed to the amide I and II group bendings, respectively. Coating of the QDs with biomaterials following their production is considered an effective methodology to modify them in order to diminish their nanotoxicity. The prepared CdTe QDs were used to image cultivated cervical cancer cells in vitro after functionalization with folic acid, and they proved to be an effective substitute or complementary tool for using traditional organic staining methods.

Cadmium sulfide (CdS) QDs are typical sulfide nanoparticles that can be synthesized by microorganisms. An easy route for synthesizing CdS QDs by the photosynthetic bacterium *Rhodopseudomonas palustris* was proposed by Bai et al. [96]. The $CdSO_4$ solution reacted with *R. palustris* biomass. After 48 h, the color of the reaction mixture changed to yellow, implying the synthesis of CdS QDs. A maximum absorption peak occurred at 425 nm, which was distinctive for quantum sized CdS particles. Moreover, XRD confirmed the crystallinity of the purified CdS QDs. TEM images showed a uniform distribution of CdS QDs with an average diameter of 0.25 ± 8.01 nm. The acidophilic bacterium of the *Acidithiobacillus* genus mediated the synthesis of CdS QDs [101]. Three *Acidithiobacillus* species, namely *A. thiooxidans*, *A. caldus*, and *A. ferrooxidans*, were exposed to sub-lethal concentrations of Cd^{2+} in presence of glutathione and cysteine. Red fluorescence was emitted by the reaction of *Acidithiobacillus* species with cadmium ions. The fluorescence of cadmium-exposed cells changed from green to red. The obtained CdS QDs displayed an absorbance band at 360 nm, as revealed by UV/Vis spectrophotometer. Once excited at 370 nm, broad emission spectrum appeared between 450 and 650 nm, which was distinctive to CdS QDs. Interestingly, microbially-fabricated QDs by acidophilic bacteria could withstand an acidic pH. These findings represent the first study to generate QDs with desired traits via extremophilic bacteria.

E. coli was tested as a bio-matrix to mediate the synthesis of CdS QDs by Yan et al. [79]. A sharp fluorescence emission peak appeared at 470 nm, indicating the presence of CdS QDs. Fluorescence inverted microscopy and TEM confirmed the presence of spherical shaped CdS QDs with a homogenous size of approximately 10 nm. The edges of the particles were lighter than the particles' center, signifying that the biosynthesized QDs might be enclosed by an enveloping capping material. Energy dispersive X-ray spectroscopy (EDX) identified the elemental structure of the particles, which were mainly made of sulfur and cadmium. XRD mapping indicated the presence of a cubic crystalline construction of CdS QDs. This microbial synthetic route resulted in the formation of uniform sized particles of CdS QDs with great fluorescence intensity. This synthesis approach had several advantages, including moderate temperature, low toxic effects, and high efficacy.

Tellurite and tellurate are the most abundant forms of Te in nature [102]. The possibility of changing them into the less toxic elemental Te (Te°) was investigated by Forootanfar et al. [103]. A bacterial strain was identified as *Pseudomonas pseudoalcaligenes* and was isolated from a hot spring. *P. pseudoalcaligenes* was capable of synthesizing tellurium nanorods (Te QNRs) [103]. An absorption peak was demonstrated at 210 nm and EDX of the purified Te QNRs displayed a characteristic Te elemental peak at 3.72 keV. Cu peaks were also detected but they were derived from the TEM copper grid. The cytotoxic effects of the biosynthesized Te QNRs on different cancer cell lines A549, HT1080, HepG2, and MCF-7 were evaluated, and the 3-(4,5-dimethylthiazol-2-yl)-2,5-diphenyltetrazolium bromide (MTT) assay was employed. The results show a direct relationship between the Te QNR applied doses and the cytotoxic effect. A low toxic effect was conducted by the rod-shaped microbially-synthesized tellurium nanostructures compared with that of Te^{4+} ions.

4.2.2. Fungal-Mediated Synthesis of QDs

F. oxysporum effectively mediated the preparation of highly fluorescent CdTe QDs at ambient conditions, as proposed by Syed and Ahmad [87]. CdTe QDs were successfully synthesized by reacting cadmium chloride and tellurium chloride with *F. oxysporum* mycelial biomass (20 g wet mycelia) for 96 h at 200 rpm. The fluorescence measurements were investigated via excitation of the reaction mixture at 400 nm. The mycosynthesized QDs demonstrated a fluorescence emission peak positioned at 475 nm, which was similar to chemically synthesized QDs. The XRD analysis showed intense peaks corresponding to (111), (220), and (311) planes. An energy dispersive X-ray (EDX) spectrum revealed signals corresponding to Cd, Te, O, and C. The C and O signals were likely due to X-ray emissions from proteins/enzymes present on the nanoparticle surfaces. X-ray photoelectron spectroscopy (XPS) illustrated the existence of Cd, Te, O, and C as the main elements. Thermogravimetric analysis (TGA) was carried out to identify the thermal properties (desorption/decomposition) of the as-prepared QDs. The TGA spectrum showed weight loss (30%) in the temperature range of 200–250 °C. The first weight loss was ascribed to the released biomolecules and water vapor, which coated the surface of the mycosynthesized QDs. Further degradation upon increasing the temperature occurred at the range of 500–700 °C. The prepared QDs exhibited antibacterial potential toward Gram-positive bacteria (e.g., *Staphylococcus aureus* NCIM 2079 and *Bacillus subtilis* NCIM 2063) and Gram-negative bacteria (e.g., *E. coli* NCIM 2065 and *Pseudomonas aeruginosa* NCIM 2200).

Selenium and lead-tolerable marine isolate *Aspergillus terreus* was used to mediate the mycosynthesis of fluorescent lead selenide QRs (PbSe QRs) [72]. An absorbance peak occurred at 872 nm, which was distinctive to quantum sized PbSe particles. A weak absorption peak at 375 nm appeared, indicating the presence of protein-capping agents. FTIR spectroscopy was used in the determination of the possible functional groups of the ligands capping the edges of the prepared QRs. The UV/Vis spectrum of the mycosynthesized PbSe QRs showed absorbance peaks at wavenumbers correlated with the presence of carboxylic and amide functional groups. Scanning electron microscope (SEM) images illustrated the presence of biogenic PbSe nanorods. The PbSe QRs crystallite size of 3.057 nm was predicted using the Scherer equation.

The phytopathogenic fungus, *Helminthosporium solani*, was incubated with an aqueous solution of cadmium chloride ($CdCl_2$) and selenium tetrachloride ($SeCl_4$) under ambient conditions. CdSe QDs were synthesized extracellularly. This was reported for the first time by Suresh [104]. The synthesized particles had 1% quantum yield and broad photoluminescence. An absorption peak at 350 nm was observed using a UV/Vis spectrophotometer, suggesting the formation of CdSe QDs. Absorption bands between 270 and 280 nm were also detected, and may have originated from proteins or peptides, thus demonstrating the possible involvement of protein or a peptide material as capping molecules. The photoluminescence properties of the mycosynthesized CdSe QDs were investigated by fluorescence spectral measurements following excitation at 380 nm. An emission band was observed at 430 nm. EDX depicted strong elemental signals for both cadmium and selenide. Particles were monodispersed, spherical in shape with an average diameter of 5.5 ± 2 nm, and a few square-shaped particles were also identified through the TEM images. The prepared particles were not in direct contact even within the aggregates, indicating powerful stabilization by the capping proteins/peptides. Particle size analysis showed that the particles had a mean diameter of 5.5 ± 2 nm. XPS was used to further validate the synthesis and assess sample purity and composition of biogenic CdSe QDs. The prominence of Cd and Se confirmed the synthesis and purification of the prepared QD particles.

The extracellular mycosynthesis of CdS QDs using the white rot fungus *Phanerochaete chrysosporium* was conducted by Chen et al. [88]. *P. chrysosporium* was incubated in cadmium nitrate tetrahydrate containing solution. The reaction solution turned yellow after 12 h, signifying the biogenic synthesis of CdS QDs. A maximum absorption peak occurred between 296 and 298 nm that was characteristic of the CdS quantum particles. The as-prepared CdS QDs emitted blue fluorescence at 458 nm. XRD confirmed the synthesis of crystalline CdS QDs with a face-centered cubic configuration. The average size of the particles was almost 2.56 nm, as calculated by the Scherer equation. TEM showed that the prepared particles were uniform in size. It was found that cysteine and proteins played a significant role during the formation and stabilization of the prepared CdS QDs.

Rhizopus stolonifer was used for the biomimetic synthesis of cadmium chalcogenide QDs [90]. The suspensions of CdTe and CdS QDs exhibited purple and greenish-blue luminescence, respectively, upon illumination via an 8 W UV lamp. Suspensions of both types of QDs remained highly stable even after four months of storage and did not exhibit any kind of aggregation. The average calculated crystallite sizes of both CdTe and CdS QDs were 7.6 and 8.8 nm, respectively, according to Scherer equation. XRD data concluded the cubic and hexagonal crystalline phases of CdTe and CdS QDs, respectively. More than 80% of the cells were viable, as confirmed by the MTT assay, using 20 µL of CdTe/CdS QDs. By loading the CdTe/CdS QDs in human breast adenocarcinoma (MCF-7) cell lines, better image contrast was obtained. These findings show that CdTe/CdS QDs could serve as alternatives to traditional organic dyes.

Increased demand of CdS QDs has led to the search for new synthesis methodologies to guarantee high production, precise control over particle size, and improved environmental friendliness within the production process, since several techniques use toxic solvents and consume high energy. In this context, Cárdenas et al. [92] used *F. oxysporum* for the preparation of hydrophilic CdS QDs. The mycelia of *F. oxysporum* were incubated with 1 mM of cadmium nitrate for 24 h at 30 °C. The biomass was then filtered. The filtrate became yellowish in color, depicting the synthesis of extracellular CdS QDs. This was confirmed by UV/Vis spectrophotometry, which revealed a characteristic band at 450 nm. Biosynthesized QDs were circular in shape with a diameter of 2.111 ± 6.116 nm and had a wurtzite crystalline structure. EDX analysis of the prepared CdS QDs showed the presence of Cd and S elements. TEM micrographs demonstrated the formation of circular-shaped agglomerated CdS QDs.

4.2.3. Yeast-Mediated Synthesis of QDs

Among the most important QDs, CdTe QDs have been extensively investigated in biomedical and industrial applications due to their distinctive properties, such as high photo-stability, controlled and narrow emission spectra, and high quantum yield, compared to conventional fluorescent dyes. CdTe QDs are promising candidates in the biological imaging of living cells. CdTe QDs with tunable fluorescence emission spectra were effectively biosynthesized using yeast cells, as demonstrated by Bao et al. [105]. This was carried out via the incubation of a type of *Saccharomyces cerevisiae*, using Cd and Te metal salt precursors. The as-prepared CdTe QDs displayed size-tunable dependent emission spectra ranging from 490 to 560 nm. A high quantum yield of ~33% was obtained. The microbially fabricated CdTe QDs were naturally surrounded with proteins and showed exceptional biocompatibility and stability. TEM revealed the presence of well-dispersed CdTe QDs with a diameter ranging between 2.0 and 3.6 nm. XRD patterns of the biosynthesized CdTe QDs showed a diffraction peak centered at $2\theta \sim 26.7°$, which corresponded to the (200) reflection for cubic CdTe QDs. The possible ligands that might have capped the as-prepared CdTe QDs were detected via FTIR spectroscopy. Two absorption peaks occurred at 1650 and 1566 cm^{-1}, which corresponded to amide I and II functional groups, respectively. The molecular mass and chemical composition of the capping proteins were further analyzed by high performance liquid chromatography (HPLC). Interestingly, HPLC revealed the presence of two types of proteins with molecular masses of 7.7 and 692 kDa with percentages of 93.4% and 6.6%, respectively. These proteins were produced from the tested yeast cell under investigation and aided in the stability and the inhibition of any possible aggregation of the prepared CdTe QDs. Accordingly, the protein-capped CdTe QDs could be tremendously beneficial in bio-labeling and -imaging applications.

ZnS QDs are among the most important and attractive semiconductor nanoparticles particularly in infrared and fast switching optical devices. Mala and Rose [83] demonstrated the microbial synthesis of ZnS QDs by *S. cerevisiae* MTCC 2918. A characteristic surface plasmon resonance (SPR) band of ZnS QDs was detected at 302.57 nm. XRD pattern showed that the nanoparticles were in the sphalerite phase. Two photoluminescence spectra were revealed at 280 and 325 nm. This suggested that the yeast had inherent sulfate-metabolizing systems that, in turn, were capable of assimilating sulfate.

5. Biomedical Applications of QDs

Semiconductor QDs are used in several applications, ranging from optoelectronic to bio-molecular, as summarized in Figure 4. They are extremely promising nanomaterials that can be applied to bio-sensing, bio-imaging, DNA detection, telecommunication, lasers, photo-detectors, and photovoltaic devices [106]. QDs are promising candidates for bio-sensing owing to their exceptional physical, chemical, and optical traits and the ability to bind different biomolecules to their surface. Innovative applications of QDs developed the current techniques for proteins and DNA detection. Moreover, QDs are beneficial in immunochemistry via molecular tracking. They represent better alternatives for the fluorescent beads that are mainly employed for studying the dynamics of neurotransmitter receptors because of their very small size (approximately 1–10 nm) compared to latex beads (approximately 500 nm) [107]. The QDs also function as biological luminescent markers capable of recognizing molecular structures. Effective multicolor cell labeling with QDs can typically be achieved through receptor-mediated diffusion or unspecific endocytosis, which provokes primitive cellular mechanisms to transfer nanoparticles via the cell membrane. Endocytosis is henceforth the least destructive mode of delivery compared with traditional approaches, such as microcapillary injection or electroporation. Endocytosis depends on injecting the target material into the cell. The inserted material becomes surrounded by a part of the cell membrane, which then buds off from the main cellular body and forms a vesicle inside the cell.

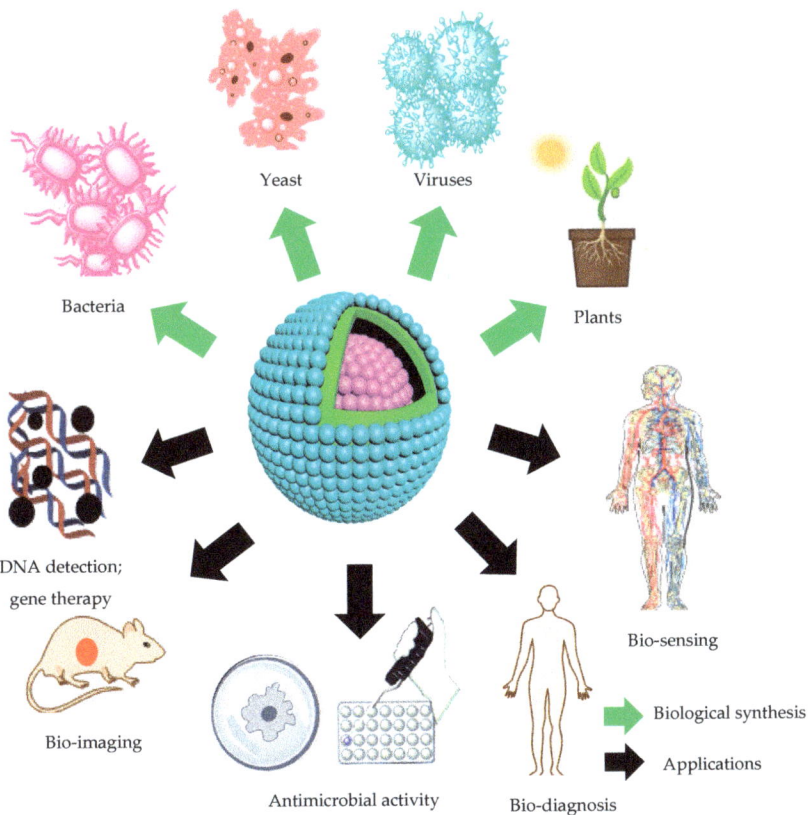

Figure 4. Scheme representing the different biological entities used for the biological fabrication of quantum dots (QDs) and their various applications.

The most commonly used fluorophores are organic fluorophores, such as genetically engineered fluorescent proteins and chemically manufactured fluorescent dyes. There are two noteworthy restrictions for using these organic fluorophores in different applications, as demonstrated by Drbohlavova et al. [108]. First, they cannot consistently fluoresce over long periods of time. Second, because of their quite broad emission spectra, they are not suited for multi-color bio-imaging applications. Accordingly, a suitable fluorescent marker should be biocompatible, highly fluorescent, and resistant to photo-bleaching.

In the late 1990s, Bruchez et al. [13] and Chan and Nie [109] published the first trials for the usage of QDs in imaging. Since then, researchers have exploited the properties of the QDs, in particular their photo-stability, luminosity, size-tunable optical and electronic features, and multiplexing potential, all of which enable QDs to be applied in a wide variety of different fields. Silver and its alloys were extensively employed for a number of years in a range of applications, including jewelry, coins, casting, and explosives [110]. Silver has also been demonstrated to have potent anti-microbial activity against various types of microbes, including bacteria, yeast, fungi, and viruses. During the last few decades, major scientific leaps in nanoscience have resulted in the development of quantum silver particles, which have biomedical potential as a result of their photoluminescence in the NIR. The use of such QDs may assist in certain biomedical applications such as photodynamic diagnostics, therapy, and in vitro imaging [111].

The silver-based QDs with potential use in biomedical applications are silver chalcogenide QDs (Ag_2X, where X = Te, Se, or S) as they exert low toxic effects in comparison with the other conventional QDs, such as CdSe, CdTe, and CdS [112]. Silver chalcogenide-based QDs also have a narrow band gap. For instance, the EBG of Ag_2S, Ag_2Se, and Ag_2Te are in the order 0.9–1.1, 0.15, and 0.67 eV, respectively [109]. The biomedical use of Ag_2X depends on the employed dopant type. For example, Ag_2Te primarily serves as a biological indicator (fluorophore), but it is rarely applied because of tellurium high toxicity. Silver chalcogenide can be employed for two main applications: high-resolution cellular imaging and in vivo tumor analysis. Jiang et al. [113], Wang et al. [114], and Zhang et al. [115] were the first researchers to report the ability to use Ag_2S QDs for the in vitro molecular imaging of living cells. Ag_2S QDs can serve as nanoprobes for cell selection in biomedical imaging applications, as they are effective in NIR II, which explains their high photoluminescence emissions. Wang and Yan [116] reported the usage of Ag_2S QDs in in vivo imaging, through which it was possible to distinguish cancer cells.

Gold QDs (AuQDs) are ideal particles for biomedical applications because they are non-toxic, inert, and biocompatible. Moreover, they exhibit ordered dispersity, and are easily functionalized. Such features make them preferable for employment in diagnosis, imaging, delivery of medical therapies to cells and tissues, control of surgical procedures, and electromagnetic radiation management [117]. It is worth noting that AuQDs possess the same characteristics as gold nanoparticles (AuNPs) but AuNPs are dissimilar from AuQDs, as they do not fluoresce. AuNPs have SPR-induced colorimetric characteristics which depend on particle size, shape, dielectrical nature, surrounding medium, and surface functionality.

5.1. Applications of QDs in Tumor Research

Despite cancer being one of the most-studied diseases worldwide, a comprehensive cure or diagnosis method has not yet been developed. Approximately 40% of the current population suffer from this deadly disease. Scientists across the world struggle to innovate new methodologies and therapies for the treatment of tumors. These techniques usually have a short lifespan due to the dynamic genetic mutations occurring in cancerous cells. As a result, there is a need for innovative technologies to help tackle cancer. Developing appropriate labels are at the main core of bio-labeling investigations. Many conservative techniques already exist which rely on the use of traditional fluorescent labels in tumor research, including organic dyes, which have some drawbacks. These include photo-bleaching effects. As a result, the application of such organic dyes as fluorescent labels is very limited. Semiconductor QDs possess attractive optical features for use in such applications, including broad emission and narrow excitation spectra, extended fluorescence lifetime, and insignificant optical bleaching. The narrow emission spectra and wide range of QD excitation wavelengths means that a single source of light can promptly stimulate different QD particles. The QD emission peaks are symmetrical, narrow, and with minor overlaps. This differs from organic fluorophores' emission peaks, which are commonly asymmetrical and with wide notable tails. This makes organic fluorophores vulnerable to disruption, so they become difficult to interpret. Furthermore, the emission wavelength of the QDs can be controlled by adjusting their crystal structure and size. QDs also possess a fluorescence intensity that is commonly greater than that of traditional dyes. QD fluorescence intensity and stability are 20–100 times greater than those of organic fluorescent dyes [118]. QDs are potent marker tools for long-term follow-up studies of vital life processes due to their anti-quenching behavior and photo-chemical stability [119]. Moreover, QDs display good optical stability in bio-imaging and a high potential to resist photo-bleaching. The photo-bleaching rate of organic fluorescent dyes makes them prone to fluorescence loss. On the contrary, the rate of photobleaching of QDs is very low [120]. CdSe/ZnS QDs possess a fluorescence intensity for almost 14 h without the signal becoming diminished [121].

As a result of these properties, QDs may become the preferred particle of choice in fluorescence imaging and labeling. Such properties may extend their use into tumor research [26]. The advantageous of photo-physical and -chemical characteristics of QDs make it feasible for QDs to be promising candidates for early diagnosis, prognosis, and monitoring of tumors. Beyond this, QDs have been used in tumor drugs delivery as well as photodynamic therapy (PDT) [122]. Among the various QD applications in tumor research, the imaging of both in vivo and in vitro cells has received worldwide attention [123]. QDs are particularly promising fluorescent labels for tumor cell imaging in vitro because of their distinctive advantages. Modifying the surface coating of QDs enhances their flexibility and labeling efficiency so that they can be used to explicitly and efficiently identify tumor cells at both subcellular and cellular levels. Immune-histochemical (IHC) and trastuzumab-conjugated QD (IHC-QD) techniques were recently developed by Miyashita et al. [124]. Images of the epidermal growth receptors in tumor tissue samples were collected from 37 breast cancer patients. The novel IHC-QD technique could overcome some of the disadvantages of the traditional IHC protocols, such as the auto-fluorescence imaging of tumor cells. Such improvements might enable a more rapid detection of HER2 expression level [124].

In case of in vivo studies, the optical probe fluorescence should effectively penetrate through tumor tissues. NIR molecular probes (700–1000 nm) are more advantageous than visible wavelength-emitting probes. Biological tissues have lower NIR absorption than visible light. This allows NIR light to penetrate deeper into the targeted tissues than visible wavelength light, thereby enabling the assessment of deeper tissues. In addition, at the NIR, there is less autofluorescence than visible wavelengths. As a result, probes emitting light at the NIR region are more suited to and preferred for in vivo imaging. Different QDs have been synthesized with fluorescence emissions ranging from UV to NIR regions, which may be used in imaging of tumor tissues, sentinel lymph nodes, and blood vessels [125]. Compared to many other imaging techniques, like computed tomography and magnetic resonance, QD imaging has the potential to provide further specific information. With respect to tumor imaging in vivo, in 2004, Voura et al. [126] developed a technique for tracking tumor cell overexpression within living animal tissues using QD labeling, emission-scanning microscopy, and multi-photon excitation. Gazouli et al. [127] investigated the labeling of vascular endothelial growth factor (VEGF) by the conjugating bevacizumab with QDs. By using this technique, non-destructive imaging of the VEGF-expressing tumor xenografts was successfully achieved in animal models. These studies suggest that QDs might be used as fluorescent trackers to label in vivo tumor cells. In addition, fluorescent indium phosphide (InP) QDs have been modified with anti-VEGF receptor 2 (anti-VEGFR2) monoclonal antibody to develop innovative chemotherapy for tumor cells. MiR-92a inhibitor induced the expression of tumor suppressor p63 [82]. Imaging and treatment of target tumor cells has been shown to be effective using VEGFR2-CD63 and the functionalized InP nanocomposite. For designing heavy metal-free QDs for in vivo imaging, Yaghini et al. [114] synthesized biocompatible QDs with strong quantum photoluminescent output that was suggested for use in mapping of lymph nodes. Furthermore, their low intrinsic toxicity has made them suitable for imaging in vivo tumors.

5.2. Applications of QDs in Drug Delivery as Drug Carriers

To date, the roles of QDs in drug delivery are chiefly divided into two main categories, in vivo fluorescent probes and drug carriers and elucidating pharmacodynamics and pharmacokinetics. QDs combined with biological molecules, such as antibodies and peptide ligands, may be used to enhance their targeting and use in drug delivery systems [128]. The first study on the use of QDs for in vivo targeted tumors diagnostic imaging and therapy was published in 2004 [129]. Since then, there has been a number of studies that have used QDs to demonstrate their potential as tools for drug delivery and tumor-targeting therapies [130]. ZnO QDs were developed as an innovative drug delivery system to control the intracellular drug release, such as doxorubicin. Following penetration into cancerous cells, controlled doses of doxorubicin would then be released due to the acidic intracellular conditions [131]. Another novel method using sulfonic-graphene (sulfonic-GQDs) was performed to target in vivo tumor cells [128]. The sulfonic-GQDs penetrated the tumor cells without any bio-ligand modification owing to the high fluid pressure in cancerous tissues. These data suggest that sulfonic-GQDs might be employed as innovative agents in drug delivery.

5.3. Applications of QDs in Photodynamic Therapy

At the end of the 1970s, PDT was introduced for the first time as a cancer treatment technique [132]. PDT depends on the use of visible light, and photosensitizers in presence of oxygen molecules. The key theory of PDT is that even after entering the body, photosensitizers appear to accumulate in damaged rather than healthy tissues. At that point, the sensitizing source of light emits the generated photosensitizers, which absorbs the photon energy and transmits it to the oxygen molecules, where photo-oxidation takes place to release reactive oxygen species (ROS). ROS interacts with cells and intracellular macromolecules, such as nucleic acids and proteins, and several subcellular organelles undergo cellular apoptosis or necrosis. Several parameters affect the success of the PDT process, including photosensitizer type and concentration in the targeted cells, laser wavelength irradiation type, duration and intensity and oxygen content of the micro-environment surrounding the targeted cells [133]. Photosensitizers are distinct molecules which can endure photochemical reactions once exposed to light. Developments in photosensitizers have led to the advancement of PDT. Fakayode et al. [134] reported the evolvement of nanoparticle self-lighting PDT (NSLPDT). NSLPDT depends on the use of QD particles, which exhibit a sustained and controlled luminescent light release after excitation. The self-luminescent particles are then conjugated with photosensitizers to synthesize a biological system that can be directly injected into patients with cancer. NSLPDT overcomes the drawbacks of weak penetration of external light sources and the poorly-induced PDT influence, which significantly improves the efficacy of the cancer therapies compared with conventional PDT.

5.4. Applications of QDs in Microbial Labeling and Tracking

5.4.1. Single-Virus Labeling and Tracking

Single-virus tracking may provide a clearer understanding of the relationship between viruses and their host cells through the imaging of the infection process, which involves attachment, entry, replication and egress. Up until now, fluorophores usually used for labeling viruses chiefly depend upon organic dyes such as cyanine5, fluorescent proteins such as green fluorescent protein, and QDs. Compared with organic dyes and fluorescent proteins, QDs retain distinctive optical characteristics, for instance, high quantum yield and photo-stability. This makes QDs very suitable for long-term tracking of single-virus with high sensitivity. Depending on the site labeled, QD labeling strategies for virus tracking are often categorized into three major groups: (i) internal components; (ii) external components; and (iii) other components.

QD possible use to identify external viral components was documented through the interaction of biotin-streptavidin with the virus particle (i.e., detected by a primary antibody) linked to a biotinylated secondary antibody and lastly distinguished by streptavidin-conjugated QDs [135]. Joo et al. [136]

succeeded in labeling retroviruses with QDs. This took place through a combination between a short acceptor peptide, which was sensitive to attachment with streptavidin-conjugated QDs. The labeling of biotin-streptavidin is an extremely simple labeling technique with improved potency and consistency. However, the labeled viruses appear to have high infectivity. Hence, it is very important to control in situ infection behaviors. To track viral infections, viruses can be labeled with several functionalized QDs by altering the viral surface coating proteins. However, the incorporation of QDs on the viral surface could have an impact on the viral ability to penetrate the host cells, thereby modifying virus-cell interactions, which would make it almost impossible to control late viral infection after losing the viral envelope [137]. Meanwhile, the encapsulation of QDs within the enveloped viral capsids may introduce an ideal solution to tackle these problems. Chemical tracking strategies have been extensively implemented. For instance, adeno virus serotype 2 was labeled successfully using QDs as reported by Joo et al. [138]. The labeling took place via amine-carboxyl crosslinking interaction to strengthen the imaging of the intracellular viral behavior within the targeted living host cells. Zhang et al. [126] suggested the encapsulation of QDs inside the core of Type I human immunodeficiency virus to facilitate viral tracking inside the living host cells. The authors revealed that conjugation of QDs with modified genomic RNAs (gRNAs) containing a viral genome sequence could be incorporated inside viruses. This further allowed the visible tracking of Type I human immunodeficiency virus infection.

Viruses are primarily comprised of DNA/RNA containing their genetic material, a protein envelop (designated as the capsid) that stores viral genetic materials, and, in certain cases, a lipid coat may exist around the protein envelope. Few viruses, however, do not have this complex structure. For instance, a prion is made up of protein rich only with histidine (His), which, in turn, provides an appropriate receptor for binding with divalent metals. Within this perspective, a modified QDs-Ni^{2+} complex of polyethylene glycol-nitrilotriacetic acid was suggested to facilitate labeling of the prion His-rich protein, and tracking the transport behavior pattern [139].

5.4.2. Bacterial Labeling

Accurate and rapid recognition of pathogenic bacteria is of a major biomedical concern [140]. In the last few years, several dyes and fluorescent polymers have been tested for bacterial labeling [141]. However, QDs have also been extensively investigated as promising bioprobe alternatives for bacterial imaging. QDs used for the labeling of bacteria are divided into four main classes; semiconductor QDs, carbon dots (CDs), silicon QDs, and polymer dots (P dots). According to Jones et al. [142], pathogenic bacteria are responsible for a large number of infectious diseases that contribute considerably to the global mortality rate. Among the bacterial species that cause several bacterial diseases are *Clostridium perfringens*, *E. coli* O157:H7, *Salmonella typhimurium*, *Listeria monocytogenes*, *Mycobacterium tuberculosis*, *S. aureus*, *Streptococcal* sp., and *Bacillus cereus* [143]. Approximately 2.8 million in the United States of America are infected annually with resistant bacterial strains, which in turn causes almost 35,000 mortalities annually [144]. It is therefore becoming increasingly critical to address effective strategies for tracking and combating pathogenic bacteria in clinical settings, food, and the environment.

Monitoring the number of pathogenic bacteria and, in particular, drug-resistant bacteria is the first and most vital step in controlling their spread. Generally, conventional plate counting and polymerase chain reaction (PCR) are two common methodologies used in bacterial detection and quantification. Although these methods are sensitive and precise, they require specific sample preparation and, as a result, they are time consuming [144]. Consequently, there is a crucial need to enhance the current methods for accelerated bacterial detection. CDs provide promising solutions to such rapid detection methods because of their outstanding fluorescence properties, potent antimicrobial potential, biocompatibility, and biosafety [145]. The efficiency of glowing CDs was proved in bacterial bioimaging and optical detection [146]. Single-walled carbon nanotubes CDs were the first source for isolating CDs. CDs involve quasi-spherical carbon QDs (CDs), graphene QDs (GQDs), and polymer QDs (PQDs) [147]. CDs and GQDs are extensively studied as anti-bacterial agents [148]. They are useful for bacterial detection and imaging as well as determining antibacterial potential. Such applications have been successful due to the interactions between the CDs and bacteria. The presence of functional groups, such as sulfhydryl, amino, hydroxyl, and carboxyl groups on CDs greatly affects their antibacterial potential. This is because of CD charged surface, unique structure, and photocatalytic features which can be easily modified by different functional groups [144]. For instance, amines and amides of N-doped-CDs play a vital role in improving the antibacterial effects [149]. The electrostatic interaction occurring between the CDs and pathogenic bacteria might be the main key factor behind CD antibacterial mechanism.

Ampicillin conjugated amino functionalized CDs (CDs-NH$_2$) were investigated as a novel approach in antibacterial therapies [150]. The resultant data showed that AMP-CDs generated ROS under visible light irradiation, which were the leading cause behind the growth inhibition of E. coli. Most E. coli strains are commensal and harmless, although there are also some pathogenic strains. Examples of such bacterial sub-strains include E. coli O157:H7 and O104:H4. Both strains are related to recent outbreaks in EU countries (Germany in 2011 and the UK in 2005 and 2009) and North America (1996 and 2006) [151]. Effective, precise, and rapid detection of E. coli is generally desired. Rapid detection of such bacteria has become a priority in a range of different disciplines, including those relating to food, water, and the environment [152]. There are a number of common methods for the detection of E. coli, including the membrane filtration, most probable number, and chromogenic enzyme-substrate techniques. Nevertheless, these techniques are either labor- or time-intensive [151]. Therefore, Yang et al. [151] documented a novel and potent approach for E. coli cell labeling with QDs. Two types of QDs were used; CdTe and CdTe/ZnS. Both QD types were used because they exhibit good biocompatibility with emission wavelengths near to the flow cytometric excitonic absorption wavelength. Hence, enabling rapid and accurate detection. Permeability was analyzed using SEM and activity measurements of the released alkaline phosphatase (PhoA) from periplasm. This method has opened a new avenue for the facile, quick, and sensitive detection of bacteria.

5.4.3. Fungal Labeling

Around 400 fungal species are identified as human pathogens, 50 of which cause several neurological disorders [153]. For instance, *F. oxysporum* is an opportunistic fungus, which becomes deadly to immunocompromised patients. It is therefore important to establish new and highly sensitive approaches for the early fungal detection. PCR can be used to detect fungi. However, PCR techniques are expensive and need a significant quantity of fungal biomass [154]. Additionally, because of the distinctive optical characteristics of semiconductor QDs and CDs, in recent years, their suitability for use in fungal labeling has been noted. Some *Candida* species are associated with superficial or invasive infections to humans, particularly in immune-depressed patients [155]. *Candida parapsilosis, C. tropicalis, C. krusei, C. glabrata,* and *C. albicans.* are among the most common *Candida* species responsible for candidiasis. The biochemical components of the cell wall in *Candida* spp. are not only responsible for their cellular integrity but also for the interaction between *Candida* sp. and the environment. The cell wall is involved in biofilm formation and interactions with the host cells [156]. Hence, any changes in the polysaccharide structural composition of the cell wall will directly affect the pathogen-associated molecular patterns [157].

5.5. MicroRNAs (miRNA) Detection

miRNAs are a category of non-coding small nucleotide RNAs and are approximately ~22–23 nucleotides long. They act as monitoring molecules in RNA silencing and are included in almost all pathological and developmental pathways in animals and humans [158]. After miRNAs were first discovered in *Caenorhabditis elegans* [159], they were subsequently identified in all living organisms and even some viruses. Nevertheless, miRNAs were not recognized as a biological regulator class until the early 2000s. Since then, more than 30,000 miRNAs have been discovered [160]. Deregulation of miRNAs can cause some illnesses, such as cancer or cardiovascular disorders [161]. Up or down alteration of miRNA expression can have deleterious consequences on cellular processes, for instance, proliferation and apoptosis. In this context, miRNAs introduce a rich biomarker base for the recognition of various illnesses.

The analysis of miRNAs is more challenging than that of other oligonucleotide targets for a number of reasons, including sequence homology among family members, short coding length, vulnerability to decomposition, and limited availability in RNA samples. Various techniques have been innovated to meet the criteria required for miRNA analysis. The identification theory is dependent upon Watson-Crick base-pairing and base-stacking. In situ hybridization, real-time PCR, Northern blotting, and various microarrays are among the most commonly used techniques for detecting miRNAs. The miRNA regeneration and reduced stability, time-consuming nature of the mentioned methodologies, and the need for separation procedures are the key drawbacks for their widespread use [162]. New techniques for the simultaneous detection of single and multiple short RNAs are being actively pursued for elucidating gene expression profiles in biological systems and for the early diagnosis of cancer and other illnesses. To increase the sensitivity, a number of new procedures have been proposed, including the nucleic acid-based locked probing assay, the ligation chain size-coded reaction, and probe-based exponential circle rolling amplification. Chen et al. [163] introduced a new electro-chemiluminescent (ECL) biosensor for detecting miRNA via the interaction of QDs conjugated with doxorubicin (DOX-QDs) with the DNA/RNA hybrids. This interaction caused the subsequent amplification of ECL emissions. The elevated ECL strength was shown to be directly correlated to the quantity of targeted miRNA in the tested samples.

6. Conclusions, Challenges and Future Prospects

Research on QDs has drawn worldwide attention from the scientific community. Growing knowledge of green chemistry principles and biological processes has led to the establishment of environmentally sustainable approaches to synthesize non-toxic and eco-friendly QDs. Unlike most of the chemical and physical QD synthesis techniques, which involve noxious chemicals and costly energy requirements, microbial synthesis of QDs is cost-effective and eco-friendly, and does not require any external reducing, capping or stabilizing agents. Microbial synthesis of QDs has therefore emerged as an attractive branch of nanobiotechnology. Furthermore, due to the high microbial diversity which mediates the fabrication of QDs, microbes are considered as potential biological factories.

This review provides a comprehensive description of the basics of QD science since their early discovery, their structural composition, their distinctive features, and the synthesis of QDs using microbial machinery, such as bacteria, yeast, and fungi. Analysis of microbially-fabricated chalcogenide QDs is discussed with regards to their structural and optical features. The structural properties have been determined in context with XRD, FTIR, SEM, and TEM analyses. Optical characterization of QDs was demonstrated via UV/Visible and fluorescence emission spectra, providing insight into the arrangements of energy levels inside the particles. In addition, the versatile biomedical applications of QDs and emerging obstacles are discussed. Despite their successful results, certain anticipated problems, such as toxicity, have also been observed; thus, the use of QDs remains controversial. Capping QDs with functional materials may result in complex lethal immune reactions. Additionally, heavy metals present in the core may be toxic to host cells. For the full determination of the cytotoxicity of QDs, an integration between the in vitro and in vivo investigations is mandatory. Toxic effects of metal chalcogenide QDs are particularly dependent on several factors including QD sizes, surface configurations, exposure routes, and agglomeration of the prepared particles. Extensive studies are therefore required to elucidate the relationship between the toxicity and the factors listed above.

The use of microbes to mediate the synthesis of QDs is remarkably significant; therefore, further studies are required for commercial development. Certain limits restrict the commercial production and applications of QDs, including: (i) the relative increasing in QD sizes during synthesis; (ii) controlling the synthesis stages such as nucleation, crystallization and crystal growth; and (iii) boosting the synthesis rate of QDs by optimizing the variables affecting the synthesis process and the downstream processing methodologies.

Future research should be directed toward: (i) identification of the exact mechanism behind the microbial synthesis of QDs; (ii) elucidation of the enzymes playing a key role during the nucleation and growth of QDs; (iii) exploration of the distribution and mode of action of QDs to promote their biomedical applications; (iv) determination of the effects of the exposure to low doses of QDs for a long time; and (v) assessment of risk management arising from QD preparation, handling, and storage procedures.

Author Contributions: Collection of data, M.A.-S., B.O., K.W. and K.-H.B.; Writing—original draft preparation, M.A.-S. and B.O.; and Writing—review and editing, K.W. and K.-H.B. All authors have read and agreed to the published version of the manuscript.

Funding: This research was funded by the Ministry of Education—the Basic Science Research Program through the National Research Foundation of Korea (NRF), grant number NRF-2019R1F1A1052625.

Conflicts of Interest: The authors declare no conflict of interest.

References

1. Xue, J.; Wang, X.; Hyun, J.; Yan, X. Fabrication, photoluminescence and applications of quantum dots embedded glass ceramics. *Chem. Eng. J.* **2020**, *383*, 123082–123115. [CrossRef]
2. Jefferson, J.H.; Häusler, W. Quantum dots and artificial atoms. *arXiv* **1997**, arXiv:cond-mat/9705012.
3. Suri, S.; Ruan, G.; Winter, J.; Schmidt, C.E. Microparticles and nanoparticles. In *Biomaterials Science*; Elsevier: Amsterdam, The Netherlands, 2013; pp. 360–388.
4. Ekimov, A.I.; Onushchenko, A.A. Quantum size effect in three-dimensional microscopic semiconductor crystals. *Jetp Lett.* **1981**, *34*, 345–349.
5. Dong, G.; Wang, H.; Chen, G.; Pan, Q.; Qiu, J. Quantum dot-doped glasses and fibers: Fabrication and optical properties. *Front. Mater.* **2015**, *2*, 1–14. [CrossRef]
6. Gao, Z.; Liu, Y.; Ren, J.; Fang, Z.; Lu, X.; Lewis, E.; Farrell, G.; Yang, J.; Wang, P. Selective doping of Ni^{2+} in highly transparent glass-ceramics containing nano-spinels $ZnGa_2O_4$ and $Zn_{1+x} Ga_{2-2x} GexO_4$ for broadband near-infrared fiber amplifiers. *Sci. Rep.* **2017**, *7*, 1783. [CrossRef]
7. Li, P.; Duan, Y.; Lu, Y.; Xiao, A.; Zeng, Z.; Xu, S.; Zhang, J. Nanocrystalline structure control and tunable luminescence mechanism of Eu-doped $CsPbBr_3$ quantum dot glass for WLEDs. *Nanoscale* **2020**, *12*, 6630–6636. [CrossRef]
8. Dong, G.; Wu, G.; Fan, S.; Zhang, F.; Zhang, Y.; Wu, B.; Ma, Z.; Peng, M.; Qiu, J. Formation, near-infrared luminescence and multi-wavelength optical amplification of PbS quantum dot-embedded silicate glasses. *J. Non-Cryst. Solids* **2014**, *383*, 192–195. [CrossRef]
9. Reed, M.A.; Randall, J.N.; Aggarwal, R.J.; Matyi, R.J.; Moore, T.M.; Wetsel, A.E. Observation of discrete electronic states in a zero-dimensional semiconductor nanostructure. *Phys. Rev. Lett.* **1988**, *60*, 535–540. [CrossRef]
10. Brus, L.E. A simple model for the ionization potential, electron affinity, and aqueous redox potentials of small semiconductor crystallites. *J. Chem. Phys.* **1983**, *79*, 5566–5571. [CrossRef]
11. Weller, H. Colloidal semiconductor q-particles: Chemistry in the transition region between solid state and molecules. *Angew. Chem. Int. Ed.* **1993**, *32*, 41–53. [CrossRef]
12. Murray, C.B.; Norris, D.J.; Bawendi, M.G. Synthesis and characterization of nearly monodisperse CdE (E = sulfur, selenium, tellurium) semiconductor nanocrystallites. *J. Am. Chem. Soc.* **1993**, *115*, 8706–8715. [CrossRef]
13. Bruchez, M.; Moronne, M.; Gin, P.; Weiss, S.; Alivisatos, A.P. Semiconductor nanocrystals as fluorescent biological labels. *Science* **1998**, *281*, 2013–2016. [CrossRef] [PubMed]
14. Yang, C.C.; Mai, Y. Size-dependent absorption properties of CdX (X = S, Se, Te) quantum dots. *Chem. Phys. Lett.* **2012**, *535*, 91–93. [CrossRef]
15. Juzenas, P.; Chen, W.; Sun, Y.-P.; Coelho, M.A.N.; Generalov, R.; Generalova, N.; Christensen, I.L. Quantum dots and nanoparticles for photodynamic and radiation therapies of cancer. *Adv. Drug Deliv. Rev.* **2008**, *60*, 1600–1614. [CrossRef]
16. Wagner, A.M.; Knipe, J.M.; Orive, G.; Peppas, N.A. Quantum dots in biomedical applications. *Acta Biomater.* **2019**, *94*, 44–63. [CrossRef]

17. Jacob, J.M.; Lens, P.N.L.; Balakrishnan, R.M. Microbial synthesis of chalcogenide semiconductor nanoparticles: A review. *Microb. Biotechnol.* **2016**, *9*, 11–21. [CrossRef]
18. Sirinakis, G.; Zhao, Z.Y.; Sevryugina, Y.; Tayi, A.; Carpenter, M. *Tailored Nanomaterials: Selective & Sensitive Chemical Sensors for Hydrocarbon Analysis*; School of NanoSciences and NanoEngineering, University at Albany, SUNY: Albany, NY, USA, 2003; Available online: http://eqs.syr.edu/documents/research (accessed on 15 June 2003).
19. Fariq, A.; Khan, T.; Yasmin, A. Microbial synthesis of nanoparticles and their potential applications in biomedicine. *J. Appl. Biomed.* **2017**, *15*, 241–248. [CrossRef]
20. Rana, A.; Yadav, K.; Jagadevan, S. A comprehensive review on green synthesis of nature-inspired metal nanoparticles: Mechanism, application and toxicity. *J. Clean. Prod.* **2020**, *272*, 122880. [CrossRef]
21. Saklani, V.; Suman, J.V.K.; Jain, K. Microbial synthesis of silver nanoparticles: A review. *J. Biotechnol. Biomater.* **2012**, *S13*, 7–10. [CrossRef]
22. Alghuthaymi, M.A.; Almoammar, H.; Rai, M.; Said-Galiev, E.; Abd-Elsalam, K.A. Myconanoparticles: Synthesis and their role in phytopathogens management. *Biotechnol. Biotechnol. Equip.* **2015**, *29*, 221–236. [CrossRef]
23. Zomorodian, K.; Pourshahid, S.; Sadatsharifi, A.; Mehryar, P.; Pakshir, K.; Rahimi, M.J.; Arabi Monfared, A. Biosynthesis and characterization of silver nanoparticles by *Aspergillus* species. *BioMed Res. Int.* **2016**, *2016*. [CrossRef] [PubMed]
24. Madakka, M.; Jayaraju, N.; Rajesh, N. Mycosynthesis of silver nanoparticles and their characterization. *MethodsX* **2018**, *5*, 20–29. [CrossRef] [PubMed]
25. Jeevanandam, J.; Chan, Y.S.; Danquah, M.K. Biosynthesis of metal and metal oxide nanoparticles. *ChemBioEng Rev.* **2016**, *3*, 55–67. [CrossRef]
26. Liu, B.; Jiang, B.; Zheng, Z.; Liu, T. Semiconductor quantum dots in tumor research. *J. Lumin.* **2019**, *209*, 61–68. [CrossRef]
27. Schiffman, J.D.; Balakrishna, R.G. Quantum dots as fluorescent probes: Synthesis, surface chemistry, energy transfer mechanisms, and applications. *Sens. Actuators B Chem.* **2018**, *258*, 1191–1214. [CrossRef]
28. Pandey, S.; Bodas, D. High-quality quantum dots for multiplexed bioimaging: A critical review. *Adv. Colloid Interface Sci.* **2020**, *278*, 102137–102153. [CrossRef]
29. Resch-Genger, U.; Grabolle, M.; Cavaliere-Jaricot, S.; Nitschke, R.; Nann, T. Quantum dots versus organic dyes as fluorescent labels. *Nat. Methods* **2008**, *5*, 763–775. [CrossRef]
30. Rizvi, S.B.; Yildirimer, L.; Ghaderi, S.; Ramesh, B.; Seifalian, A.M.; Keshtgar, M. A novel POSS-coated quantum dot for biological application. *Int. J. Nanomed.* **2012**, *7*, 3915–3927. [CrossRef]
31. Smith, A.M.; Duan, H.; Mohs, A.M.; Nie, S. Bioconjugated quantum dots for in vivo molecular and cellular imaging. *Adv. Drug Deliv. Rev.* **2008**, *60*, 1226–1240. [CrossRef]
32. Reshma, V.G.; Mohanan, P.V. Quantum dots: Applications and safety consequences. *J. Lumin.* **2019**, *205*, 287–298. [CrossRef]
33. Zhang, Y.; Clapp, A. Overview of stabilizing ligands for biocompatible quantum. *Sensors* **2011**, *11*, 11036–11055. [CrossRef] [PubMed]
34. Martynenko, I.V.; Litvin, A.P.; Purcell-Milton, F.; Baranov, A.V.; Fedorov, A.V.; Gun'Ko, Y.K. Application of semiconductor quantum dots in bioimaging and biosensing. *J. Mater. Chem. B* **2017**, *5*, 6701–6727. [CrossRef] [PubMed]
35. Credi, A. *Photoactive Semiconductor Nanocrystal Quantum Dots: Fundamentals and Applications*; Springer: Gewerbestrasse, Swizterland, 2017; ISBN 3319511920.
36. Reiss, P.; Protiere, M.; Li, L. Core/shell semiconductor nanocrystals. *Small* **2009**, *5*, 154–168. [CrossRef] [PubMed]
37. Ziaudeen, S.A.; Gaddam, R.R.; Pallapothu, P.K.; Sugumar, M.K.; Rangarajan, J. Supra gap excitation properties of differently confined PbS-nano structured materials studied with opto-impedance spectroscopy. *J. Nanophotonics* **2013**, *7*, 73075–73088. [CrossRef]
38. Mews, A.; Eychmüller, A.; Giersig, M.; Schooss, D.; Weller, H. Preparation, characterization, and photophysics of the quantum dot quantum well system cadmium sulfide/mercury sulfide/cadmium sulfide. *J. Phys. Chem.* **1994**, *98*, 934–941. [CrossRef]
39. Cao, Y.; Banin, U. Synthesis and characterization of InAs/InP and InAs/CdSe core/shell nanocrystals. *Angew. Chem. Int. Ed.* **1999**, *38*, 3692–3694. [CrossRef]

40. Battaglia, D.; Li, J.J.; Wang, Y.; Peng, X. Colloidal two-dimensional systems: CdSe quantum shells and wells. *Angew. Chem. Int. Ed.* **2003**, *42*, 5035–5039. [CrossRef]
41. Zhong, X.; Xie, R.; Zhang, Y.; Basche, T.; Knoll, W. High-quality violet-to red-emitting ZnSe/CdSe core/shell nanocrystals. *Chem. Mater.* **2005**, *17*, 4038–4042. [CrossRef]
42. Nadagouda, M.N.; Varma, R.S. A greener synthesis of core (Fe, Cu)-shell (Au, Pt, Pd, and Ag) nanocrystals using aqueous vitamin C. *Cryst. Growth Des.* **2007**, *7*, 2582–2587. [CrossRef]
43. Yang, Z.; Gao, M.; Wu, W.; Yang, X.; Sun, X.W.; Zhang, J.; Wang, H.-C.; Liu, R.-S.; Han, C.-Y.; Yang, H. Recent advances in quantum dot-based light-emitting devices: Challenges and possible solutions. *Mater. Today* **2019**, *24*, 69–93. [CrossRef]
44. Arya, H.; Kaul, Z.; Wadhwa, R.; Taira, K.; Hirano, T.; Kaul, S.C. Quantum dots in bio-imaging: Revolution by the small. *Biochem. Biophys. Res. Commun.* **2005**, *329*, 1173–1177. [CrossRef] [PubMed]
45. Rizvi, S.B.; Ghaderi, S.; Keshtgar, M.; Seifalian, A.M. Semiconductor quantum dots as fluorescent probes for in vitro and in vivo bio-molecular and cellular imaging. *Nano Rev.* **2010**, *1*, 5161. [CrossRef] [PubMed]
46. Perini, G.; Palmieri, V.; Ciasca, G.; De Spirito, M.; Papi, M. Unravelling the potential of graphene quantum dots in biomedicine and neuroscience. *Int. J. Mol. Sci.* **2020**, *21*, 3712. [CrossRef] [PubMed]
47. Singh, A.K.; Pal, P.; Gupta, V.; Yadav, T.P.; Gupta, V.; Singh, S.P. Green synthesis, characterization and antimicrobial activity of zinc oxide quantum dots using *Eclipta alba*. *Mater. Chem. Phys.* **2018**, *203*, 40–48. [CrossRef]
48. Michalet, X.; Pinaud, F.F.; Bentolila, L.A.; Tsay, J.M.; Doose, S.; Li, J.J.; Sundaresan, G.; Wu, A.M.; Gambhir, S.S.; Weiss, S. Quantum dots for live cells, in vivo imaging, and diagnostics. *Science* **2005**, *307*, 538–544. [CrossRef]
49. Medintz, I.L.; Uyeda, H.T.; Goldman, E.R.; Mattoussi, H. Quantum dot bioconjugates for imaging, labeling and sensing. *Nat. Mater.* **2005**, *4*, 435–446. [CrossRef]
50. Demchenko, A.P. *Introduction to Fluorescence Sensing*; Springer Science & Business Media: Heidelberg, Germany, 2008; ISBN 140209003X.
51. Yaghini, E.; Seifalian, A.M.; MacRobert, A.J. in vivo applications of quantum dot nanoparticles for optical diagnostics and therapy. *Nanomedicine* **2011**, *2*, 21–40.
52. Ducheyne, P. *Comprehensive Biomaterials*, 2nd ed.; Elsevier: Amsterdam, The Netherlands, 2017; ISBN 0081006926.
53. Cordones, A.A.; Leone, S.R. Mechanisms for charge trapping in single semiconductor nanocrystals probed by fluorescence blinking. *Chem. Soc. Rev.* **2013**, *42*, 3209–3221. [CrossRef]
54. Mahler, B.; Spinicelli, P.; Buil, S.; Quelin, X.; Hermier, J.; Dubertret, B.; Brossel, L.K.; Umr, C.; Pierre, U.; Cedex, P.; et al. Towards non-blinking quantum dots: The effect of thick shell. *Proc. SPIE* **2009**, *7189*, 1–9. [CrossRef]
55. Mingqian, T.; Aiguo, W. *Nanomaterials for Tumor Targeting Theranostics: A Proactive Clinical Perspective*; World Scientific: Rosewood Drive, SC, USA, 2016; ISBN 981463543X.
56. Stokes, G.G. On the change of refrangibility of light. *Philos. Trans. R. Soc.* **1852**, *142*, 463–562. [CrossRef]
57. McCartney, M.; Whitaker, A.; Wood, A. *George Gabriel Stokes: Life, Science and Faith*; Oxford University Press: New York, NY, USA, 2019; ISBN 0198822863.
58. Zhou, R.; Lu, X.; Yang, Q.; Wu, P. Nanocrystals for large Stokes shift-based optosensing. *Chin. Chem. Lett.* **2019**, *30*, 1843–1848. [CrossRef]
59. Vollmer, F.; Rettig, W.; Birckner, E. Photochemical mechanisms producing large fluorescence stokes shifts. *J. Fluoresc.* **1994**, *4*, 65–69. [CrossRef] [PubMed]
60. Rizvi, S.B.; Rouhi, S.; Taniguchi, S.; Yang, S.Y.; Green, M.; Keshtgar, M.; Seifalian, A.M. Near-infrared quantum dots for HER2 localization and imaging of cancer cells. *Int. J. Nanomed.* **2014**, *9*, 1323–1337. [CrossRef]
61. Jung, J.; Lin, C.H.; Yoon, Y.J.; Malak, S.T.; Zhai, Y.; Thomas, E.L.; Vardeny, V.; Tsukruk, V.V.; Lin, Z. Crafting core/graded shell–shell quantum dots with suppressed re-absorption and tunable Stokes shift as high optical gain materials. *Angew. Chem. Int. Ed.* **2016**, *55*, 5071–5075. [CrossRef]
62. Chen, M.; Chen, R.; Shi, Y.; Wang, J.; Cheng, Y.; Li, Y.; Gao, X.; Yan, Y.; Sun, J.Z.; Qin, A. Malonitrile-functionalized tetraphenylpyrazine: Aggregation-induced emission, ratiometric detection of hydrogen sulfide, and mechanochromism. *Adv. Funct. Mater.* **2018**, *28*, 1704689–1704699. [CrossRef]
63. Lu, X.; Zhang, J.; Xie, Y.-N.; Zhang, X.; Jiang, X.; Hou, X.; Wu, P. Ratiometric phosphorescent probe for thallium in serum, water, and soil samples based on long-lived, spectrally resolved, Mn-doped ZnSe quantum dots and carbon dots. *Anal. Chem.* **2018**, *90*, 2939–2945. [CrossRef]

64. Ca, N.X.; Hien, N.T.; Luyen, N.T.; Lien, V.T.K.; Thanh, L.D.; Do, P.V.; Bau, N.Q.; Pham, T.T. Photoluminescence properties of CdTe/CdTeSe/CdSe core/alloyed/shell type-II quantum dots. *J. Alloys Compd.* **2019**, *787*, 823–830. [CrossRef]
65. Aikens, C.M. Electronic and geometric structure, optical properties, and excited state behavior in atomically precise thiolate-stabilized noble metal nanoclusters. *Acc. Chem. Res.* **2018**, *51*, 3065–3073. [CrossRef]
66. John, M.S.; Nagoth, J.A.; Ramasamy, K.P.; Mancini, A.; Giuli, G.; Natalello, A.; Ballarini, P.; Miceli, C.; Pucciarelli, S. Synthesis of bioactive silver nanoparticles by a *Pseudomonas* strain associated with the antarctic psychrophilic protozoon *Euplotes focardii*. *Mar. Drugs* **2020**, *18*, 38. [CrossRef]
67. Omran, B.A.; Nassar, H.N.; Younis, S.A.; Fatthallah, N.A.; Hamdy, A.; El-Shatoury, E.H.; El-Gendy, N.S. Physiochemical properties of *Trichoderma longibrachiatum* DSMZ 16517-synthesized silver nanoparticles for the mitigation of halotolerant sulphate-reducing bacteria. *J. Appl. Microbiol.* **2019**, *126*, 138–154. [CrossRef]
68. Khanna, P.; Kaur, A.; Goyal, D. Algae-based metallic nanoparticles: Synthesis, characterization and applications. *J. Microbiol. Methods* **2019**, *163*, 105656–105680. [CrossRef] [PubMed]
69. Hefferon, K.L. Repurposing plant virus nanoparticles. *Vaccines* **2018**, *6*, 11. [CrossRef]
70. Seifipour, R.; Nozari, M.; Pishkar, L. Green synthesis of silver nanoparticles using *Tragopogon collinus* leaf extract and study of their antibacterial effects. *J. Inorg. Organomet. Polym. Mater.* **2020**. [CrossRef]
71. Omran, B.A.; Nassar, H.N.; Fatthallah, N.A.; Hamdy, A.; El-Shatoury, E.H.; El-Gendy, N.S. Waste upcycling of *Citrus sinensis* peels as a green route for the synthesis of silver nanoparticles. *Energy Sources Part A Recover. Util. Environ. Eff.* **2017**, *40*, 1–10. [CrossRef]
72. Jacob, J.M.; Balakrishnan, R.M.; Kumar, U.B. Biosynthesis of lead selenide quantum rods in marine *Aspergillus terreus*. *Mater. Lett.* **2014**, *124*, 279–281. [CrossRef]
73. Mi, C.; Wang, Y.; Zhang, J.; Huang, H.; Xu, L.; Wang, S.; Fang, X.; Fang, J.; Mao, C.; Xu, S. Biosynthesis and characterization of CdS quantum dots in genetically engineered *Escherichia coli*. *J. Biotechnol.* **2011**, *153*, 125–132. [CrossRef] [PubMed]
74. Yang, Z.; Lu, L.; Berard, V.F.; He, Q.; Kiely, C.J.; Berger, B.W.; McIntosh, S. Biomanufacturing of CdS quantum dots. *Green Chem.* **2015**, *17*, 3775–3782. [CrossRef]
75. Yue, L.; Wang, J.; Zhang, Y.; Qi, S.; Xin, B. Controllable biosynthesis of high-purity lead-sulfide (PbS) nanocrystals by regulating the concentration of polyethylene glycol in microbial system. *Bioprocess Biosyst. Eng.* **2016**, *39*, 1839–1846. [CrossRef]
76. Bruna, N.; Collao, B.; Tello, A.; Caravantes, P.; Díaz-Silva, N.; Monrás, J.P.; Órdenes-Aenishanslins, N.; Flores, M.; Espinoza-Gonzalez, R.; Bravo, D.; et al. Synthesis of salt-stable fluorescent nanoparticles (quantum dots) by polyextremophile halophilic bacteria. *Sci. Rep.* **2019**, *9*, 1953. [CrossRef]
77. Oliva-Arancibia, B.; Órdenes-Aenishanslins, N.; Bruna, N.; Ibarra, P.S.; Zacconi, F.C.; Pérez-Donoso, J.M.; Poblete-Castro, I. Co-synthesis of medium-chain-length polyhydroxyalkanoates and CdS quantum dots nanoparticles in *Pseudomonas putida* KT2440. *J. Biotechnol.* **2017**, *264*, 29–37. [CrossRef]
78. Venegas, F.A.; Saona, L.A.; Monrás, J.P.; Órdenes-Aenishanslins, N.; Giordana, M.F.; Ulloa, G.; Collao, B.; Bravo, D.; Pérez-Donoso, J.M. Biological phosphorylated molecules participate in the biomimetic and biological synthesis of cadmium sulphide quantum dots by promoting H$_2$S release from cellular thiols. *RSC Adv.* **2017**, *7*, 40270–40278. [CrossRef]
79. Yan, Z.Y.; Du, Q.Q.; Qian, J.; Wan, D.Y.; Wu, S.M. Eco-friendly intracellular biosynthesis of CdS quantum dots without changing *Escherichia coli*'s antibiotic resistance. *Enzym. Microb. Technol.* **2017**, *96*, 96–102. [CrossRef] [PubMed]
80. Kominkova, M.; Milosavljevic, V.; Vitek, P.; Polanska, H.; Cihalova, K.; Dostalova, S.; Hynstova, V.; Guran, R.; Kopel, P.; Richtera, L.; et al. Comparative study on toxicity of extracellularly biosynthesized and laboratory synthesized CdTe quantum dots. *J. Biotechnol.* **2017**, *241*, 193–200. [CrossRef] [PubMed]
81. Ashengroph, M.; Khaledi, A.; Bolbanabad, E.M. Extracellular biosynthesis of cadmium sulphide quantum dot using cell-free extract of *Pseudomonas chlororaphis* CHR05 and its antibacterial activity. *Process Biochem.* **2020**, *89*, 63–70. [CrossRef]
82. Wu, Y.-Z.; Sun, J.; Zhang, Y.; Pu, M.; Zhang, G.; He, N.; Zeng, X. Effective integration of targeted tumor imaging and therapy using functionalized InP QDs with VEGFR2 monoclonal antibody and miR-92a inhibitor. *ACS Appl. Mater. Interfaces* **2017**, *9*, 13068–13078. [CrossRef]
83. Sandana Mala, J.G.; Rose, C. Facile production of ZnS quantum dot nanoparticles by *Saccharomyces cerevisiae* MTCC 2918. *J. Biotechnol.* **2014**, *170*, 73–78. [CrossRef]

84. Brooks, J.; Lefebvre, D.D. Optimization of conditions for cadmium selenide quantum dot biosynthesis in *Saccharomyces cerevisiae*. *Appl. Microbiol. Biotechnol.* **2017**, *101*, 2735–2745. [CrossRef]
85. Al-Shalabi, Z.; Doran, P.M. Biosynthesis of fluorescent CdS nanocrystals with semiconductor properties: Comparison of microbial and plant production systems. *J. Biotechnol.* **2016**, *223*, 13–23. [CrossRef]
86. Cao, K.; Chen, M.M.; Chang, F.Y.; Cheng, Y.Y.; Tian, L.J.; Li, F.; Deng, G.Z.; Wu, C. The biosynthesis of cadmium selenide quantum dots by *Rhodotorula mucilaginosa* PA-1 for photocatalysis. *Biochem. Eng. J.* **2020**, *156*, 107497. [CrossRef]
87. Syed, A.; Ahmad, A. Extracellular biosynthesis of CdTe quantum dots by the fungus *Fusarium oxysporum* and their anti-bacterial activity. *Spectrochim. Acta Part A Mol. Biomol. Spectrosc.* **2013**, *106*, 41–47. [CrossRef]
88. Chen, G.; Yi, B.; Zeng, G.; Niu, Q.; Yan, M.; Chen, A.; Du, J.; Huang, J.; Zhang, Q. Facile green extracellular biosynthesis of CdS quantum dots by white rot fungus *Phanerochaete chrysosporium*. *Colloids Surf. B Biointerfaces* **2014**, *117*, 199–205. [CrossRef] [PubMed]
89. Borovaya, M.; Pirko, Y.; Krupodorova, T.; Naumenko, A.; Blume, Y.; Yemets, A. Biosynthesis of cadmium sulphide quantum dots by using *Pleurotus ostreatus* (Jacq.) P. Kumm. *Biotechnol. Biotechnol. Equip.* **2015**, *29*, 1156–1163. [CrossRef]
90. Mareeswari, P.; Brijitta, J.; Etti, S.H.; Meganathan, C.; Kaliaraj, G.S. Rhizopus stolonifer mediated biosynthesis of biocompatible cadmium chalcogenide quantum dots. *Enzym. Microb. Technol.* **2016**, *95*, 225–229. [CrossRef] [PubMed]
91. Tian, L.; Zhou, N.; Liu, X.; Zhang, X.; Zhu, T.; Li, L. Fluorescence dynamics of the biosynthesized CdSe quantum dots in *Candida utilis*. *Sci. Rep.* **2017**, *7*, 2048–2054. [CrossRef]
92. Cárdenas, S.; Issell, D.; Gomez-Ramirez, M.; Rojas-Avelizapa, N.G.; Vidales-Hurtado, M.A. Synthesis of cadmium sulfide nanoparticles by biomass of *Fusarium oxysporum* f. sp. *lycopersici*. *J. Nano Res.* **2017**, *46*, 179–191. [CrossRef]
93. Jacob, J.M.; Rajan, R.; Kurup, G.G. Biologically synthesized ZnS quantum dots as fluorescent probes for lead (II) sensing. *Luminescence* **2020**, 1–10. [CrossRef]
94. Jacob, J.M.; Rajan, R.; Aji, M.; Kurup, G.G.; Pugazhendhi, A. Bio-inspired ZnS quantum dots as efficient photo catalysts for the degradation of methylene blue in aqueous phase. *Ceram. Int.* **2019**, *45*, 4857–4862. [CrossRef]
95. Qin, Z.; Yue, Q.; Liang, Y.; Zhang, J.; Zhou, L.; Hidalgo, O.B.; Liu, X. Extracellular biosynthesis of biocompatible cadmium sulfide quantum dots using *Trametes versicolor*. *J. Biotechnol.* **2018**, *284*, 52–56. [CrossRef]
96. Bai, H.J.; Zhang, Z.M.; Guo, Y.; Yang, G.E. Biosynthesis of cadmium sulfide nanoparticles by photosynthetic bacteria *Rhodopseudomonas palustris*. *Colloids Surf. B Biointerfaces* **2009**, *70*, 142–146. [CrossRef]
97. Pandian, S.R.K.; Deepak, V.; Kalishwaralal, K.; Gurunathan, S. Biologically synthesized fluorescent CdS NPs encapsulated by PHB. *Enzym. Microb. Technol.* **2011**, *48*, 319–325. [CrossRef]
98. Raouf Hosseini, M.; Nasiri Sarvi, M. Recent achievements in the microbial synthesis of semiconductor metal sulfide nanoparticles. *Mater. Sci. Semicond. Process.* **2015**, *40*, 293–301. [CrossRef]
99. Ahmad, A.; Senapati, S.; Khan, M.I.; Kumar, R.; Sastry, M. Extracellular biosynthesis of monodisperse gold nanoparticles by a novel extremophilic actinomycete, *Thermomonospora* sp. *Langmuir* **2003**, *19*, 3550–3553. [CrossRef]
100. Bao, H.; Lu, Z.; Cui, X.; Qiao, Y.; Guo, J.; Anderson, J.M.; Li, C.M. Extracellular microbial synthesis of biocompatible CdTe quantum dots. *Acta Biomater.* **2010**, *6*, 3534–3541. [CrossRef] [PubMed]
101. Ulloa, G.; Collao, B.; Araneda, M.; Escobar, B.; Álvarez, S.; Bravo, D.; Pérez-Donoso, J.M. Use of acidophilic bacteria of the genus *Acidithiobacillus* to biosynthesize CdS fluorescent nanoparticles (quantum dots) with high tolerance to acidic pH. *Enzym. Microb. Technol.* **2016**, *95*, 217–224. [CrossRef] [PubMed]
102. Baesman, S.M.; Bullen, T.D.; Dewald, J.; Zhang, D.; Curran, S.; Islam, F.S.; Beveridge, T.J.; Oremland, R.S. Formation of tellurium nanocrystals during anaerobic growth of bacteria that use Te oxyanions as respiratory electron acceptors. *Appl. Environ. Microbiol.* **2007**, *73*, 2135–2143. [CrossRef] [PubMed]
103. Forootanfar, H.; Amirpour-Rostami, S.; Jafari, M.; Forootanfar, A.; Yousefizadeh, Z.; Shakibaie, M. Microbial-assisted synthesis and evaluation the cytotoxic effect of tellurium nanorods. *Mater. Sci. Eng. C* **2015**, *49*, 183–189. [CrossRef] [PubMed]
104. Suresh, A.K. Extracellular bio-production and characterization of small monodispersed CdSe quantum dot nanocrystallites. *Spectrochim. Acta Part A Mol. Biomol. Spectrosc.* **2014**, *130*, 344–349. [CrossRef] [PubMed]

105. Bao, N.; Yang, Y.; Commonwealth, T. Biosynthesis of biocompatible cadmium telluride quantum dots using yeast cells. *Nano Res.* **2010**, *3*, 481–489. [CrossRef]
106. Naik, V.; Zantye, P.; Gunjal, D.; Gore, A.; Anbhule, P.; Kowshik, M.; Bhosale, S.V.; Kolekar, G. Nitrogen-doped carbon dots via hydrothermal synthesis: Naked eye fluorescent sensor for dopamine and used for multicolor cell imaging. *ACS Appl. Bio Mater.* **2019**, *2*, 2069–2077. [CrossRef]
107. Ramalingam, G.; Saravanan, K.V.; Vizhi, T.K.; Rajkumar, M.; Baskar, K. Synthesis of water-soluble and bio-taggable CdSe@ZnS quantum dots. *RSC Adv.* **2018**, *8*, 8516–8527. [CrossRef]
108. Drbohlavova, J.; Adam, V.; Kizek, R.; Hubalek, J. Quantum dots—Characterization, preparation and usage in biological systems. *Int. J. Mol. Sci.* **2009**, *10*, 656–673. [CrossRef]
109. Chan, W.C.W.; Nie, S. Quantum dot bioconjugates for ultrasensitive nonisotopic detection. *Science* **1998**, *281*, 2016–2018. [CrossRef] [PubMed]
110. Granada-Ramírez, D.A.; Arias-Cerón, J.S.; Rodriguez-Fragoso, P.; Vázquez-Hernández, F.; Luna-Arias, J.P.; Herrera-Perez, J.L.; Mendoza-Álvarez, J.G. Quantum dots for biomedical applications. In *Nanobiomaterials*; Narayan, R., Ed.; Woodhead Publishing Elsevier: Chennai, India, 2017; pp. 411–436.
111. Gui, R.; Jin, H.; Wang, Z.; Zhang, F.; Xia, J.; Yang, M.; Bi, S.; Xia, Y. Room-temperature phosphorescence logic gates developed from nucleic acid functionalized carbon dots and graphene oxide. *Nanoscale* **2015**, *7*, 8289–8293. [CrossRef] [PubMed]
112. Prasad, P.N. *Introduction to Nanomedicine and Nanobioengineering*; John Wiley & Sons: New York, NY, USA, 2012; Volume 7, ISBN 1118093437.
113. Jiang, P.; Zhu, C.-N.; Zhang, Z.-L.; Tian, Z.-Q.; Pang, D.-W. Water-soluble Ag_2S quantum dots for near-infrared fluorescence imaging in vivo. *Biomaterials* **2012**, *33*, 5130–5135. [CrossRef] [PubMed]
114. Wang, C.; Wang, Y.; Xu, L.; Zhang, D.; Liu, M.; Li, X.; Sun, H.; Lin, Q.; Yang, B. Facile aqueous-phase synthesis of biocompatible and fluorescent Ag_2S nanoclusters for bioimaging: Tunable photoluminescence from red to near infrared. *Small* **2012**, *8*, 3137–3142. [CrossRef]
115. Zhang, Y.; Hong, G.; Zhang, Y.; Chen, G.; Li, F.; Dai, H.; Wang, Q. Ag_2S quantum dot: A bright and biocompatible fluorescent nanoprobe in the second near-infrared window. *ACS Nano* **2012**, *6*, 3695–3702. [CrossRef]
116. Wang, Y.; Yan, X.P. Fabrication of vascular endothelial growth factor antibody bioconjugated ultrasmall near-infrared fluorescent Ag_2S quantum dots for targeted cancer imaging in vivo. *Chem. Commun.* **2013**, *49*, 3324–3326. [CrossRef]
117. Kim, D.; Jeong, Y.Y.; Jon, S. A drug-loaded aptamer-gold nanoparticle bioconjugate for combined CT imaging and therapy of prostate cancer. *ACS Nano* **2010**, *4*, 3689–3696. [CrossRef]
118. Gao, X.; Yang, L.; Petros, J.A.; Marshall, F.F.; Simons, J.W.; Nie, S. in vivo molecular and cellular imaging with quantum dots. *Curr. Opin. Biotechnol.* **2005**, *16*, 63–72. [CrossRef]
119. Erdem, T.; Nizamoglu, S.; Demir, H.V. Computational study of power conversion and luminous efficiency performance for semiconductor quantum dot nanophosphors on light-emitting diodes. *Opt. Express* **2012**, *20*, 3275–3295. [CrossRef]
120. Zhao, M.X.; Zeng, E.Z. Application of functional quantum dot nanoparticles as fluorescence probes in cell labeling and tumor diagnostic imaging. *Nanoscale Res. Lett.* **2015**, *10*, 1–9. [CrossRef] [PubMed]
121. Jaiswal, J.K.; Mattoussi, H.; Mauro, J.M.; Simon, S.M. Long-term multiple color imaging of live cells using quantum dot bioconjugates. *Nat. Biotechnol.* **2003**, *21*, 47–51. [CrossRef] [PubMed]
122. Li, Z.; Wang, D.; Xu, M.; Wang, J.; Hu, X.; Anwar, S.; Tedesco, A.C.; Morais, P.C.; Bi, H. Fluorine-containing graphene quantum dots with a high singlet oxygen generation applied for photodynamic therapy. *J. Mater. Chem. B* **2020**, *8*, 2598–2606. [CrossRef] [PubMed]
123. Koutsogiannis, P.; Thomou, E.; Stamatis, H.; Gournis, D.; Rudolf, P. Advances in fluorescent carbon dots for biomedical applications. *Adv. Phys. X* **2020**, *5*, 178592–178630. [CrossRef]
124. Miyashita, M.; Gonda, K.; Tada, H.; Watanabe, M.; Kitamura, N.; Kamei, T.; Sasano, H.; Ishida, T.; Ohuchi, N. Quantitative diagnosis of HER2 protein expressing breast cancer by single-particle quantum dot imaging. *Cancer Med.* **2016**, *5*, 2813–2824. [CrossRef]
125. Trapiella-Alfonso, L.; Pons, T.; Lequeux, N.; Leleu, L.; Grimaldi, J.; Tasso, M.; Oujagir, E.; Seguin, J.; d'Orlyé, F.; Girard, C. Clickable-zwitterionic copolymer capped-quantum dots for in vivo fluorescence tumor imaging. *ACS Appl. Mater. Interfaces* **2018**, *10*, 17107–17116. [CrossRef]

126. Voura, E.B.; Jaiswal, J.K.; Mattoussi, H.; Simon, S.M. Tracking metastatic tumor cell extravasation with quantum dot nanocrystals and fluorescence emission-scanning microscopy. *Nat. Med.* **2004**, *10*, 993–998. [CrossRef]
127. Gazouli, M.; Bouziotis, P.; Lyberopoulou, A.; Ikonomopoulos, J.; Papalois, A.; Anagnou, N.P.; Efstathopoulos, E.P. Quantum dots-bevacizumab complexes for in vivo imaging of tumors. in vivo **2014**, *28*, 1091–1095.
128. Yao, J.; Li, P.; Li, L.; Yang, M. Biochemistry and biomedicine of quantum dots: From biodetection to bioimaging, drug discovery, diagnostics, and therapy. *Acta Biomater.* **2018**, *74*, 36–55. [CrossRef]
129. Gao, X.; Cui, Y.; Levenson, R.M.; Chung, L.W.K.; Nie, S. in vivo cancer targeting and imaging with semiconductor quantum dots. *Nat. Biotechnol.* **2004**, *22*, 969–976. [CrossRef]
130. Olerile, L.D.; Liu, Y.; Zhang, B.; Wang, T.; Mu, S.; Zhang, J.; Selotlegeng, L.; Zhang, N. Near-infrared mediated quantum dots and paclitaxel co-loaded nanostructured lipid carriers for cancer theragnostic. *Colloids Surf. B Biointerfaces* **2017**, *150*, 121–130. [CrossRef]
131. Cai, X.; Luo, Y.; Zhang, W.; Du, D.; Lin, Y. pH-Sensitive ZnO quantum dots–doxorubicin nanoparticles for lung cancer targeted drug delivery. *ACS Appl. Mater. Interfaces* **2016**, *8*, 22442–22450. [CrossRef]
132. Penjweini, R.; Liu, B.; Kim, M.M.; Zhu, T.C. Explicit dosimetry for 2-(1-hexyloxyethyl)-2-devinyl pyropheophorbide-a-mediated photodynamic therapy: Macroscopic singlet oxygen modeling. *J. Biomed. Opt.* **2015**, *20*, 128003–128013. [CrossRef]
133. Gallardo-Villagrán, M.; Leger, D.Y.; Liagre, B.; Therrien, B. Photosensitizers used in the photodynamic therapy of rheumatoid arthritis. *Int. J. Mol. Sci.* **2019**, *20*, 3339. [CrossRef]
134. Fakayode, O.J.; Tsolekile, N.; Songca, S.P.; Oluwafemi, O.S. Applications of functionalized nanomaterials in photodynamic therapy. *Biophys. Rev.* **2018**, *10*, 49–67. [CrossRef]
135. Agrawal, A.; Tripp, R.A.; Anderson, L.J.; Nie, S. Real-time detection of virus particles and viral protein expression with two-color nanoparticle probes. *J. Virol.* **2005**, *79*, 8625–8628. [CrossRef]
136. Joo, K.-I.; Lei, Y.; Lee, C.-L.; Lo, J.; Xie, J.; Hamm-Alvarez, S.F.; Wang, P. Site-specific labeling of enveloped viruses with quantum dots for single virus tracking. *ACS Nano* **2008**, *2*, 1553–1562. [CrossRef]
137. Wen, L.; Lin, Y.; Zheng, Z.-H.; Zhang, Z.-L.; Zhang, L.-J.; Wang, L.-Y.; Wang, H.-Z.; Pang, D.-W. Labeling the nucleocapsid of enveloped baculovirus with quantum dots for single-virus tracking. *Biomaterials* **2014**, *35*, 2295–2301. [CrossRef]
138. Yoo, J.Y.; Kim, J.; Kim, J.; Huang, J.; Zhang, S.N.; Kang, Y.; Kim, H.; Yun, C. Short hairpin RNA-expressing oncolytic adenovirus-mediated inhibition of IL-8: Effects on antiangiogenesis and tumor growth inhibition. *Gene Ther.* **2008**, *16*, 635–651. [CrossRef]
139. Luo, K.; Li, S.; Xie, M.; Wu, D.; Wang, W.; Chen, R.; Huang, L.; Huang, T.; Pang, D.; Xiao, G. Real-time visualization of prion transport in single live cells using quantum dots. *Biochem. Biophys. Res. Commun.* **2010**, *394*, 493–497. [CrossRef]
140. Cheng, D.; Yu, M.; Fu, F.; Han, W.; Li, G.; Xie, J.; Song, Y.; Swihart, M.T.; Song, E. Dual recognition strategy for specific and sensitive detection of bacteria using aptamer-coated magnetic beads and antibiotic-capped gold nanoclusters. *Anal. Chem.* **2016**, *88*, 820–825. [CrossRef]
141. Jia, H.-R.; Zhu, Y.-X.; Chen, Z.; Wu, F.-G. Cholesterol-assisted bacterial cell surface engineering for photodynamic inactivation of Gram-positive and Gram-negative bacteria. *ACS Appl. Mater. Interfaces* **2017**, *9*, 15943–15951. [CrossRef]
142. Jones, K.E.; Patel, N.G.; Levy, M.A.; Storeygard, A.; Balk, D.; Gittleman, J.L.; Daszak, P. Global trends in emerging infectious diseases. *Nature* **2008**, *451*, 990–993. [CrossRef]
143. Chen, J.; Andler, S.M.; Goddard, J.M.; Nugen, S.R.; Rotello, V.M. Integrating recognition elements with nanomaterials for bacteria sensing. *Chem. Soc. Rev.* **2017**, *46*, 1272–1283. [CrossRef]
144. Cui, F.; Ye, Y.; Ping, J.; Sun, X. Carbon dots: Current advances in pathogenic bacteria monitoring and prospect applications. *Biosens. Bioelectron.* **2020**, *156*, 112085–112098. [CrossRef]
145. Qu, J.H.; Wei, Q.; Sun, D.W. Carbon dots: Principles and their applications in food quality and safety detection. *Crit. Rev. Food Sci. Nutr.* **2018**, *58*, 2466–2475. [CrossRef]
146. Shi, X.; Wei, W.; Fu, Z.; Gao, W.; Zhang, C.; Zhao, Q.; Deng, F.; Lu, X. Review on carbon dots in food safety applications. *Talanta* **2019**, *194*, 809–821. [CrossRef]

147. Xu, X.; Ray, R.; Gu, Y.; Ploehn, H.J.; Gearheart, L.; Raker, K.; Scrivens, W.A. Electrophoretic analysis and purification of fluorescent single-walled carbon nanotube fragments. *J. Am. Chem. Soc.* **2004**, *126*, 12736–12737. [CrossRef]
148. Devi, P.; Thakur, A.; Bhardwaj, S.K.; Saini, S.; Rajput, P.; Kumar, P. Metal ion sensing and light activated antimicrobial activity of *Aloe vera* derived carbon dots. *J. Mater. Sci. Mater. Electron.* **2018**, *29*, 17254–17261. [CrossRef]
149. Travlou, N.A.; Giannakoudakis, D.A.; Algarra, M.; Labella, A.M.; Rodríguez-Castellón, E.; Bandosz, T.J. S- and N-doped carbon quantum dots: Surface chemistry dependent antibacterial activity. *Carbon* **2018**, *135*, 104–111. [CrossRef]
150. Jijie, R.; Barras, A.; Bouckaert, J.; Dumitrascu, N.; Szunerits, S.; Boukherroub, R. Enhanced antibacterial activity of carbon dots functionalized with ampicillin combined with visible light triggered photodynamic effects. *Colloids Surf. B Biointerfaces* **2018**, *170*, 347–354. [CrossRef] [PubMed]
151. Yang, C.; Xie, H.; Li, Y.; Zhang, J.K.; Su, B.L. Direct and rapid quantum dots labeling of *Escherichia coli* cells. *J. Colloid Interface Sci.* **2013**, *393*, 438–444. [CrossRef]
152. Bae, P.K.; So, H.M.; Kim, K.N.; You, H.S.; Choi, K.S.; Kim, C.H.; Park, J.K.; Lee, J.O. Simple route for the detection of *Escherichia coli* using quantum dots. *Biochip J.* **2010**, *4*, 129–133. [CrossRef]
153. Gao, G.; Jiang, Y.-W.; Sun, W.; Wu, F.-G. Fluorescent quantum dots for microbial imaging. *Chin. Chem. Lett.* **2018**, *29*, 1475–1485. [CrossRef]
154. Rispail, N.; De Matteis, L.; Santos, R.; Miguel, A.S.; Custardoy, L.; Testillano, P.S.; Risueño, M.C.; Pérez-de-Luque, A.; Maycock, C.; Fevereiro, P. Quantum dot and superparamagnetic nanoparticle interaction with pathogenic fungi: Internalization and toxicity profile. *ACS Appl. Mater. Interfaces* **2014**, *6*, 9100–9110. [CrossRef]
155. Wang, X.; van de Veerdonk, F.L.; Netea, M.G. Basic genetics and immunology of *Candida* infections. *Infect. Dis. Clin. N. Am.* **2016**, *30*, 85–102. [CrossRef]
156. Oliveira, W.F.; Cabrera, M.P.; Santos, N.R.M.; Napoleão, T.H.; Paiva, P.M.G.; Neves, R.P.; Silva, M.V.; Santos, B.S.; Coelho, L.C.B.B.; Cabral Filho, P.E.; et al. Evaluating glucose and mannose profiles in *Candida* species using quantum dots conjugated with Cramoll lectin as fluorescent nanoprobes. *Microbiol. Res.* **2020**, *230*, 126330–126336. [CrossRef]
157. Singh, G.; Lakhi, K.S.; Sil, S.; Bhosale, S.V.; Kim, I.; Albahily, K.; Vinu, A. Biomass derived porous carbon for CO_2 capture. *Carbon* **2019**, *148*, 164–186. [CrossRef]
158. Ha, M.; Kim, V.N. Regulation of microRNA biogenesis. *Nat. Rev. Mol. Cell Biol.* **2014**, *15*, 509–524. [CrossRef]
159. Feinbaum, R.; Ambros, V.; Lee, R. The *C. elegans* heterochronic gene lin-4 encodes small RNAs with antisense complementarity to lin-14. *Cell* **2004**, *116*, 843–854. [CrossRef]
160. O'Brien, J.; Hayder, H.; Zayed, Y.; Peng, C. Overview of microRNA biogenesis, mechanisms of actions, and circulation. *Front. Endocrinol.* **2018**, *9*, 1–12. [CrossRef] [PubMed]
161. Goryacheva, O.A.; Mishra, P.K.; Goryacheva, I.Y. Luminescent quantum dots for miRNA detection. *Talanta* **2018**, *179*, 456–465. [CrossRef] [PubMed]
162. Zhuang, J.; Tang, D.; Lai, W.; Chen, G.; Yang, H. Immobilization-free programmable hairpin probe for ultrasensitive electronic monitoring of nucleic acid based on a biphasic reaction mode. *Anal. Chem.* **2014**, *86*, 8400–8407. [CrossRef] [PubMed]
163. Chen, Y.; Xiang, Y.; Yuan, R.; Chai, Y. Intercalation of quantum dots as the new signal acquisition and amplification platform for sensitive electrochemiluminescent detection of microRNA. *Anal. Chim. Acta* **2015**, *891*, 130–135. [CrossRef] [PubMed]

© 2020 by the authors. Licensee MDPI, Basel, Switzerland. This article is an open access article distributed under the terms and conditions of the Creative Commons Attribution (CC BY) license (http://creativecommons.org/licenses/by/4.0/).

Article

Protective Effects of Traditional Polyherbs on Cisplatin-Induced Acute Kidney Injury Cell Model by Inhibiting Oxidative Stress and MAPK Signaling Pathway

VinayKumar Dachuri [1,2,†], Phil Hyun Song [3,†], Young Woo Kim [4], Sae-Kwang Ku [1,2,*] and Chang-Hyun Song [1,2,*]

1. Department of Anatomy and Histology, College of Korean Medicine, Daegu Haany University, Gyeongsan 38610, Korea; dachurivinay@dhu.ac.kr
2. Research Center for Herbal Convergence on Liver Disease, College of Korean Medicine, Daegu Haany University, Gyeongsan 38610, Korea
3. Department of Urology, College of Medicine, Yeungnam University, Daegu 42415, Korea; sph04@hanmail.net
4. School of Korean Medicine, Dongguk University, Gyeongju 38066, Korea; ywk@dongguk.ac.kr
* Correspondence: gucci200@hanmail.net (S.-K.K.); dvmsong@hotmail.com (C.-H.S.); Tel.: +82-53-819-1549 (S.-K.K.); +82-53-819-1822 (C.-H.S.)
† These authors contributed equally to this work.

Academic Editor: Nagaraj Basavegowda
Received: 19 October 2020; Accepted: 29 November 2020; Published: 30 November 2020

Abstract: Acute kidney injury (AKI) is a disease caused by sudden renal dysfunction, which is an important risk factor for chronic renal failure. However, there is no effective treatment for renal impairment. Although some traditional polyherbs are commercially available for renal diseases, their effectiveness has not been reported. Therefore, we examined the nephroprotective effects of polyherbs and their relevant mechanisms in a cisplatin-induced cell injury model. Rat NRK-52E and human HK-2 subjected to cisplatin-induced AKI were treated with four polyherbs, Injinhotang (IJ), Ucha-Shinki-Hwan (US), Yukmijihwang-tang (YJ), and Urofen[TM] (Uro) similar with Yondansagan-tang, for three days. All polyherbs showed strong free radical scavenging activities, and the treatments prevented cisplatin-induced cell death in both models, especially at 1.2 mg/mL. The protective effects involved antioxidant effects by reducing reactive oxygen species and increasing the activities of superoxide dismutase and catalase. The polyherbs also reduced the number of annexin V-positive apoptotic cells and the expression of cleaved caspase-3, along with inhibited expression of mitogen-activated protein kinase-related proteins. These findings provide evidence for promoting the development of herbal formulas as an alternative therapy for treating AKI.

Keywords: acute kidney injury; acute renal failure; renal cell injury; polyherb; cisplatin; nephroprotective; antioxidant; antiapoptosis; MAPK

1. Introduction

Acute kidney injury (AKI), also known as acute renal failure, is a disease caused by sudden abruption in renal function with severe tubular damage, which is considered a global health problem that accounts for 9.5% of in-hospital mortality [1]. Its etiology includes infections, sepsis, failure of renal cell repair, and nephrotoxic drug insults. Cisplatin is one of the most effective chemotherapeutic agents for treating various cancers of the ovary, head and neck, lung, breast, and bladder; however, its application is limited by the development of nephrotoxicity [2]. Indeed, AKI occurs in 20–30% of patients treated with cisplatin [3]. The main treatments for AKI focus on symptomatic therapies including saline hydration and diuresis, and discontinuation of cisplatin and other specific drugs until renal function is recovered. Considering that AKI increases the risk of chronic or end-stage renal disease with an exceptionally high mortality rate, the socioeconomic importance is increasing [4]. End-stage renal disease can only be treated with dialysis and kidney replacement. Therefore, therapeutic strategies are urgently needed to reduce the risk of developing cisplatin-induced nephrotoxicity.

The pathological progress of AKI involves oxidative stress, apoptosis, and inflammation [5]. In particular, oxidative stress is closely associated with cisplatin-induced AKI; reactive oxygen species (ROS) overwhelming the endogenous antioxidant defense system induce lipid peroxidation and cell damage, resulting in a reduction in the glomerular filtration rate [6]. The overproduced ROS serves as intracellular signaling molecules to activate nuclear factor (NF)-κB [7] and mitogen-activated protein kinase (MAPK) signaling [8]. Indeed, cisplatin-induced renal tubular damage involves the activation of signaling proteins such as p38, extracellular signal-regulated kinase (ERK), and c-Jun N-terminal kinase (JNK) [9]. There have been many efforts to develop pharmacological agents for treating renal impairment, with growing interest in the beneficial effects of various antioxidants, such as vitamins C (ascorbic acid) and E, curcumin, selenium, and bixin [10]. Among these, ascorbic acid is a powerful antioxidant and free radical scavenger, which ameliorates renal failure and tubular cell damage in the cisplatin-induced AKI model [11,12]. Furthermore, several dietary phytochemicals containing polyphenols and flavonoids also appear to be efficient in AKI animal models [13]. However, most antioxidant agents have shown inconsistencies between preclinical and clinical studies despite their positive effects in the AKI model [10].

Conversely, there are traditional polyherb formulae used for treating renal deficiency since ancient times in East Asia, including Korea, China, and Japan. Some traditional polyherbs have been approved as pharmaceuticals and they are commercially available, based on long-term clinical records rather than scientific evidence; however, their effectiveness in different renal diseases needs to be clarified. Furthermore, their use should be carefully considered, because some herbal components (i.e., aristolochic acid, croton, podophyllotoxin, and other plant alkaloids) can induce severe nephrotoxicity [14]. There have been recent reports supporting the nephroprotective effects of *Boerhaavia diffusa*, *Rheum emodi*, *Nelumbo nucifera*, and *Crataeva nurvala*, as well as active compounds including curcumin and red ginseng extracts [15–17]. However, because most traditional herbal medicines are used as combination formulae consisting of various herbs for synergic effects, extensive studies are needed to prove their effectiveness for clinical use. Although mechanistic studies on individual herbs are difficult, recent systems pharmacology can provide an understanding of the nature of traditional medicines and their mechanisms [18,19].

We have attempted to screen the nephroprotective effects of polyherbs used for renal diseases in traditional Korean medicine [20], and from these, we selected four polyherbs showing strong antioxidant activities for this study. These polyherbs included the traditional medicines Injinhotang (IJ), Ucha-Shinki-Hwan (US; *Gochajinkigan* in Japanese), Yukmijihwang-tang (YJ; *Liuweidihuang-tang* in Chinese; *Rokumijio-to* in Japanese), and Urofen^{TM} (Uro) comprising ingredients similar to Yondansagan-tang (*Longdanxiegan-tang* in Chinese; *Ryutanshakan-to* in Japanese). The nephroprotective effects and relevant mechanisms were examined in a cisplatin-induced renal cell injury model.

2. Results

2.1. Free Radical Scavenging Activities of Traditional Polyherbs

In the results of the 1,1-diphenyl-2-picrylhydrazyl (DPPH) assay, one-way analysis of variance (ANOVA) showed significant differences among the groups (F = 121.7; $p < 0.01$, Figure 1). The post-hoc tests versus the vehicle control revealed significant decreases by 15.2%, 37.9%, 29.8%, and 28.3% in the IJ, Uro, US, and YJ groups, respectively ($p < 0.01$). The value was also decreased by 38.7% in the ascorbic acid (AA) group ($p < 0.01$).

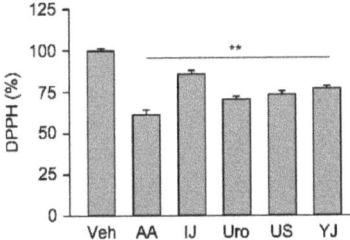

Figure 1. Free radical scavenging activity of polyherbs. Antioxidant activity of four polyherbs, IJ, Uro, US, and YJ, was assessed using the 1,1-diphenyl-2-picrylhydrazyl (DPPH) assay. Ascorbic acid (AA) was used as a positive control. The results were compared with the vehicle control (Veh), and were expressed as a percentage of the control. Values are means ± standard deviation (SD) from three independent experiments. **: $p < 0.01$ versus the Veh group.

2.2. Effects on Cell Viabilities in Cisplatin-Induced Nephrotoxicity

Compared to the normal control without cisplatin, the cell viabilities were 51.7% and 48.8% in the vehicle-treated (Veh) group of normal rat kidney (NRK) and human kidney (HK)-2 epithelial cells after cisplatin induction, respectively ($p < 0.01$), approximately reaching to the half-maximal inhibitory concentration (IC_{50}) values (Figure 2). Two-way ANOVA showed significant main effects for the group in the NRK ($F = 187.0; p < 0.01$) and HK-2 ($F = 101.1; p < 0.01$), as well as for the dose in the NRK ($F = 122.6; p < 0.01$) and HK-2 ($F = 119.2; p < 0.01$). There were significant interactions between the group and dose in the NRK ($F = 122.2; p < 0.01$) and HK-2 ($F = 250.7; p < 0.01$), indicating dose-dependent differences among the groups. Compared to the Veh group in the AKI model, the cell viabilities significantly increased in treatments with IJ at 0.6 and 1.2 mg/mL, Uro at 0.3–2.4 mg/mL, US at 1.2 and 2.4 mg/mL, YJ at 0.3–2.4 mg/mL, and AA at 35–18 µg/mL in the NRK model ($p < 0.01$). The increases were found in all the treatments at 0.6 and 1.2 mg/mL in the HK-2 ($p < 0.01$). The cell viabilities were increased more with the polyherbs at 1.2 mg/mL and AA at 35 µg/mL compared to the other doses in both models, and the treatments were, thus, used at these doses for further experiments.

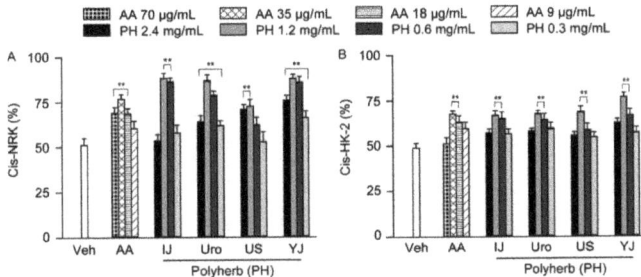

Figure 2. Protective effects on cisplatin-induced cell injuries. Renal cell injury was induced by cisplatin in normal rat kidney (Cis-NRK, (**A**)) and human kidney-2 (Cis-HK-2, (**B**)) cells. The cells were treated with AA, Injinhotang (IJ), UrofenTM (Uro), Ucha-Shinki-Hwan (US), or Yukmijihwang-tang (YJ) at the indicated doses for 3 days. The results are expressed as a percentage of the cell viability in the cisplatin non-treated normal control. Values are means ± SD from three independent experiments. **: $p < 0.01$ versus the vehicle group (Veh).

2.3. Effects on ROS Production and Antioxidant Activities

The cells stained with dichlorofluorescein diacetate (H2DCFDA) for ROS levels were evidently more in the cisplatin-treated Veh group of the NRK and HK-2; however, these seemed to be fewer in the treatments with polyherbs (Figure 3). Few stained cells were observed in the normal control of the NRK and HK-2 cells without cisplatin treatment, regardless of the groups. One-way ANOVA showed significant differences among the groups for ROS in the NRK (F = 39.4; $p < 0.01$) and HK-2 models (F = 53.1; $p < 0.01$). The levels of ROS in the NRK and HK-2 increased 2.5- and 3.1-fold in the cisplatin-treated Veh group, respectively, compared to that in the corresponding cisplatin non-treated controls ($p < 0.01$). However, the post-hoc tests versus the cisplatin-treated Veh group revealed significant decreases of 16.6%, 13.9%, 21.5%, 26.5%, and 21.9% in IJ, Uro, US, YJ, and AA groups in the NRK, and of 25.0%, 14.4%, 26.4%, 33.6%, and 34.1% in the HK-2, respectively ($p < 0.01$). In addition, significant differences were also observed among the groups for superoxide dismutase (SOD) activity in the NRK (F = 181.9; $p < 0.01$) and HK-2 (F = 130.3; $p < 0.01$), and for catalase activity in the NRK (F = 70.9; $p < 0.01$) and HK-2 (F = 102.3; $p < 0.01$). Compared to the cisplatin non-treated control, the cisplatin-treated Veh group showed significant decreases in the activities of SOD and catalase ($p < 0.01$, Figure 3): the SOD activity decreased by 39.5% and 41.2% and the catalase activity decreased by 39.9% and 38.6%, respectively in the NRK and HK-2 models. Compared to the cisplatin-treated Veh group, SOD activity significantly increased by 1.4 folds in the Uro, US, YJ, and AA groups, and 1.3 folds in the IJ in NRK, and by 1.4 folds in the US and 1.3 folds in the other groups in HK-2 ($p < 0.01$). Catalase activity increased by 1.4 folds in the IJ, US, and AA, and 1.3 folds in the Uro and YJ in NRK, and by 1.3 folds in the IJ and AA, and 1.2 folds in the other groups in HK-2 ($p < 0.01$). However, there were no significant differences among the groups in the cisplatin non-treated conditions.

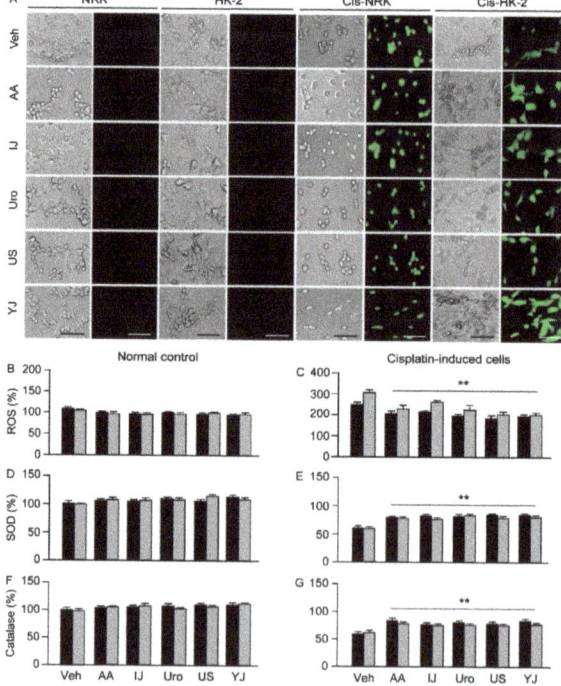

Figure 3. Effects on reactive oxygen species production and antioxidant activities. The cisplatin-induced NRK and HK-2 cells (Cis-NRK and Cis-HK-2, respectively) and the corresponding normal controls (NRK and HK-2) were treated with AA, IJ, Uro, US, or YJ for 3 days. Representative images of cells stained with dichlorofluorescein diacetate for reactive oxygen species (ROS) are shown in (**A**). Scale bars indicate 50 μm. Then, the levels of reactive oxygen species (ROS, (**A–C**)) and activities of the superoxide dismutase (SOD, (**D,E**)) and catalase (**F,G**) were assessed. Graphs show the results as percentages of the cisplatin-nontreated control. Values are expressed as means ± SD from three independent experiments. **: $p < 0.01$ versus the cisplatin-treated Veh group.

2.4. Effects on Apoptotic Changes

Flow cytometric analysis showed that the number of annexin V-positive cells was low in the cisplatin non-treated NRK and HK-2, whereas they were higher in the cisplatin-treated Veh control of both cell models (Figure 4). However, the number of annexin V-positive cells tended to be lower in both cell models treated with the polyherbs. There were significant differences among the groups in NRK (F = 130.5; $p < 0.01$) and HK-2 (F = 197.9; $p < 0.01$). The post-hoc tests versus the cisplatin-treated Veh control revealed significant decreases by 37.6%, 53.9%, 46.5%, 64.3%, and 44.9% in the IJ, Uro, US, YJ, and AA groups, respectively, in the NRK, and by 24.6%, 47.6%, 45.4%, 54.0%, and 48.7% in the HK-2 ($p < 0.01$). However, there were no significant differences in the number of cells immunostained for annexin-V only.

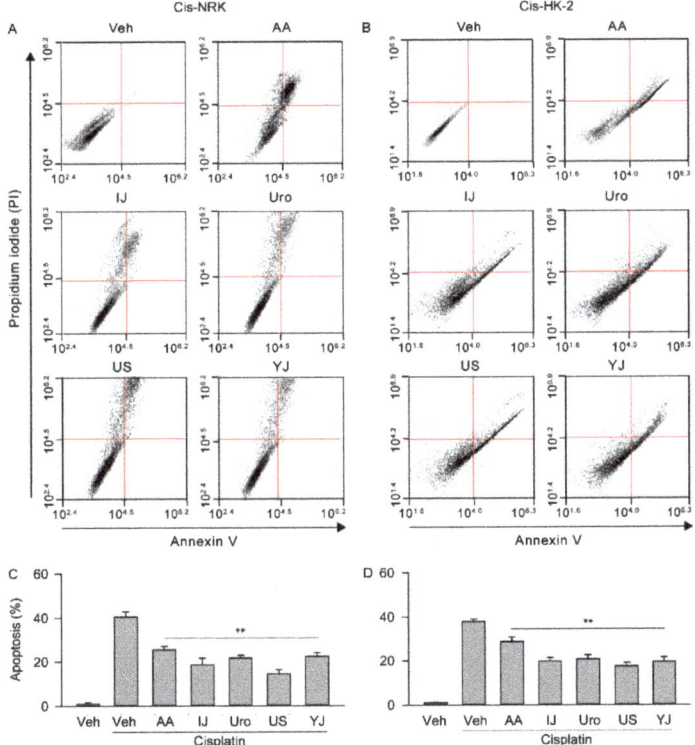

Figure 4. Effects on apoptotic changes. The cisplatin-induced NRK (Cis-NRK, (**A**)) and HK-2 (Cis-HK-2, (**B**)) cells were treated with AA, IJ, Uro, US, or YJ, and apoptosis was measured by the annexin V- and PI-double positive cells by flow cytometry. The results are expressed as percentages of the cisplatin non-treated normal control (**C,D**). Values are means ± SD from three independent experiments. **: $p < 0.01$ versus the cisplatin-treated Veh group.

2.5. Effects on Cell Proliferation

Cell proliferation was assessed in the cisplatin non-treated normal cells after treatment with the polyherbs for three days under serum-free conditions (Figure 5). Ki-67-positive cells showed no differences among the groups in the NRK; however, they were significantly different in the HK-2 (F = 16.6; $p < 0.01$). The post-hoc tests versus the normal Veh group showed significant increases by 1.2 folds in US and YJ ($p < 0.01$). Consistently, the 3-(4,5-Dimethylthiazol-2-yl)-2,5-diphenyltetrazolium bromide (MTT) assay for cell growth also showed that cells treated with US and YJ increased by 1.3 and 1.4 folds, respectively ($p < 0.01$).

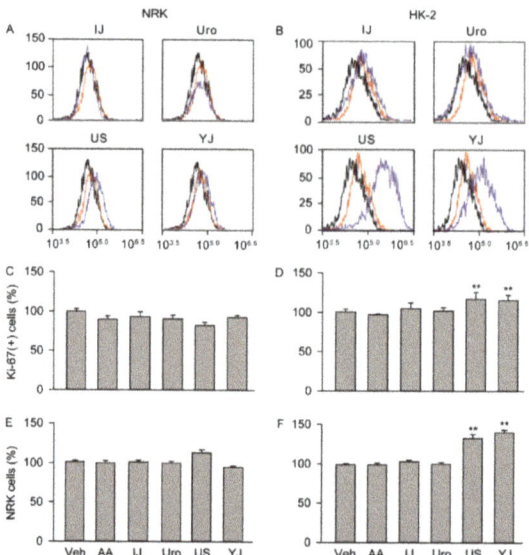

Figure 5. Effects on cell proliferation. Cells were treated with AA, IJ, Uro, US, or YJ under serum-free conditions for 3 days, and Ki-67-positive proliferative cells were measured in the NRK (**A**) and HK-2 (**B**) using flow cytometry. Blue and red lines indicate the polyherb treatment groups and vehicle control (Veh), respectively. Black indicates the experimental control omitting the primary antibody. Results are expressed as a percentage of the Veh group (**C,D**). In addition, the cell growth was assessed and expressed as a percentage of the Veh group (**E,F**). Values are means ± SD from five independent experiments. **: $p < 0.01$ versus the Veh group.

2.6. Effects on Expression of Caspase-3 and MAPK Signaling Proteins

The expression levels of cleaved caspase-3 and the phosphorylated forms of p-38 (p38-α) and JNK (JNK2) increased upon cisplatin treatment (Figure 6). For cleaved caspase-3, the cisplatin-treated Veh group showed a significant increase by 3.0 folds compared to that in the cisplatin non-treated normal control ($p < 0.01$). There were significant differences among the groups in NRK (F = 97.8, $p < 0.01$) and HK-2 (F = 75.7, $p < 0.01$). The post-hoc tests versus the cisplatin-treated Veh group showed significant decreases by 49.4%, 27.3%, 45.1%, 30.0%, and 43.4% in the IJ, Uro, US, YJ, and AA groups, respectively, in the NRK, and by 45.4%, 24.4%, 37.2%, 44.9%, and 25.2% in the HK-2 ($p < 0.01$). Furthermore, compared to the cisplatin non-treated Veh group, the cisplatin-treated Veh group showed significant increases by 3.7 and 3.1 folds in the expression of p38-α and JNK2, respectively, in the NRK, and by 3.4 and 3.7 folds in the HK-2 models ($p < 0.01$). There were significant differences among the groups in the expression of p38-α in NRK (F = 116.0, $p < 0.01$) and HK-2 (F = 55.1, $p < 0.01$), as well as JNK2 in NRK (F = 77.7, $p < 0.01$) and the HK-2 (F = 92.9, $p < 0.01$). The expression of p38-α decreased by 15.1%, 24.5%, 19.7%, 28.1%, and 25.8% in the IJ, Uro, US, YJ, and AA groups, and by 16.2%, 19.4%, 21.7%, 27.2%, and 22.2%, respectively in NRK and HK-2 ($p < 0.01$). The JNK2 decreased by 24.2%, 31.4%, 26.1%, 38.1%, and 37.9% in the IJ, Uro, US, YJ, and AA groups, respectively, in the NRK, and by 18.1%, 31.4%, 36.2%, 24.8%, and 30.1% in the HK-2 ($p < 0.01$).

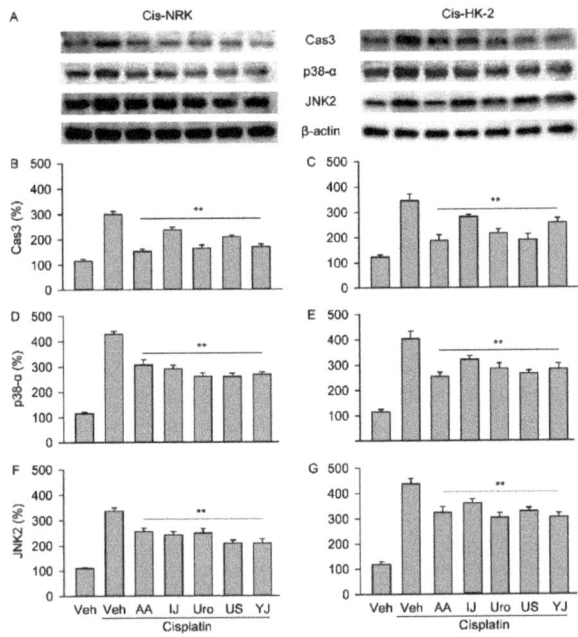

Figure 6. Effects on the expression of cleaved caspase-3 and activated mitogen-activated protein kinase (MAPK)-related proteins. The cisplatin-induced cells were treated with AA, IJ, Uro, US, or YJ for 3 days, and the expression of cleaved caspase 3 (Cas3) and phosphorylated MAPK proteins, p-38α and c-Jun N-terminal kinase (JNK2), was assessed using western-blotting (**A**). The expression was normalized to the levels of β-actin. The results are expressed as percentages of the cisplatin non-treated control (**B–G**). Values are means ± SD from three independent experiments. **: $p < 0.01$ versus the cisplatin-treated Veh group.

3. Discussion

Similar to other studies [21–23], we found that cisplatin showed IC_{50} values at 20 μM in NRK and 16 μM in HK-2, and ascorbic acid was the most efficient at 35 μg/mL (250 μM) in both cisplatin-induced cell injury models. The cell models showed increased ROS production and reduced activities of antioxidant enzymes, SOD and catalase, due to responses to the oxidative stress [24]. There have been accumulated pieces of evidence that cisplatin-induced nephrotoxicity involves oxidative damage to renal tubular cells and tissues [6,25]. Cisplatin reacts with an endogenous antioxidant, glutathione, and produces reactive electrophiles, which deteriorate mitochondrial function, and disrupt the electron transport chain, leading to increased ROS production [5,25]. The overproduced ROS reacts with lipids, proteins, and nucleic acids, causing oxidative stress and damage [26]. Here, the polyherbs showed significant free radical scavenging activities, and the treatments prevented cisplatin-induced cell death in both NRK and HK-2 models, especially at 1.2 mg/mL. Furthermore, the polyherbs reduced the ROS levels, which might increase the activities of SOD and catalase probably by the reduced consumption to the oxidative stress. These results were similar to those of the treatment of ascorbic acid used as a strong antioxidant. It is likely that the antioxidant properties of the polyherbs might contribute to reduce the oxidative stress and conserve antioxidant enzymes, resulting in protective effects against cisplatin-induced cell injuries. Many animal studies have shown nephroprotective effects via antioxidant activity [6,27], suggesting that antioxidant polyherbs can have therapeutic potential in AKI.

Our previous study has shown the nephroprotective effects of four other polyherbs including Bojungikki-tang, Palmijihwang-tang, Oryeong-san, and Wiryeong-tang [23]. The polyherbs are main traditional medicines for treating renal diseases; however, the antioxidant effects were little in treatments with Oryeong-san or lower in treatments with others. Given that the pathogenesis of AKI involves tubular oxidative stress, we selected the current polyherbs showing stronger antioxidant properties than those of Bojungikki-tang and Wiryeong-tang. However, because all the polyherbs used previously and currently inhibited cisplatin-induced apoptosis and cell death, the relevant pathway was further examined. Increased ROS is known to activate the transcription factor, NF-κB, which plays a key role in inflammatory progress [28]. Activation of NF-κB increases the levels of proinflammatory cytokines (i.e., tumor necrosis factor-α and interleukin-6) and pro-apoptotic proteins (Bcl-2 family), and mediates cell survival and differentiation [29]. The MAPK signaling pathway comprising p38, ERK, and JNK is an upstream component of NF-κB [30]. Furthermore, ROS also activates JNK in proximal tubular epithelial cells, and oxidative stress with activated JNK accelerates MAPK signaling, which induces renal cell apoptosis [31–33]. Indeed, activation of MAPK proteins has been associated with renal cell apoptosis and inflammation, leading to renal dysfunction [30,34]. In this context, the MAPK pathway has a significant correlation with the regulation of oxidative stress, apoptosis, and inflammation, and agents inhibiting oxidative stress and the MAPK pathway can have therapeutic potential for cisplatin-induced AKI. Here, a cisplatin-induced cell model showed increased expression of activated caspase-3 and apoptotic cell death as a result of the mitochondrial apoptotic cascade [35]. However, treatment with the polyherbs resulted in antiapoptotic effects, accompanied by inhibition of activation of MAPK proteins (p-38 and JNK). This indicates that the nephroprotective effects of the polyherbs involve the inhibition of oxidative stress and MAPK signaling.

In traditional Korean medicine, IJ is prescribed for treating inflammation-related diseases, especially in the liver, and Uro, US, and YJ, are used for treating kidney deficiency [20]. YJ, consisting of six herbs, is the most common polyherb used for treating renal diseases, and many reports have shown its preventive effects on renal hypertension and ischemic acute renal failure [36,37], along with antioxidant and anti-tumor effects [36,38,39]. US consists of 10 herbs, including one in YJ. The *Alismatis rhizome* and *Dioscoreae rhizome* contained in the YJ and US have been shown to improve blood flow [40] and to exert anti-inflammatory effects [41–43]. In addition, the main herb (*Artemisia capillaris*) of IJ and its compounds (dapillarisin) have shown strong antioxidant and anti-inflammatory effects [44,45]; Yondansagan-tang, which contains components similar to those of Uro, is used to treat hepatorenal syndrome with hepatic inflammation and dysuresia [46]. In addition, the single herbs of *Rhubarb* and *Gardenia* fruit included in IJ and Uro; *Cinnamon* in Uro and US; *Rehmannia* Root in Uro, US, and YJ; *Moutan* root bark in US and YJ; and *Cornus* fruit in YJ have been reported to exert nephroprotective effects in animal models [47,48]. However, because of a lack of mechanism studies, it is difficult to speculate which herbs or combinations can have nephroprotective effects. The present study is the first to report the nephroprotective effects of IJ, Uro, US, and YJ, and their relevant mechanisms. Because traditional polyherbal formulae are used in various herbal combinations based on accumulated clinical experiences, they may have synergic effects via multi-targeting. Furthermore, YJ possessing antitumor effects can be used in combination with chemotherapeutic agents for treating renal cancer-related AKI. Future studies are, thus, needed to clarify the exact mechanisms by which polyherbal combinations enhance their beneficial effects.

Traditional polyherbs have advantages in that they can be applied to clinical patients for a long time with few side effects. Here, polyherb-related cell death was not observed in the NRK and HK-2 cells; rather, cell proliferation was observed in HK-2 cells treated with US or YJ. As mentioned above, US and YJ share six herbs, and their specific components could contribute to their proliferative effects on human tubular cells; however, these components are unclear. Here, we demonstrated that antioxidant polyherbs might exert nephroprotective effects by regulating MAPK signaling. The polyherbs used were approved by the Korea Food and Drug Administration (FDA); however, their use has generally ceased in hospitalized patients with AKI. These results support scientific evidence for the effectiveness of polyherbs and their relevant mechanisms, which provide useful information for their clinical application in AKI.

4. Materials and Methods

4.1. Preparation of Traditional Polyherbs

The four polyherbs used here have been approved by the Korea FDA as general pharmaceuticals, and are commercially available. These were IJ (Panparu™, Hanpoong Pharmaceutical Co. Ltd., Daejeon, Korea), Uro (Urofen™, Hanpoong Pharmaceutical Co. Ltd.), US (Bosinji™, Jeil Pharmaceutical Co. Ltd., Seoul, Korea), and YJ (Yunbohwan™, Kyoungbang Pharmacy, Incheon, Korea). Their individual ingredients are listed in Table 1. The polyherbs were dissolved in absolute dimethyl sulfoxide (DMSO; Sigma-Aldrich, St. Louis, MO, USA), and then diluted with the cell culture medium at a final concentration of 2.4 mg/mL with 0.5% DMSO as a vehicle. They were filtered through a pore size of 0.22 μm, and stored at 4 °C in the dark until use. The effects of the polyherbs were compared with those of AA (Sigma-Aldrich) as a positive control.

Table 1. Individual herbs composing the polyherbs used in this study.

	Ingredients
IJ	Artemisiae Capillaris Herba 2 g, Gardenia Fruit 1 g, Rhubarb 0.67 g
Uro	Akebiae Caulis 416.7 mg, Alisma Rhizome 250 mg, Angelica Gigas Root 416.7 mg, Cinnamon Bark 16.7 mg, Ephedra Herb 16.7 mg, Forsythia Fruit 16.7 mg, Gardenia Fruit 125 mg, Gentian Root 125 mg, Glycyrrhiza 125 mg, Plantago Seed 250 mg, Rhubarb 16.7 mg, Scutellaria Root 250 mg, Raw Ginger 16.7 mg, Rehmannia Root 416.7 mg
US	Achyranthes Root 3.0 mg, Alisma Rhizome 3.0 mg, Cinnamon Bark 1 g, Cornus Fruit 3.0 mg, Dioscorea Rhizome 3.0 mg, Hoelen 3 g, Moutan Root Bark 3.0 mg, Psyllium Husk 3 g, Pulvis Aconiti Tuberis Purificatum 1.0 mg, Rehmannia Root 5.0 mg
YJ	Alisma Rhizome 240 mg, Cornus Fruit 320 mg, Dioscorea Rhizome 320 mg, Hoelen 240 mg, Moutan Root Bark 240 mg, Steamed Rehmannia Root 640 mg

4.2. Free Radical Scavenging Activity

Antioxidant activities of polyherbs were assessed using the DPPH (Sigma-Aldrich) assay. Briefly, the polyherbal solution was incubated with 0.4 mM DPPH solution in methanol at a final concentration of 1 mg/mL for 30 min in the dark. Distilled water containing 0.5% DMSO was used as the vehicle control. Absorbance was measured at 517 nm using an automated microplate reader (BIO-TEK, Winooski, VT, USA) and antioxidant activity was calculated using the following formula:

$$\text{Inhibition (\%)} = 1 - \frac{(\text{Absorbance of control} - \text{Absorbance of polyherb})}{\text{Absorbance of control}} \times 100$$

4.3. Cell Culture

Kidney proximal tubular epithelial cell lines, rat NRK-52E (NRK) and human HK-2, were obtained from the American Type Culture Collection (URL www.atcc.org). NRK and HK-2 were cultured in Dulbecco's modified medium–high glucose (Hyclone, Logan, UT, USA) and Roswell Park Memorial Institute 1640 (Gibco, Grand Island, NY, USA), respectively. The media were supplemented with 100 U/mL penicillin/streptomycin and 10% fetal bovine serum (FBS; Gibco). The cells were maintained at 37 °C in a humidifying incubator with 5% CO_2.

4.4. Cisplatin-Induced Acute Kidney Cell Model and Treatments

When cells were grown to 80–90% confluence, they were seeded in 96-well (1×10^4 cells/well) and 6-well (1×10^6 cells/well) plates. The AKI cell model was induced by cisplatin (Sigma-Aldrich) at 20 μM in NRK and at 16 μM in HK-2, as the IC_{50} under the serum-free conditions. The cells were then treated with the polyherbs for three days, and the results were compared with those of the negative control treated with vehicle alone (Veh).

4.5. Cell Viability Assay

Cell viability was assessed using a MTT (TCI Chemicals, Tokyo, Japan) assay. MTT solution at 0.5 mg/mL in distilled water was added to the treated cells and incubated for 1 h at 37 °C. The cells were then lysed in DMSO, and the absorbance was measured at 570 nm using a microplate reader (BIO-TEK). Viability was represented as a percentage of the Veh group.

4.6. Assessment of ROS Levels

Cells (1×10^4 cells/well in a black 96-well flat-bottom plate) or cell suspensions (5×10^4 cells/well in a black 96-well V-bottom plate) were incubated with 5 µM H2DCFDA (Invitrogen, Waltham, MA, USA) at 37 °C for 15 min. The stained cells were observed using an inverted fluorescence microscope, and the fluorescence intensities of the cell suspensions were measured at Ex-495nm and Em-520nm using a microplate reader (BIO-TEK).

4.7. Assessment of Activities of Antioxidant Enzymes

Activities of antioxidant enzymes, SOD and catalase, were measured using commercial enzyme-linked immunosorbent assay kits (#706002 for SOD and #707002 for catalase, Cayman, Ann Arbor, MI, USA), according to the manufacturer's instructions. For SOD, cells were sonicated in 20 mM HEPES buffer (pH 7.2) containing 1 mM ethylene glycol tetraacetic acid, 210 mM mannitol, and 70 mM sucrose. The cell lysates were centrifuged at $1500 \times g$ for 5 min at 4 °C, and the supernatants were reacted with tetrazolium salt. For catalase, cells were sonicated in 50 mM potassium phosphate buffer (pH 7.0) containing 1 mM ethylenediaminetetraacetic acid. The lysates were centrifuged at $10,000 \times g$ for 15 min at 4 °C, and the supernatants were reacted with formaldehyde. The absorbance of the reactions was measured at 450 and 540 nm for SOD and catalase, respectively, under the standard curves using a microplate reader (BIO-TEK).

4.8. Flow Cytometric Analysis

Apoptotic changes and cell proliferation were measured using flow cytometry, as described previously [23]. Briefly, for apoptosis, cell samples were incubated with a rabbit anti-annexin V–FITC (1:1000, #14085, Abcam, Cambridge, UK) for 30 min, followed by propidium iodide at 50 µg/mL (Life Technologies, Carlsbad, CA, USA) for 10 min. For determining cell proliferation, cells were fixed in 4% formaldehyde solution, and cell membranes were permeated with saponin at 1 mg/mL (TCI chemicals). The cells were incubated with a rabbit anti-Ki-67 antibody (1:100, #15580, Abcam), and then with an Alexa 488-conjugated goat anti-rabbit IgG antibody (1:1000, #11008, Life Technologies) for 30 min each. All steps were performed on ice, and cells were washed three times with phosphate-buffered saline containing 2% FBS between each step. The cells omitting the primary antibodies were used as negative controls. The immunopositive cells were analyzed on a BD Accuri C6-Plus flow cytometer (BD Bioscience, San Jose, CA, USA).

4.9. Immunoblotting

Cells were centrifuged at $10,000 \times g$ for 10 min at 4 °C, and the cell pellet was lysed in RIPA buffer (Rock Land, Pottstown, PA, USA) with 1 mg/mL protease inhibitor (Leupeptin^{TM}, Roche, Mannheim, Germany) for 30 min. After measuring the amount of total protein using the BCA assay (Thermo-Fisher, Rockford, IL, USA), the lysates were mixed with sodium dodecylsulfate (SDS)-gel loading buffer (Biorad, Hercules, CA, USA), and boiled for 10 min. Equal amounts of samples were electrophoresed on 10% SDS-polyacrylamide gel electrophoresis gels, and transferred onto nitrocellulose membranes using semi-dry blot transfer (Bio-Rad). The membrane was blocked with 5% skim milk in tris-buffered saline with 0.1% tween (TBST), and then incubated overnight at 4 °C with the following primary antibodies: mouse anti-cleaved caspase-3 (1:100, #9668, Cell Signaling, Danvers, MA, USA), mouse anti-p38-α (1:500, #8691, R&D systems, Minneapolis, MN, USA), rabbit anti-JNK2 (1:1000, #178953, Abcam), and mouse anti-β-actin antibodies (1:2000, Abcam). The next day, the cells were incubated with anti-mouse and anti-rabbit horseradish peroxidase-conjugated secondary antibodies (1:1000, #1706515 and #1706516, respectively, Biorad) for 1 h. The membrane was washed with TBST five times for 30 min after each incubation. The expression was visualized using WESTARηC2.0 (Cyanagen, Bologna, Italy), and analyzed using a ChemiDoc instrument (Bio-Rad). The expression levels were normalized to those of β-actin.

4.10. Statistical Analysis

Data are expressed as the means ± standard deviation in each experiment performed independently at least three times. The homogeneity of variance was examined by the Levene test. As it was not significant, multi-comparison ANOVA was examined, followed by Tukey post-hoc tests. Dose-dependent effects of polyherbs on the cell viabilities in the AKI model were examined by two-way ANOVA with main factors for the group and the dose, and the others were examined by one-way ANOVA. Multi-comparison was described to be significant in the treatment group compared to the Veh group. A p-value of less than 0.05 was considered statistically significant.

Author Contributions: Conceptualization, methodology, and validation, V.D., P.H.S., Y.W.K., and C.-H.S.; software, formal analysis, investigation, resources, and data curation, V.D. C.-H.S., and S.-K.K.; writing—original draft preparation, V.D. and P.H.S.; visualization, V.D., and Y.W.K.; writing—review and editing and supervision, C.-H.S., and S.-K.K.; project administration and funding acquisition, P.H.S. and C.-H.S. All authors have read and agreed to the published version of the manuscript.

Funding: This work was supported by the National Research Foundation of Korea (NRF) grant funded by the Korea government (MSIT) (grant number 2018R1A5A2025272) and the Basic Science Research Program through the NRF funded by the Ministry of Education (grant numbers 2017R1D1A3B03031470 and 2017R1D1A3B03031498).

Acknowledgments: V.D. and P.H.S. contributed equally to this work. We appreciate Hanpoong, Jeil, and Kyoungbang Pharmaceutical companies, and the related persons for providing the polyherbs.

Conflicts of Interest: The authors declare no conflict of interest.

References

1. Bellomo, R.; Kellum, J.A.; Ronco, C. Acute kidney injury. *Lancet* **2012**, *380*, 756–766. [CrossRef]
2. Dasari, S.; Tchounwou, P.B. Cisplatin in cancer therapy: Molecular mechanisms of action. *Eur. J. Pharm.* **2014**, *740*, 364–378. [CrossRef] [PubMed]
3. Nematbakhsh, M.; Nasri, H. The effects of vitamin E and selenium on cisplatin-induced nephrotoxicity in cancer patients treated with cisplatin-based chemotherapy: A randomized, placebo-controlled study. *J. Res. Med. Sci.* **2013**, *18*, 625.
4. Coca, S.G.; Singanamala, S.; Parikh, C.R. Chronic kidney disease after acute kidney injury: A systematic review and meta-analysis. *Kidney Int.* **2012**, *81*, 442–448. [CrossRef] [PubMed]
5. Ozkok, A.; Edelstein, C.L. Pathophysiology of cisplatin-induced acute kidney injury. *Biomed. Res. Int.* **2014**, *2014*, 967826. [CrossRef]
6. Song, K.I.; Park, J.Y.; Lee, S.; Lee, D.; Jang, H.J.; Kim, S.N.; Ko, H.; Kim, H.Y.; Lee, J.W.; Hwang, G.S.; et al. Protective effect of tetrahydrocurcumin against cisplatin-induced renal damage: In vitro and in vivo studies. *Planta Med.* **2015**, *81*, 286–291. [CrossRef]
7. Asehnoune, K.; Strassheim, D.; Mitra, S.; Kim, J.Y.; Abraham, E. Involvement of reactive oxygen species in Toll-like receptor 4-dependent activation of NF-kappa B. *J. Immunol.* **2004**, *172*, 2522–2529. [CrossRef]
8. Nathan, C. Specificity of a third kind: Reactive oxygen and nitrogen intermediates in cell signaling. *J. Clin. Investig.* **2003**, *111*, 769–778. [CrossRef]
9. Arany, I.; Megyesi, J.K.; Kaneto, H.; Price, P.M.; Safirstein, R.L. Cisplatin-induced cell death is EGFR/src/ERK signaling dependent in mouse proximal tubule cells. *Am. J. Physiol. Ren. Physiol.* **2004**, *287*, F543–F549. [CrossRef]
10. Dennis, J.M.; Witting, P.K. Protective Role for Antioxidants in Acute Kidney Disease. *Nutrients* **2017**, *9*, 718. [CrossRef]

11. Maliakel, D.M.; Kagiya, T.V.; Nair, C.K.K. Prevention of cisplatin-induced nephrotoxicity by glucosides of ascorbic acid and α-tocopherol. *Exp. Toxicol. Pathol.* **2008**, *60*, 521–527. [CrossRef] [PubMed]
12. Abdel-Daim, M.M.; Abushouk, A.I.; Donia, T.; Alarifi, S.; Alkahtani, S.; Aleya, L.; Bungau, S.G. The nephroprotective effects of allicin and ascorbic acid against cisplatin-induced toxicity in rats. *Environ. Sci. Pollut. Res.* **2019**, *26*, 13502–13509. [CrossRef] [PubMed]
13. Chatterjee, P.K. Novel pharmacological approaches to the treatment of renal ischemia-reperfusion injury: A comprehensive review. *Naunyn Schmiedebergs Arch. Pharm.* **2007**, *376*, 43. [CrossRef] [PubMed]
14. Yang, B.; Xie, Y.; Guo, M.; Rosner, M.H.; Yang, H.; Ronco, C. Nephrotoxicity and Chinese Herbal Medicine. *Clin. J. Am. Soc. Nephrol.* **2018**, *13*, 1605–1611. [CrossRef] [PubMed]
15. Quan, H.Y.; Kim, D.Y.; Chung, S.H. Korean red ginseng extract alleviates advanced glycation end product-mediated renal injury. *J. Ginseng Res.* **2013**, *37*, 187–193. [CrossRef]
16. Chen, H.; Busse, L.W. Novel Therapies for Acute Kidney Injury. *Kidney Int. Rep.* **2017**, *2*, 785–799. [CrossRef]
17. Sharma, S.; Baboota, S.; Amin, S.; Mir, S.R. Ameliorative effect of a standardized polyherbal combination in methotrexate-induced nephrotoxicity in the rat. *Pharm. Biol.* **2020**, *58*, 184–199. [CrossRef]
18. Li, Y.; Zhang, J.; Zhang, L.; Chen, X.; Pan, Y.; Chen, S.S.; Zhang, S.; Wang, Z.; Xiao, W.; Yang, L.; et al. Systems pharmacology to decipher the combinational anti-migraine effects of Tianshu formula. *J. Ethnopharmacol.* **2015**, *174*, 45–56. [CrossRef]
19. Luo, Y.; Wang, Q.; Zhang, Y. A systems pharmacology approach to decipher the mechanism of danggui-shaoyao-san decoction for the treatment of neurodegenerative diseases. *J. Ethnopharmacol.* **2016**, *178*, 66–81. [CrossRef]
20. Huh, J. *Dong Ui Bo Gam*; Dong Ui Bo Gam Publisher: Seoul, Korea, 2005.
21. Jeong, J.J.; Park, N.; Kwon, Y.J.; Ye, D.J.; Moon, A.; Chun, Y.J. Role of annexin A5 in cisplatin-induced toxicity in renal cells: Molecular mechanism of apoptosis. *J. Biol. Chem.* **2014**, *289*, 2469–2481. [CrossRef]
22. Dogra, S.; Bandi, S.; Viswanathan, P.; Gupta, S. Arsenic trioxide amplifies cisplatin toxicity in human tubular cells transformed by HPV-16 E6/E7 for further therapeutic directions in renal cell carcinoma. *Cancer Lett.* **2015**, *356*, 953–961. [CrossRef] [PubMed]
23. Dachuri, V.; Song, P.H.; Ku, S.K.; Song, C.H. Protective Effects of Traditional Herbal Formulas on Cisplatin-Induced Nephrotoxicity in Renal Epithelial Cells via Antioxidant and Antiapoptotic Properties. *Evid. Based Complement. Altern. Med.* **2020**, *2020*, 5807484. [CrossRef] [PubMed]
24. Baliga, R.; Ueda, N.; Walker, P.D.; Shah, S.V. Oxidant mechanisms in toxic acute renal failure. *Drug Metab. Rev.* **1999**, *31*, 971–997. [CrossRef] [PubMed]
25. Yao, X.; Panichpisal, K.; Kurtzman, N.; Nugent, K. Cisplatin nephrotoxicity: A review. *Am. J. Med. Sci.* **2007**, *334*, 115–124. [CrossRef] [PubMed]
26. Pabla, N.; Dong, Z. Cisplatin nephrotoxicity: Mechanisms and renoprotective strategies. *Kidney Int.* **2008**, *73*, 994–1007. [CrossRef] [PubMed]
27. Al-Majed, A.A.; Sayed-Ahmed, M.M.; Al-Yahya, A.A.; Aleisa, A.M.; Al-Rejaie, S.S.; Al-Shabanah, O.A. Propionyl-L-carnitine prevents the progression of cisplatin-induced cardiomyopathy in a carnitine-depleted rat model. *Pharm. Res.* **2006**, *53*, 278–286. [CrossRef]
28. Benedetti, G.; Fredriksson, L.; Herpers, B.; Meerman, J.; van de Water, B.; de Graauw, M. TNF-α-mediated NF-κB survival signaling impairment by cisplatin enhances JNK activation allowing synergistic apoptosis of renal proximal tubular cells. *Biochem. Pharmacol.* **2013**, *85*, 274–286. [CrossRef]
29. Guerrero-Beltrán, C.E.; Mukhopadhyay, P.; Horváth, B.; Rajesh, M.; Tapia, E.; García-Torres, I.; Pedraza-Chaverri, J.; Pacher, P. Sulforaphane, a natural constituent of broccoli, prevents cell death and inflammation in nephropathy. *J. Nutr. Biochem.* **2012**, *23*, 494–500. [CrossRef]
30. Ma, X.; Dang, C.; Kang, H.; Dai, Z.; Lin, S.; Guan, H.; Liu, X.; Wang, X.; Hui, W. Saikosaponin-D reduces cisplatin-induced nephrotoxicity by repressing ROS-mediated activation of MAPK and NF-κB signalling pathways. *Int. Immunopharmacol.* **2015**, *28*, 399–408. [CrossRef]
31. Cui, X.-L.; Douglas, J.G. Arachidonic acid activates c-jun N-terminal kinase through NADPH oxidase in rabbit proximal tubular epithelial cells. *Proc. Natl. Acad. Sci. USA* **1997**, *94*, 3771–3776. [CrossRef]
32. Tsuruya, K.; Tokumoto, M.; Ninomiya, T.; Hirakawa, M.; Masutani, K.; Taniguchi, M.; Fukuda, K.; Kanai, H.; Hirakata, H.; Iida, M. Antioxidant ameliorates cisplatin-induced renal tubular cell death through inhibition of death receptor-mediated pathways. *Am. J. Physiol. Ren. Physiol.* **2003**, *285*, F208–F218. [CrossRef] [PubMed]

33. Bragado, P.; Armesilla, A.; Silva, A.; Porras, A. Apoptosis by cisplatin requires p53 mediated p38α MAPK activation through ROS generation. *Apoptosis* **2007**, *12*, 1733–1742. [CrossRef] [PubMed]
34. Kim, H.Y.; Okubo, T.; Juneja, L.R.; Yokozawa, T. The protective role of amla (Emblica officinalis Gaertn.) against fructose-induced metabolic syndrome in a rat model. *Br. J. Nutr.* **2010**, *103*, 502–512. [CrossRef] [PubMed]
35. Liu, X.; Huang, Z.; Zou, X.; Yang, Y.; Qiu, Y.; Wen, Y. Panax notoginseng saponins attenuates cisplatin-induced nephrotoxicity via inhibiting the mitochondrial pathway of apoptosis. *Int. J. Clin. Exp. Pathol.* **2014**, *7*, 8391.
36. Cai, B.; Jiang, T. Study on preventive and curative effects of liu wei di huang tang on tumors. *J. Tradit. Chin. Med. Chung I Tsa Chih Ying Wen Pan* **1994**, *14*, 207–211. [PubMed]
37. Kang, D.G.; Sohn, E.J.; Moon, M.K.; Mun, Y.J.; Woo, W.H.; Kim, M.K.; Lee, H.S. Yukmijihwang-tang ameliorates ischemia/reperfusion-induced renal injury in rats. *J. Ethnopharmacol.* **2006**, *104*, 47–53. [CrossRef]
38. Kim, J.S.; Na, C.S.; Pak, S.C.; Kim, Y.G. Effects of yukmi, an herbal formula, on the liver of senescence accelerated mice (SAM) exposed to oxidative stress. *Am. J. Chin. Med.* **2000**, *28*, 343–350. [CrossRef]
39. Wu, C.-T.; Tsai, Y.-T.; Lin, J.-G.; Fu, S.-l.; Lai, J.-N. Chinese herbal products and the reduction of risk of breast cancer among females with type 2 diabetes in Taiwan: A case–control study. *Medicine* **2018**, *97*, e11600. [CrossRef]
40. Suzuki, Y.; Goto, K.; Ishige, A.; Komatsu, Y.; Kamei, J. Effect of Gosha-jinki-gan, a Kampo medicine, on enhanced platelet aggregation in streptozotocin-induced diabetic rats. *Jpn. J. Pharm.* **1998**, *78*, 87–91. [CrossRef]
41. Tian, T.; Chen, H.; Zhao, Y.Y. Traditional uses, phytochemistry, pharmacology, toxicology and quality control of Alisma orientale (Sam.) Juzep: A review. *J. Ethnopharmacol.* **2014**, *158*, 373–387. [CrossRef]
42. Fu, P.K.; Yang, C.Y.; Tsai, T.H.; Hsieh, C.L. Moutan cortex radicis improves lipopolysaccharide-induced acute lung injury in rats through anti-inflammation. *Phytomedicine* **2012**, *19*, 1206–1215. [CrossRef]
43. Liao, J.C.; Deng, J.S.; Chiu, C.S.; Hou, W.C.; Huang, S.S.; Shie, P.H.; Huang, G.J. Anti-Inflammatory Activities of Cinnamomum cassia Constituents In Vitro and In Vivo. *Evid. Based Complement. Altern. Med.* **2012**, *2012*, 429320. [CrossRef]
44. Seo, H.-C.; Suzuki, M.; Ohnishi-Kameyama, M.; Oh, M.-J.; Kim, H.-R.; Kim, J.-H.; Nagata, T. Extraction and identification of antioxidant components from Artemisia capillaris herba. *Plant Foods Hum. Nutr.* **2003**, *58*, 1–12. [CrossRef]
45. Han, K.-H.; Jeon, Y.-J.; Athukorala, Y.; Choi, K.-D.; Kim, C.-J.; Cho, J.-K.; Sekikawa, M.; Fukushima, M.; Lee, C.-H. A water extract of Artemisia capillaris prevents 2, 2'-azobis (2-amidinopropane) dihydrochloride-induced liver damage in rats. *J. Med. Food* **2006**, *9*, 342–347. [CrossRef] [PubMed]
46. Cheng, H.Y.; Huang, H.H.; Yang, C.M.; Lin, L.T.; Lin, C.C. The in vitro anti-herpes simplex virus type-1 and type-2 activity of Long Dan Xie Gan Tan, a prescription of traditional Chinese medicine. *Chemotherapy* **2008**, *54*, 77–83. [CrossRef] [PubMed]
47. Lee, S.; Jung, K.; Lee, D.; Lee, S.R.; Lee, K.R.; Kang, K.S.; Kim, K.H. Protective effect and mechanism of action of lupane triterpenes from Cornus walteri in cisplatin-induced nephrotoxicity. *Bioorg. Med. Chem. Lett.* **2015**, *25*, 5613–5618. [CrossRef] [PubMed]
48. Li, J.; Tan, Y.J.; Wang, M.Z.; Sun, Y.; Li, G.Y.; Wang, Q.L.; Yao, J.C.; Yue, J.; Liu, Z.; Zhang, G.M.; et al. Loganetin protects against rhabdomyolysis-induced acute kidney injury by modulating the toll-like receptor 4 signalling pathway. *Br. J. Pharm.* **2019**, *176*, 1106–1121. [CrossRef] [PubMed]

Publisher's Note: MDPI stays neutral with regard to jurisdictional claims in published maps and institutional affiliations.

© 2020 by the authors. Licensee MDPI, Basel, Switzerland. This article is an open access article distributed under the terms and conditions of the Creative Commons Attribution (CC BY) license (http://creativecommons.org/licenses/by/4.0/).

Article

Impact of Ag/ZnO Reinforcements on the Anticancer and Biological Performances of CA@Ag/ZnO Nanocomposite Materials

Nadiyah Alahmadi [1,*] and Mahmoud A. Hussein [2,3,*]

1. Department of Chemistry, College of Science, University of Jeddah, Jeddah 21959, Saudi Arabia
2. Department of Chemistry, Faculty of Science, King Abdulaziz University, Jeddah 21589, Saudi Arabia
3. Chemistry Department, Facul1ty of Science, Assiut University, Assiut 71516, Egypt
* Correspondence: nalahmadi@uj.edu.sa (N.A.); mahussein74@yahoo.com or maabdo@kau.edu.sa (M.A.H.)

Abstract: In this study, an unpretentious, non-toxic, and cost-effective dissolution casting method was utilized to synthesize a group of anticancer and biologically active hybrid nanocomposite materials containing biopolymer cellulose acetate. Pristine ZnO and $Ag_{(0.01,\ 0.05,\ 0.1)}$/ZnO hybrid nanofillers based on variable Ag NP loadings were prepared via green procedures in the presence of gum arabic (GA). The chemical structures and the morphological features of the designed nanocomposite materials were investigated by PXRD, TEM, SEM, FTIR, TGA, and XPS characterization techniques. The characterization techniques confirmed the formation of CA@$Ag_{(0.01,\ 0.05,\ 0.1)}$/ZnO hybrid nanocomposite materials with an average crystallite size of 15 nm. All investigated materials showed two degradation steps. The thermal stability of the fabricated samples was ranked in the following order: CA/ZnO < CA@$Ag_{(0.01)}$/ZnO < CA@$Ag_{(0.05)}$/ZnO = CA@$Ag_{(0.1)}$/ZnO. Hence, the higher Ag doping level slightly enhanced the thermal stability. The developed nanocomposites were tested against six pathogens and were used as the target material to reduce the number of cancer cells. The presence of Ag NPs had a positive impact on the biological and the anticancer activities of the CA-reinforced Ag/ZnO composite materials. The CA@$Ag_{(0.1)}$/ZnO hybrid nanocomposite membrane had the highest antimicrobial activity in comparison to the other fabricated materials. Furthermore, the developed CA@$Ag_{(0.1)}$/ZnO hybrid nanocomposite material effectively induced cell death in breast cancer.

Keywords: cellulose acetate; Ag-doped ZnO; nanocomposite materials; green synthesis; antibacterial activity; anticancer activity

Citation: Alahmadi, N.; Hussein, M.A. Impact of Ag/ZnO Reinforcements on the Anticancer and Biological Performances of CA@Ag/ZnO Nanocomposite Materials. *Molecules* 2023, 28, 1290. https://doi.org/10.3390/molecules28031290

Academic Editors: Nagaraj Basavegowda and Kwang-Hyun Baek

Received: 29 December 2022
Revised: 23 January 2023
Accepted: 25 January 2023
Published: 29 January 2023

Copyright: © 2023 by the authors. Licensee MDPI, Basel, Switzerland. This article is an open access article distributed under the terms and conditions of the Creative Commons Attribution (CC BY) license (https://creativecommons.org/licenses/by/4.0/).

1. Introduction

Owing to their inherently advantageous structural, electrical, and mechanical characteristics, polymer nanocomposites have experienced rapid growth in popularity and advancement over the past few generations. The use of nanofiller in a polymer host could be beneficial for a number of purposed, such as in biosensors, energy storage devices, photocatalysts, drug delivery, and other applications [1]. Biopolymers differ from the traditional polymers. They are created or obtained from living creatures, such as plants and microbes [2]. Biopolymers may be natural or synthetic in nature [1]. The use of biopolymers could produce material with unique properties, including biodegradability, biocompatibility, and sustainability [3]. Biodegradable polymers, or biopolymers, are synthetic materials that can be decomposed by microorganisms such as bacteria and fungi [1,3]. As a result, they do not harm the environment in any way [1]. Under aerobic conditions, biopolymers degrade into CO_2, H_2O, and biomass, whereas under anaerobic conditions, they decompose into methane, hydrocarbons, and biomass [3]. Therefore, biopolymers are promising substitutes for petroleum-based materials. Biopolymers can be categorized into three groups depending on their sources. Biomass biopolymers, such as polysaccharides, proteins, and lipids, are extracted from biomass. Biopolymers such as polylactic acid (PLA) can be chemically synthesized from biomass. Microorganism biopolymers, such as microbial polysaccharides and microbial polyesters, are produced by microorganisms [3]. Improvements in their mechanical and thermal characters can be achieved through nanoparticle inclusion into the polymeric matrices. A number of nanoparticles, including metallic-based materials and their oxides, as well as nanoclays, have been utilized to enhance the properties of polymers [4,5]. The development of biopolymers containing nanocomposites and the many fields to which they have been applied have made possible new avenues of study. Additionally, these types of nanocomposites are an emerging category of biohybrid composites that typically combine biopolymer matrices with nanoscale reinforcing elements [6]. Such eco-friendly NCs are well-rounded, so they should improve compatibility, recycling, and output rates [3]. Moreover, biopolymers containing nanocomposites have a highly flexible range of significant industrial uses in modern technology, such as for automotive parts, environmentally sound packaging utilities, biomedical applications, smart electronics, and a wide range of other applications [6].

Zinc oxide nanoparticles (ZnO NPs) constitute a multifunctional metal oxide because of their unique electronic structure. They are also classified as semiconductor materials (n-type), owing to the wide range of their bandgap (3.37 eV). Moreover, as non-toxic materials, they have potential in environmental and biological applications. ZnO NPs are applied in a wide range of applications, for instance, gas sensors, optoelectronic devices, dye-sensitized solar cells, and photocatalysts [7,8]. ZnO NPs have been incorporated into biopolymers for target applications. Althomali et al. modified glassy carbon electrodes with polyaniline@dialdehyde carboxymethyl cellulose/ZnO nanocomposites (PANI/D-CMC/ZnO) for the detection of H_2O_2 [9]. Akshaykranth et al. fabricated a novel nanocomposite that consisted of polylactic acid (PLA) and curcumin–ZnO to investigate its optical and antibacterial properties [10]. Kotharangannagari et al. applied hybrid nanocomposites of starch/lysine@ZnO NPs in food packing applications [5]. Because of their antibacterial properties, a lot of attention has been paid to silver nanoparticles (Ag NPs). They are widely used in biological applications, such as biosensors, forensic science, burn treatment, wound healing, biomolecule diagnostics, and chemotherapeutic processes [11–14]. It has been demonstrated that well-known oxide materials can be combined with biopolymers such as polysaccharides and their derivatives to form nanocomposites that either have a novel functionality or improve upon an existing function. Ail et al. used Ag/ZnO/chitosan (Ag/ZnO/Cs) ternary bionanocomposites to improve the antibacterial, optical, and photocatalytic properties of the metals [7]. Zare et al. combined Ag/Zn NPs with blended poly(3-hydroxybutyrate-co-3-hydroxyvalerate)-chitosan (PHBVCS) biopolymers to make food packaging that promotes longer shelf life [15]. Trandafilović et al. investigated the photocatalytic and antimicrobial activity of alginate-ZnO/Ag nanocomposites [16]. Shi et al. constructed Ag–ZnO/cellulose nanocomposites as effective photocatalysts for the photodegradation of methyl orange (MO) [17]. Peng et al. reported decorating cellulose/chitosan with Ag/Ag$_2$O/ZnO (AZ@CC) to explore the resulting photocatalytic and antimicrobial activity [18]. Cellulose is the natural biopolymer that is found in the greatest abundance on the planet [17]. Among the most significant cellulose derivatives is cellulose acetate (CA) [19], which is extracted and synthesized from renewable and natural resources [20,21]. CA has attracted considerable attention, as it is a biodegradable polymer with hydrophilic features, excellent chemical and mechanical stability, and low toxicity [22,23]. Because of its advantages, CA has the potential to be broadly utilized in industrial and biomedical applications, including drug delivery, wound dressing, and separation membrane technology [24–26]. The number of studies related to CA have increased [27]. In addition, gum arabic is one of the cellulose derivatives that could be applied as a stabilizer in the synthesis of nanoparticles, as it is commercially available at low cost [28–34]. The main strategy employed in the present work was to estimate the anticancer and biological performances of cellulose acetate biopolymer membrane-reinforced Ag/ZnO hybrid nanomaterials. A green synthesis procedure was utilized to prepare the Ag/ZnO hybrid reinforcement agent in the presence of gum arabic. We also focused on structural investigations of the developed hybrid nanocomposite materials combined with variations in Ag/ZnO dopant concentrations referring to the silver loadings using PXRD, FTIR, TEM, SEM, XPS, and TGA techniques; morphological studies were also undertaken.

2. Experimental

2.1. Reagents and Materials

Silver nitrate (purity > 99%), zinc (II) nitrate hexahydrate (purity > 98%), cellulose acetate, sodium hydroxide pellets, ethanol, and acetone (analytical grade) were obtained from Sigma-Aldrich, Fisher Chemical, and Techno PharmChem India, respectively. Gum arabic (GA) was available from a commercial market. Deionized water was used to prepare all aqueous solutions. All chemicals were used as obtained.

2.2. Preparation of Pristine ZnO Nanoparticles

Freen synthesis of pristine ZnO NPs was carried out as follows: an aqueous solution of Zn^{2+} (50 mL, 0.1 M) and GA (40 mL, 1% (w/v)) was stirred for 30 min, followed by an adjustment of the pH to 10. The stirring was continued for an additional 180 min. For a period of 24 h, the suspended solution was aged at room temperature. Before being dried in a furnace, the emulsion was first centrifuged; then, any residue that remained was thoroughly cleaned with ethanol and deionized water. The ZnO nanoparticles were then subjected to a calcination process for 1 h at 400 °C in an oven.

2.3. Green Synthesis of $Ag_{(0.01,\ 0.05,\ 0.1)}$/ZnO Hybrid Nanomaterials

A green synthesis for the Ag/ZnO hybrid nanomaterials based on variable Ag loading was carried out as follows. Equal amounts of Ag^+ and Zn^{2+} and 40 mL of aqueous solution GA (1% (w/v)) were stirred together for 30 min. The pH value was adjusted to pH = 10, and the aqueous solution was left at room temperature for 24 h. After the colloidal was centrifuged, it was washed with a small amount of ethanol and deionized water. The suspended solids were then centrifuged again. The precipitate was then dried in the oven and subjected to a calcination process for 1 h at 400 °C in a furnace. The concentration of Zn^{2+} was kept constant at 0.1 M, and three Ag^+ concentrations were used, i.e., 0.01, 0.05, and 0.1 M. Scheme 1 is an illustration of the preparation of Ag/ZnO hybrid nanomaterials.

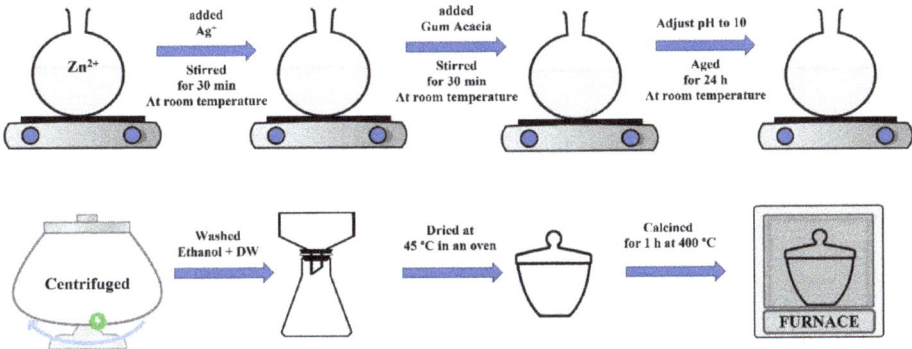

Scheme 1. An illustration of the green synthesis of $Ag_{(0.01,\ 0.05,\ 0.1)}$/ZnO hybrid nanomaterials.

2.4. CA/ZnO and CA@Ag/ZnO Hybrid Membrane Fabrication Procedures

The dissolution casting technique was utilized to fabricate the developed CA/ZnO and CA@Ag/ZnO hybrid membranes. A solution of 25 mL of acetone consisting of dissolved cellulose acetate powder (1 g) and a 10% fixed loading of pure ZnO nanoparticles was used to prepare the CA/ZnO hybrid membrane. The solution was stirred for 2 h, then sonicated for 30 min. The casting procedure was then carried out, and the homogeneous solution was poured into a glass petri dish. To avoid contamination from environmental particles, the dishes were wrapped in aluminum foil and dried for 24 h. Similar procedures were applied for the fabrication of the CA@Ag$_{(0.01-0.1)}$/ZnO hybrid membranes using three different loadings of Ag (0.01, 0.05, and 0.1) each time.

2.5. Utilized Instrumentation

The properties of the CA biopolymer, the prepared green nanomaterials, and the developed CA/ZnO and CA@Ag/ZnO hybrid membranes were studied using a number of characterization techniques, such as PXRD, FTIR, TEM, SEM, TGA, and XPS. The instrumentation used in this study was as follows. The PXRD pattern from 5° to 80° was recorded with a Bruker D8 Advance X-ray diffractometer using Cu K (=1.5406A) radiation at 40 kV and 20 mA. The FTIR spectra were recorded by (FT/IR-4100, JASCO, Japan) in the range 400–4000 cm^{-1}. The produced hybrid materials were analyzed using a TGA-50 Shimadzu Thermo gravimetric analyzer. We investigated the morphology and chemical makeup using a JEMF200 multipurpose electron microscope and a JSM-7610F Plus Schottky field emission scanning electron microscope. X-ray photoelectron spectra were captured using a Thermo Scientific K-Alpha™ spectrometer (XPS).

2.6. Biological Screening

The biological activities of CA biopolymer and the developed CA/ZnO and CA@Ag/ZnO hybrid membranes were displayed via the agar diffusion technique. They were tested against a wide range of bacterial species, Gram-negative bacteria strains, Gram-positive bacteria strains, and fungi; for example, Serratia marcescens ATCC 21074 and Escherichia coli ATCC 35218; Bacillus Cereus ATCC 14579 and Staphylococcus aureus ATCC 29213; and Candida albicans ATCC 76615 and Aspergillus flavus ATCC 9643. All microorganisms were supplied by King Abdulaziz University (microbiology lab located at King Fahad Medical Center) in Jeddah, Saudi Arabia. Previous studies revealed the methodology used to evaluate antibacterial efficacy of the new nanocomposites [35]. Table 1 displays the results of measuring the size of the growth inhibition zone.

Table 1. Antimicrobial activities of CA/ZnO and CA@Ag$_{(0.01, 0.05, 0.1)}$/ZnO hybrid materials.

Symbol	Microorganism Species/Inhibition Zone (mm)					
	S. aureus	B. subtilis	E. coli	S. marcescens	A. flavus	C. albicans
CA/ZnO	8	3	-	7	-	2
CA@Ag$_{(0.01)}$/ZnO	12	11	6	8	-	4
CA@Ag$_{(0.05)}$/ZnO	14	13	9	8	-	6
CA@Ag$_{(0.1)}$/ZnO	15	16	12	10	-	9

2.7. In Vitro MCF7 Anticancer Activity

2.7.1. Cell Culturing

The breast cancer cell lines (MCF-7) were cultured in Dulbecco's Modified Eagle Medium (DMEM), which was complemented with 10% fetal bovine serum (FBS), 100 g/mL streptomycin, and 100 units/mL penicillin. After the cell lines had developed to their full potential, they were subjected to an incubation period at a temperature of 40 degrees Celsius. The compounds that were analyzed were suspected of having undergone many doses of testing with MCF-7.

2.7.2. Experimental (Cell Count and Cell Viability) Investigations

The CA@Ag$_{(0.1)}$/ZnO nanocomposite was chosen for this investigation because it had previously been explored for its antimicrobial activity. In the previous investigation, it was shown to demonstrate a modest level of antimicrobial activity against the bacteria and fungus that were being studied. Over the course of 48 h, MCF-7 breast cancer cell lines were cultivated in two different environments: in the absence of CA@Ag$_{(0.1)}$/ZnO nanocomposite and in its presence at several concentrations (0, 0.5, 1, 2, and 3 mg/mL). Following the incubation time, microscopic pictures were captured using inverted microscopy. Thereafter, MCF-7 cell lines were harvested and quantified utilizing a hemocytometer [36–38].

In addition, an MTT test was carried out to determine cell viability according to the recommendations provided by the manufacturer (Invitrogen, Carlsbad, CA, USA) [39]. This test assessed the viability of cells by evaluating the metabolic activity of the cells that were found to be viable. In this regard, MCF-7 cell lines were implanted in 96-well, flat-bottomed plates, and cells were plated in the 96-multiwell plate (104 cells/well) for 24 h prior to treatment with the material to enable cell adhesion to the plate wall. Experiments were carried out for forty-eight hours, first in the absence of the CA@Ag$_{(0.1)}$/ZnO nanocomposite and then in its presence at varied concentrations, namely 0, 0.5, 1, 2, and 3 mg/mL.

3. Results and Discussion

3.1. Chemical Structure Evaluations of CA/ZnO and CA@Ag/ZnO Hybrid Membranes

The chemical structure of the developed CA/ZnO, as well as that of CA@Ag/ZnO in hybrid membranes with variable silver loadings, was evaluated by eco-friendly and cost-effective dissolution casting methods. Prior to the fabrication process, a green preparation of pristine ZnO and hybrid Ag-doped ZnO NPs was carried out in the presence of gum arabic. The solution's immediate color change from colorless to dark gray confirmed the reduction in silver ions. GA is a water-soluble polysaccharide-based biomaterial. The advantage of gum arabic in this application is its ability to stabilize the desired hybrid nanocomposite materials.

The PXRD features of CA, CA/ZnO, and CA@Ag$_{(0.01-0.1)}$/ZnO hybrid composite membranes are shown in Figure 1. This figure shows that pristine CA had a typical diffraction pattern, including two separate diffraction peaks at 18 and 22° in the 2 direction, as previously reported in the literature [40]. The crystalline nature of CA@ZnO was demonstrated by a hexagonal structure with a P 63 mc space group, which is consistent with the reference data for the material (JCPDS no. 36-1451). CA@Ag$_{(0.01, 0.05, 0.1)}$/ZnO hybrid nanocomposites had a face-centered cubic crystalline structure, as reflected by their respective PXRD patterns, i.e., space group Fm-3m. These peaks were well-matched with the reference peaks of silver (JCPDS no. 04-0783). The refraction peaks at 32.79 and 54.9° indicated the existence of Ag NPs. Figure 1 confirms the doping of the ZnO NPs with Ag NPs. Moreover, the ZnO NPs maintained their crystal structure. Scherrer's formula was applied to the FWHM of the prominent peaks and their location to derive the crystallite size (D) of the produced nanocomposite (nm). The pristine CA crystallite size (D) was 3.1 nm, whereas the average crystallite size of the fabricated nanocomposite membranes was approximately 15 nm.

Figure 2 displays a TEM image of CA@Ag$_{(0.1)}$/ZnO hybrid membrane (a) as a selected example, and its related particle size distribution histogram is shown in Figure 2b. It can be seen from the image that the prepared sample had a spherical shape, and the average size of nanoparticles was 15 nm. The Ag/ZnO hybrid showed improved compatibility and good distribution of nanoparticles into the CA polymer matrix.

Figure 3 displays SEM micrographs of pristine CA (a) CA@Ag$_{(0.1)}$/ZnO hybrid material (b), a high-resolution image of the CA@Ag$_{(0.1)}$/ZnO hybrid material (c), and the EDX signals and their percentage composition of the CA@Ag$_{(0.1)}$/ZnO hybrid material (d). Figure 3c shows that after the surface of CA was changed with a Ag$_{(0.1)}$/ZnO hybrid membrane, a variety of accumulated bubbles of different shapes appeared. Energy-dispersive X-ray (EDX) was used to examine the elemental composition of the prepared sample, as seen in Figure 3d. There were two peaks at 1.2 and 9.00 keV denoting Zn. The peak at 0.5 keV denoted O. The peak at 3.0 keV corresponded to silver. Moreover, an energy peak of C linked to carbon in CA was also present. These findings confirm the formation of a CA@Ag$_{(0.1)}$/ZnO hybrid nanocomposite membrane.

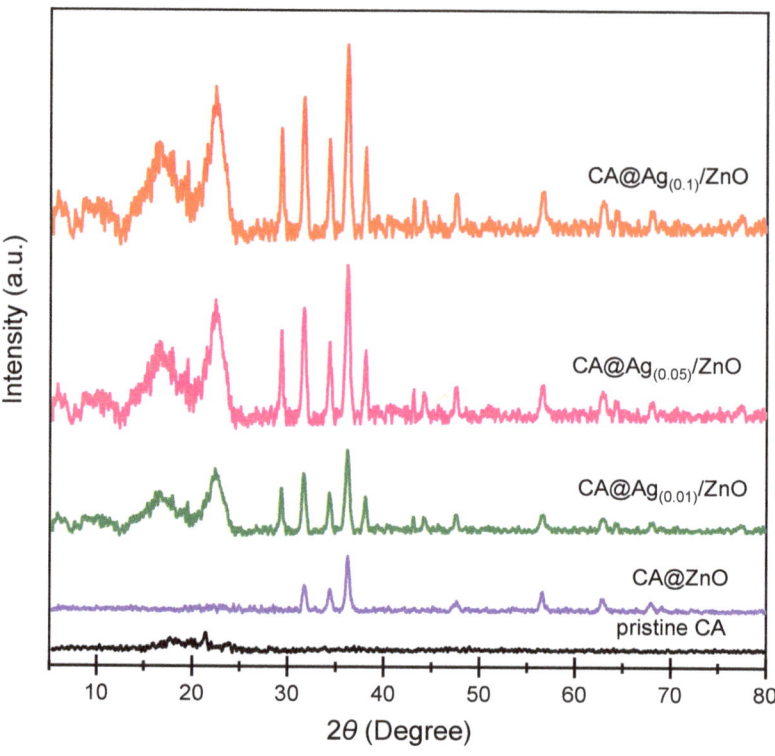

Figure 1. PXRD diffractograms of pristine CA, CA/ZnO, and CA@Ag$_{(0.01, 0.05, 0.1)}$/ZnO hybrid membranes.

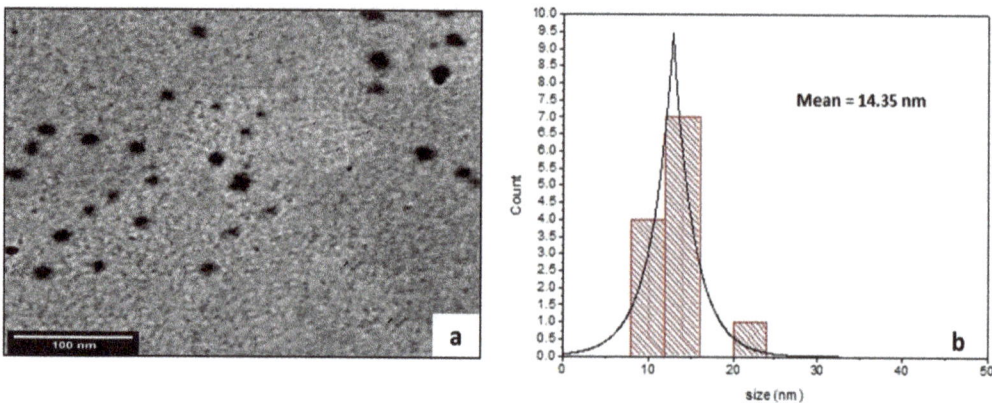

Figure 2. TEM image of CA@Ag$_{(0.1)}$/ZnO hybrid membrane (**a**) and a histogram of its particle size distribution (**b**).

Figure 4 shows the FTIR analyses of the developed hybrid materials in the range of 500–4000 cm^{-1}. FTIR analysis clarified the material functionality and nanocomposite formation of the hybrid nanocomposite. Figure 4 shows the FTIR spectra of pristine CA (black line), CA/ZnO composite material (purple line), and CA@Ag$_{(0.01, 0.05, 0.1)}$/ZnO hybrid materials as final targeted products (green, pink, and red lines, respectively). The FTIR spectrum of pure CA showed an absorption-stretching broadband around 3400 cm^{-1}, representing the presence of hydroxyl groups. The C-H group was obtained from the foundations of various peaks centered at 2973 and 1373 cm^{-1}. The peak at 1061 cm^{-1} was attributed to the C-O group. The presence of the peak around 1500 cm^{-1} was attributed to the stretching of the CA composition of the C=O ester carbonyl group. The peak at 1061 cm^{-1} was attributed to the 1120 cm^{-1} (acetate C-C-O stretching) and 1016 cm^{-1} (C-O stretching), which are characteristic of CA [41,42]. The FTIR spectra of CA@ZnO hybrid materials (purple line) display a characteristic peak of CA. In addition, a new peak around 477 cm^{-1} was assigned to ZnO vibration, which further confirmed the formation of ZnO [42]. It can be seen that after the formation of of silver nanoparticles on the surface of the zinc oxide nanoparticles, the intensity of the ZnO peaks decreased in CA@Ag$_{(0.01, 0.05, 0.1)}$/ZnO (green, pink, and red lines, respectively) [43].

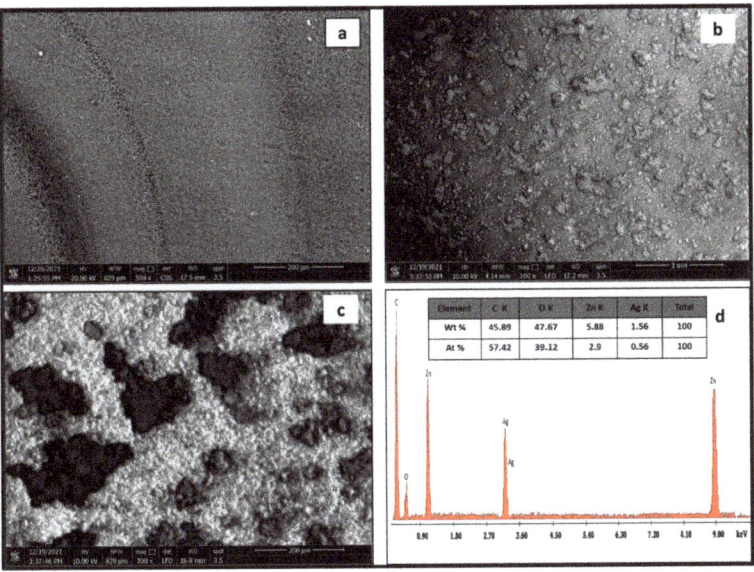

Figure 3. SEM micrographs of pristine CA (**a**) CA@Ag$_{(0.1)}$/ZnO hybrid material at low (**b**) and high magnifications of (**c**) and EDX analysis of CA@Ag$_{(0.1)}$/ZnO hybrid material (**d**).

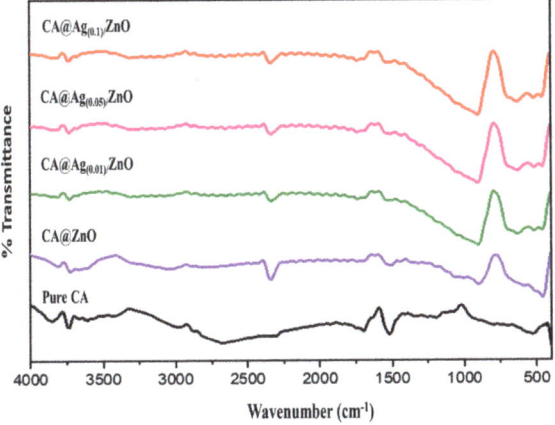

Figure 4. FTIR of pristine CA, CA/ZnO, and CA@Ag$_{(0.01, 0.05, 0.1)}$/ZnO hybrid materials.

Figure 5 illustrates the TGA patterns of pristine CA, CA/ZnO, and CA@Ag$_{(0.01, 0.05, 0.1)}$/ZnO hybrid materials. An initial weight loss amounting to ~10% occurred at around 100 °C in all samples as a result of the removal of H$_2$O molecules and/or trapped solvents on the surface. The TGA thermogram for the pristine CA (black line) displayed single-step weight losses. This TGA thermogram also exhibited a sharp reduction in weight; 50% of the weight loss was observed at temperatures less than 400 °C. The thermal degradation of pristine CA was complete at 600 °C. In the fabricated CA/ZnO and CA@Ag$_{(0.01, 0.05, 0.1)}$/ZnO hybrid materials, the TGA curves mainly displayed two-step weight losses. The CA/ZnO thermogram showed lower thermal degradation in the first step compared to the CA, which meant that the addition of ZnO accelerated the thermal decomposition of the pristine CA. Meanwhile, it displayed a significant increase in thermal stability in the second step. The first step was rapid, starting at around 217 °C and ending at around 385 °C, whereas the second step was slow and was completed at around 492 °C. Furthermore, CA@Ag$_{(0.01, 0.05, 0.1)}$/ZnO hybrid materials showed identical decomposition patterns consisting of two main decomposition stages. The TGA curves of CA@Ag$_{(0.05)}$/ZnO and CA@Ag$_{(0.1)}$/ZnO were nearly identical (a tiny shift was observed, as illustrated in the subfigure of Figure 5). The first step was rapid, starting at around 264 °C and ending at around 385 °C. The second step was complete at around 500 °C. However, there was a noticeable shift in the TGA curve of CA@Ag$_{(0.01)}$/ZnO. This shift was clearly noted between 250 and 350 °C. The first step was rapid, starting at 200 °C and ending at around 386°C. The second step was complete at around 425 °C. An almost 75% weight loss was observed in all the fabricated hybrid materials at around 450 °C. The thermal stability was highly affected by the silver loading in the fabricated hybrid materials. The thermal stabilities for those developed materials were detected in the following order: CA/ZnO < CA@Ag$_{(0.01)}$/ZnO < CA@Ag$_{(0.05)}$/ZnO = CA@Ag$_{(0.1)}$/ZnO.

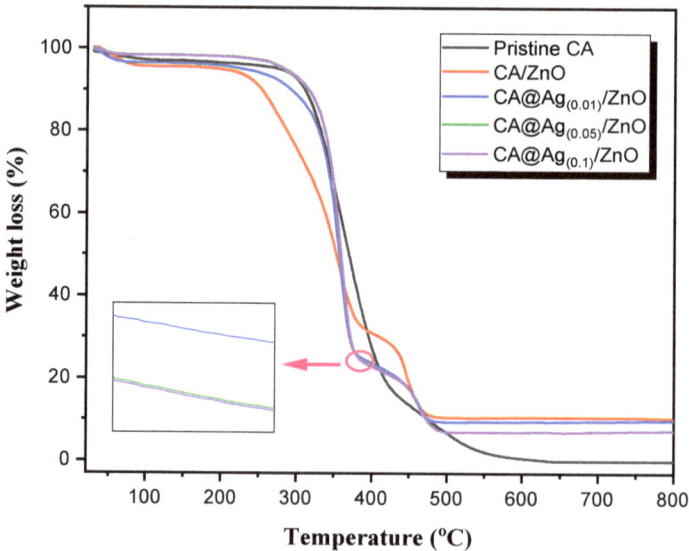

Figure 5. TGA thermograms of pristine CA, CA/ZnO and CA@Ag$_{(0.01, 0.05, 0.1)}$/ZnO hybrid materials.

Figure 6 displays the high-resolution XPS spectra of the fabricated CA@Ag$_{(0.1)}$/ZnO hybrid nanocomposite. Figure 6a presents two observable peaks at 1045.00 eV and 1022.01 eV that are characteristic of Zn 2p$_{1/2}$ and Zn 2p$_{3/2}$, respectively. This observation confirms the presence of Zn^{2+} ions within the fabricated hybrid nanocomposite. Figure 6b presents the high-resolution Ag 3d spectrum giving rise to two peaks at 367.91 and 373.85 eV, corresponding to the Ag 3d$_{5/2}$ and Ag 3d$_{3/2}$ orbitals typical of Ag, respectively. Figure 6c shows a distinct peak realized at 532.84 eV, which could be correlated to O1s. This observation endorses the presence of an oxide lattice phase within the hybrid nanocomposite [44–46].

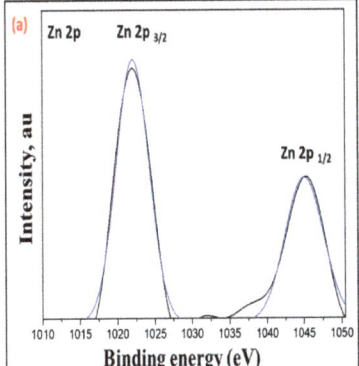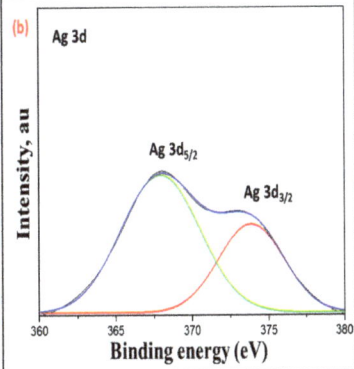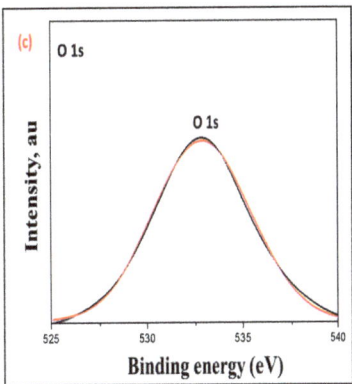

Figure 6. High-resolution XPS spectra of CA@Ag$_{(0.1)}$/ZnO hybrid material, the Zn 2p (**a**), Ag 3d (**b**), and O1s (**c**).

3.2. Antimicrobial Activities

The fabricated CA/ZnO and CA@Ag$_{(0.01, 0.05, 0.1)}$/ZnO hybrid materials were biologically screened against some selected bacterial and fungal microorganisms; the obtained results are illustrated in Table 1 and Figure 7. The biological characters showed that the presence of Ag NPs influenced the antibacterial capabilities of the developed nanocomposite materials. Table 1 records the effect of each tested material shown in the obtained inhibition zones (mm) of the bacterial species and fungi, whereas Figure 7 shows a screening illustration of CA/ZnO and CA@Ag$_{(0.01, 0.05, 0.1)}$/ZnO hybrid materials against the same investigated microorganisms. A disk diffusion method was applied to evaluate the antimicrobial activity assay of prepared samples. It can be seen from Table 1 and Figure 7 that all fabricated hybrid composite materials demonstrated significant biological performance against the majority of the tested bacteria and fungi, except A. flavus. Moreover, there was a proportional relation between the concentration of Ag NPs in the prepared samples and the inhibition zone (mm). The replication process and the growth of microorganisms were noticeably affected by the increase in Ag NP loading. Thus, the CA@Ag$_{(0.1)}$/ZnO hybrid nanocomposite membrane had the highest antimicrobial activity among the prepared samples. The biological properties showed that the presence of Ag NPs influenced the antibacterial capabilities of the nanocomposite.

Many different mechanisms have been proposed for the inhibition and destruction of bacterial cells [18,47,48]. According to some studies, electrostatic interaction between nanomaterials and microorganisms would release a positive ion that would lead to membrane breakdown or permeability disruption; for instance, ZnO NPs and Ag NPs would produce Zn^{2+} and Ag^+ ions, respectively, which would lead to cell death [7]. Such ions may have the potential to interfere with the negatively charged functional groups (such as –NH, –COOH, and –SH) that are found in proteins and nucleic acids, which may lead to the suppression of DNA replication and the rupture of cell walls. Moreover, it has been reported in the literature that the formation of reactive oxidation species (ROS), such as hydroxyl radical (·OH) and hydrogen peroxide (H_2O_2), superoxide radical (O_2^-), and singlet oxygen (O_2), can cause bacterial cell death. These species exhibited different levels of activity and dynamics with ZnO NPs and Ag NPs. ROS caused oxidative stress in bacteria, resulting in cell death. Moreover, the presence of Ag would reduce the recombination rate of the electron–hole pair, enhancing the production of ROS. It has been suggested that the cause of cell death may be the direct interaction between various NPs and the bacterial membrane [18,47,48].

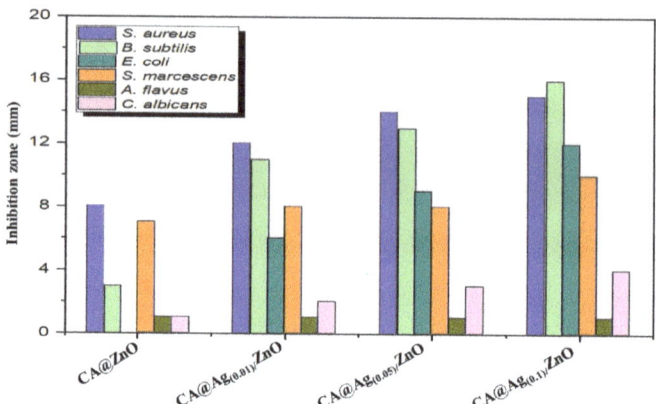

Figure 7. Antimicrobial activities of CA/ZnO and CA@Ag$_{(0.01, 0.05, 0.1)}$/ZnO hybrid materials in the presence of selected microorganisms.

3.3. In Vitro MCF7 Anticancer Activity

Figure 8 shows images of the MCF7 cell used as a control (a) and that in the presence of variable concentrations of CA@Ag$_{(0.1)}$/ZnO hybrid composite material (0.50, 1, 2, and 3 mg/mL) (b-e) at a magnification of X = 600. Figure 8 also shows the relationship between cellular uptake and the cytotoxicity of the hybrid nanocomposite. In addition, the results in Figure 8 indicate the positive effect of the hybrid nanocomposite on the health of cells as depicted by their more flattened appearance. Because of this effect, the CA@Ag$_{(0.1)}$/ZnO hybrid nanocomposite induces cell death in breast cancer. This result is consistent with the findings reported in the abovementioned figure.

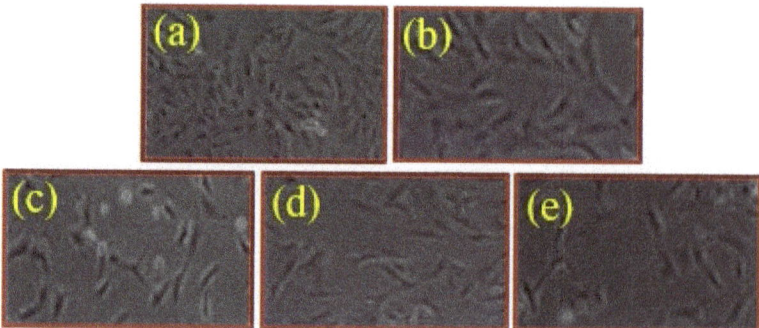

Figure 8. Images of MCF7 cells for the control (**a**) and in the presence of variable concentrations of CA@Ag$_{(0.1)}$/ZnO nanocomposite 0.50, 1, 2, and of 3 mg/mL (**b–e**) at a magnification of X = 600.

Furthermore, Figure 9 displays the MCF7 cell counts at variable concentrations of CA@Ag$_{(0.1)}$/ZnO nanocomposites, namely 0.50, 1, 2, and 3 mg/mL. It can be concluded that the cytotoxicity potential associated with the nanocomposites occurred in a concentration-dependent manner. The hybrid nanocomposite caused 50% cell growth inhibition at a concentration of 1 mg/Ml. Moreover, Figure 10 displays the MCF7 cell viability at variable concentrations of CA@Ag$_{(0.1)}$/ZnO hybrid material, namely 0.50, 1, 2, and 3 mg/mL.

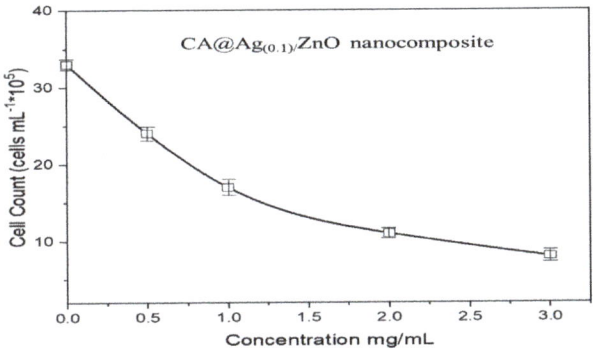

Figure 9. MCF7 cell counts at variable concentrations of CA@Ag$_{(0.1)}$/ZnO hybrid material (0.50, 1, 2, and 3 mg/Ml).

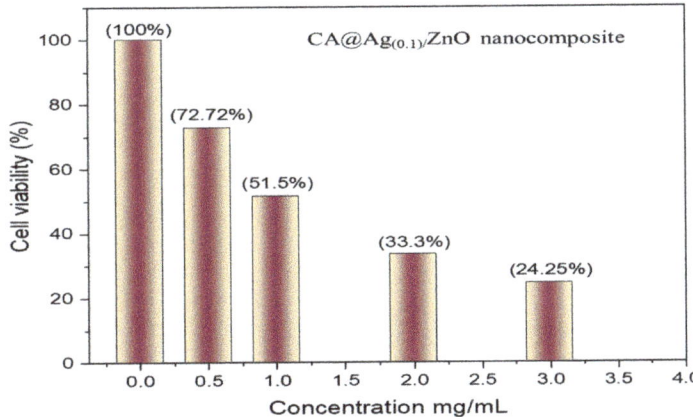

Figure 10. MCF7 cell viability at variable concentrations of CA@Ag$_{(0.1)}$/ZnO hybrid material (0.50, 1, 2, and 3 mg/mL).

4. Conclusions

In order to manufacture a series of biologically active hybrid composite materials depending on cellulose acetate-reinforced hybrid Ag/ZnO nanomaterials, an easy, non-toxic, and cost-effective casting approach was efficiently utilized. A green synthesis method was used to prepare ZnO NPs and Ag-doped ZnO in three different concentrations. The chemical structure and morphologies of the developed composite materials were characterized using a number of techniques, such as PXRD, TEM, SEM, FTIR, TGA, and XPS. The average crystallite size (nm) for such hybrid materials was 15 nm. The doping of the ZnO NPs with Ag NPs and Ag$_2$O NPs was confirmed. The hybrid nanocomposite membranes were spherical, exhibiting thermal stability at temperatures over 400 °C. The biological properties of fabricated materials were tested against selected bacterial and fungal species using a common biological tool. The results showed that the fabricated materials did not demonstrate any antimicrobial activities against *A. flavus*. The growth of the tested microorganisms was noticeably affected by the increase in the concentration of Ag NP loading. Moreover, hybrid nanocomposite membranes were used as target materials to reduce the number of MCF7 cancer cell lines. In addition, the cell count and cell viability in the presence of Ag NPs were analyzed through an MTT test. The biological properties showed that the presence of Ag NPs influenced the antibacterial capabilities of the nanocomposite.

Author Contributions: Conceptualization, M.A.H.; methodology, N.A. and M.A.H.; validation, N.A. and M.A.H.; formal analysis, N.A. and M.A.H.; investigation, N.A. and M.A.H.; writing—original draft preparation, N.A.; writing—review and editing, M.A.H.; funding acquisition, N.A. All authors have read and agreed to the published version of the manuscript.

Funding: This work funded by the University of Jeddah, Saudi Arabia under grant No. UJ-20-146-DR. Therefore, the authors acknowledge, with thanks, the University technical and financial support.

Institutional Review Board Statement: Not applicable.

Informed Consent Statement: Not applicable.

Data Availability Statement: Not applicable.

Conflicts of Interest: The authors declare no conflict of interest.

Sample Availability: Samples of the compounds are not available from the authors.

References

1. Sharma, B.; Malik, P.; Jain, P. Biopolymer Reinforced Nanocomposites: A Comprehensive Review. *Mater. Today Commun.* **2018**, *16*, 353–363. [CrossRef]
2. Christian, S.J. 5—*Natural Fibre-Reinforced Noncementitious Composites (Biocomposites)*; Harries, K.A., Sharma, B., Eds.; Woodhead Publishing: Sawston, UK, 2016; pp. 111–126. [CrossRef]
3. Brandelli, A.; Lopes, N.A. Chapter 9—*Nanocomposite Antimicrobial Films Based on Biopolymers*; Rai, M., dos Santos, C.A., Eds.; Elsevier: Amsterdam, The Netherlands, 2021; pp. 149–170. [CrossRef]
4. Shankar, S.; Teng, X.; Rhim, J.-W. Properties and Characterization of Agar/CuNP Bionanocomposite Films Prepared with Different Copper Salts and Reducing Agents. *Carbohydr. Polym.* **2014**, *114*, 484–492. [CrossRef] [PubMed]
5. Kotharangannagari, V.K.; Krishnan, K. Biodegradable Hybrid Nanocomposites of Starch/Lysine and ZnO Nanoparticles with Shape Memory Properties. *Mater. Des.* **2016**, *109*, 590–595. [CrossRef]
6. Bai, H.; Liang, Z.; Wang, D.; Guo, J.; Zhang, S.; Ma, P.; Dong, W. Biopolymer Nanocomposites with Customized Mechanical Property and Exceptionally Antibacterial Performance. *Compos. Sci. Technol.* **2020**, *199*, 108338. [CrossRef]
7. Ali, H.; Ismail, A.M.; Menazea, A.A. Multifunctional Ag/ZnO/Chitosan Ternary Bio-Nanocomposites Synthesized via Laser Ablation with Enhanced Optical, Antibacterial, and Catalytic Characteristics. *J. Water Process Eng.* **2022**, *49*, 102940. [CrossRef]
8. Viorica, G.P.; Musat, V.; Pimentel, A.; Calmeiro, T.R.; Carlos, E.; Baroiu, L.; Martins, R.; Fortunato, E. Hybrid (Ag)ZnO/Cs/PMMA Nanocomposite Thin Films. *J. Alloys Compd.* **2019**, *803*, 922–933. [CrossRef]
9. Althomali, R.H.; Alamry, K.A.; Hussein, M.A.; Guedes, R.M. Hybrid PANI@dialdehyde Carboxymethyl Cellulose/ZnO Nanocomposite Modified Glassy Carbon Electrode as a Highly Sensitive Electrochemical Sensor. *Diam. Relat. Mater.* **2022**, *122*, 108803. [CrossRef]
10. Akshaykranth, A.; Jayarambabu, N.; Kumar, A.; Venkatappa Rao, T.; Kumar, R.R.; Srinivasa Rao, L. Novel Nanocomposite Polylactic Acid Films with Curcumin-ZnO: Structural, Thermal, Optical and Antibacterial Properties. *Curr. Res. Green Sustain. Chem.* **2022**, *5*, 100332. [CrossRef]
11. Sudhakar, K.; Won, S.Y.; Han, S.S. Gelatin Stabilized Silver Nanoparticles for Wound Healing Applications. *Mater. Lett.* **2022**, *325*, 132851. [CrossRef]
12. Nakamura, S.; Sato, M.; Sato, Y.; Ando, N.; Takayama, T.; Fujita, M.; Ishihara, M. Synthesis and Application of Silver Nanoparticles (Ag NPs) for the Prevention of Infection in Healthcare Workers. *Int. J. Mol. Sci.* **2019**, *20*, 3620. [CrossRef]
13. Zhang, Y.-W.; Wang, L.-K.; Fang-Zhou, L.; Yuan, B.-H.; Zou, X.-M.; Wang, R.-T. Synthesis and Characterization of Silver Nanoparticles Green-Formulated by Allium Stipitatum and Treat the Colorectal Cancer as a Modern Chemotherapeutic Supplement. *Inorg. Chem. Commun.* **2022**, *143*, 109781. [CrossRef]
14. Bharathi, D.; Diviya Josebin, M.; Vasantharaj, S.; Bhuvaneshwari, V. Biosynthesis of Silver Nanoparticles Using Stem Bark Extracts of Diospyros Montana and Their Antioxidant and Antibacterial Activities. *J. Nanostructure Chem.* **2018**, *8*, 83–92. [CrossRef]
15. Zare, M.; Namratha, K.; Ilyas, S.; Hezam, A.; Mathur, S.; Byrappa, K. Smart Fortified PHBV-CS Biopolymer with ZnO–Ag Nanocomposites for Enhanced Shelf Life of Food Packaging. *ACS Appl. Mater. Interfaces* **2019**, *11*, 48309–48320. [CrossRef] [PubMed]
16. Trandafilović, L.V.; Whiffen, R.K.; Dimitrijević-Branković, S.; Stoiljković, M.; Luyt, A.S.; Djoković, V. ZnO/Ag Hybrid Nanocubes in Alginate Biopolymer: Synthesis and Properties. *Chem. Eng. J.* **2014**, *253*, 341–349. [CrossRef]
17. Shi, C.; Zhang, L.; Bian, H.; Shi, Z.; Ma, J.; Wang, Z. Construction of Ag–ZnO/Cellulose Nanocomposites via Tunable Cellulose Size for Improving Photocatalytic Performance. *J. Clean. Prod.* **2021**, *288*, 125089. [CrossRef]
18. Peng, Y.; Zhou, H.; Wu, Y.; Ma, Z.; Zhang, R.; Tu, H.; Jiang, L. A New Strategy to Construct Cellulose-Chitosan Films Supporting Ag/Ag2O/ZnO Heterostructures for High Photocatalytic and Antibacterial Performance. *J. Colloid Interface Sci.* **2022**, *609*, 188–199. [CrossRef] [PubMed]
19. Qiu, X.; Hu, S. "Smart" Materials Based on Cellulose: A Review of the Preparations, Properties, and Applications. *Materials* **2013**, *6*, 738–781. [CrossRef] [PubMed]

20. Bhansali, M.; Dabholkar, N.; Swetha, P.; Dubey, S.K.; Singhvi, G. *Chapter 18—Solid Oral Controlled-Release Formulations*; Academic Press: Cambridge, MA, USA, 2021; pp. 313–331. [CrossRef]
21. Fischer, S.; Thümmler, K.; Volkert, B.; Hettrich, K.; Schmidt, I.; Fischer, K. Properties and Applications of Cellulose Acetate. *Macromol. Symp.* **2008**, *262*, 89–96. [CrossRef]
22. Miao, X.; Lin, J.; Bian, F. Utilization of Discarded Crop Straw to Produce Cellulose Nanofibrils and Their Assemblies. *J. Bioresour. Bioprod.* **2020**, *5*, 26–36. [CrossRef]
23. Wei, D.W.; Wei, H.; Gauthier, A.C.; Song, J.; Jin, Y.; Xiao, H. Superhydrophobic Modification of Cellulose and Cotton Textiles: Methodologies and Applications. *J. Bioresour. Bioprod.* **2020**, *5*, 1–15. [CrossRef]
24. Vatankhah, E.; Prabhakaran, M.P.; Jin, G.; Mobarakeh, L.G.; Ramakrishna, S. Development of Nanofibrous Cellulose Acetate/Gelatin Skin Substitutes for Variety Wound Treatment Applications. *J. Biomater. Appl.* **2013**, *28*, 909–921. [CrossRef] [PubMed]
25. Madaeni, S.S.; Derakhshandeh, K.; Ahmadi, S.; Vatanpour, V.; Zinadini, S. Effect of Modified Multi-Walled Carbon Nanotubes on Release Characteristics of Indomethacin from Symmetric Membrane Coated Tablets. *J. Memb. Sci.* **2012**, *389*, 110–116. [CrossRef]
26. Zugenmaier, P. 4. Characteristics of Cellulose Acetates 4.1 Characterization and Physical Properties of Cellulose Acetates. *Macromol. Symp.* **2004**, *208*, 81–166. [CrossRef]
27. Vatanpour, V.; Pasaoglu, M.E.; Barzegar, H.; Teber, O.O.; Kaya, R.; Bastug, M.; Khataee, A.; Koyuncu, I. Cellulose Acetate in Fabrication of Polymeric Membranes: A Review. *Chemosphere* **2022**, *295*, 133914. [CrossRef] [PubMed]
28. Roque, A.C.A.; Bicho, A.; Batalha, I.L.; Cardoso, A.S.; Hussain, A. Biocompatible and Bioactive Gum Arabic Coated Iron Oxide Magnetic Nanoparticles. *J. Biotechnol.* **2009**, *144*, 313–320. [CrossRef] [PubMed]
29. Wilson, O.C.; Blair, E.; Kennedy, S.; Rivera, G.; Mehl, P. Surface Modification of Magnetic Nanoparticles with Oleylamine and Gum Arabic. *Mater. Sci. Eng. C* **2008**, *28*, 438–442. [CrossRef]
30. Williams, D.N.; Gold, K.A.; Holoman, T.R.P.; Ehrman, S.H.; Wilson, O.C. Surface Modification of Magnetic Nanoparticles Using Gum Arabic. *J. Nanoparticle Res.* **2006**, *8*, 749–753. [CrossRef]
31. Balu, S.; Palanisamy, S.; Velusamy, V.; Yang, T.C.K. Ultrasonics—Sonochemistry Sonochemical Synthesis of Gum Guar Biopolymer Stabilized Copper Oxide on Exfoliated Graphite: Application for Enhanced Electrochemical Detection of H_2O_2 in Milk and Pharmaceutical Samples. *Ultrason. Sonochemistry* **2019**, *56*, 254–263. [CrossRef]
32. Pauzi, N.; Zain, N.M.; Yusof, N.A.A. Gum Arabic as Natural Stabilizing Agent in Green Synthesis of ZnO Nanofluids for Antibacterial Application. *J. Environ. Chem. Eng.* **2020**, *8*, 103331. [CrossRef]
33. Li, Y.F.; Gan, W.P.; Zhou, J.; Lu, Z.Q.; Yang, C.; Ge, T.T. Hydrothermal Synthesis of Silver Nanoparticles in Arabic Gum Aqueous Solutions. *Trans. Nonferrous Met. Soc. China* **2015**, *25*, 2081–2086. [CrossRef]
34. Barik, P.; Bhattacharjee, A.; Roy, M. Preparation, Characterization and Electrical Study of Gum Arabic/ZnO Nanocomposites. *Bull. Mater. Sci.* **2015**, *38*, 1609–1616. [CrossRef]
35. Ge, B.; Wang, F.; Sjölund-Karlsson, M.; McDermott, P.F. Antimicrobial Resistance in Campylobacter: Susceptibility Testing Methods and Resistance Trends. *J. Microbiol. Methods* **2013**, *95*, 57–67. [CrossRef] [PubMed]
36. Liu, T.; van den Berk, L.; Wondergem, J.A.J.; Tong, C.; Kwakernaak, M.C.; Braak, B.T.; Heinrich, D.; van de Water, B.; Kieltyka, R.E. Squaramide-Based Supramolecular Materials Drive HepG2 Spheroid Differentiation. *Adv. Healthc. Mater.* **2021**, *10*, e2001903. [CrossRef] [PubMed]
37. Pocasap, P.; Weerapreeyakul, N.; Junhom, C.; Phiboonchaiyanan, P.P.; Srisayam, M.; Nonpunya, A.; Siriwarin, B.; Khamphio, M.; Nanok, C.; Thumanu, K.; et al. FTIR Microspectroscopy for the Assessment of Mycoplasmas in HepG2 Cell Culture. *Appl. Sci.* **2020**, *10*, 3766. [CrossRef]
38. Skehan, P.; Storeng, R.; Scudiero, D.; Monks, A.; McMahon, J.; Vistica, D.; Warren, J.T.; Bokesch, H.; Kenney, S.; Boyd, M.R. New Colorimetric Cytotoxicity Assay for Anticancer-Drug Screening. *J. Natl. Cancer Inst.* **1990**, *82*, 1107–1112. [CrossRef] [PubMed]
39. Md, S.; Alhakamy, N.A.; Akhter, S.; Awan, Z.A.Y.; Aldawsari, H.M.; Alharbi, W.S.; Haque, A.; Choudhury, H.; Sivakumar, P.M. Development of Polymer and Surfactant Based Naringenin Nanosuspension for Improvement of Stability, Antioxidant, and Antitumour Activity. *J. Chem.* **2020**, *2020*, 3489393. [CrossRef]
40. Nasiri Khalil Abad, S.; Mozammel, M.; Moghaddam, J.; Mostafaei, A.; Chmielus, M. Highly Porous, Flexible and Robust Cellulose Acetate/Au/ZnO as a Hybrid Photocatalyst. *Appl. Surf. Sci.* **2020**, *526*, 146237. [CrossRef]
41. Wang, D.; Yang, J.; Yang, H.; Zhao, P.; Shi, Z. Fe-Complex Modified Cellulose Acetate Composite Membrane with Excellent Photo-Fenton Catalytic Activity. *Carbohydr. Polym.* **2022**, *296*, 119960. [CrossRef] [PubMed]
42. Anitha, S.; Brabu, B.; John Thiruvadigal, D.; Gopalakrishnan, C.; Natarajan, T.S. Optical, Bactericidal and Water Repellent Properties of Electrospun Nano-Composite Membranes of Cellulose Acetate and ZnO. *Carbohydr. Polym.* **2013**, *97*, 856–863. [CrossRef]
43. Fouladi-Fard, R.; Aali, R.; Mohammadi-Aghdam, S.; Mortazavi-derazkola, S. The Surface Modification of Spherical ZnO with Ag Nanoparticles: A Novel Agent, Biogenic Synthesis, Catalytic and Antibacterial Activities. *Arab. J. Chem.* **2022**, *15*, 103658. [CrossRef]
44. Ding, W.; Zhao, L.; Yan, H.; Wang, X.; Liu, X.; Zhang, X.; Huang, X.; Hang, R.; Wang, Y.; Yao, X.; et al. Bovine Serum Albumin Assisted Synthesis of $Ag/Ag_2O/ZnO$ Photocatalyst with Enhanced Photocatalytic Activity under Visible Light. *Colloids Surfaces A Physicochem. Eng. Asp.* **2019**, *568*, 131–140. [CrossRef]

45. Cao, W.; An, Y.; Chen, L.; Qi, Z. Visible-Light-Driven Ag_2MoO_4/Ag_3PO_4 Composites with Enhanced Photocatalytic Activity. *J. Alloys Compd.* **2017**, *701*, 350–357. [CrossRef]
46. Alahmadi, N.; Amin, M.S.; Mohamed, R.M. Superficial Visible-Light-Responsive Pt@ZnO Nanorods Photocatalysts for Effective Remediation of Ciprofloxacin in Water. *J. Nanoparticle Res.* **2020**, *22*, 230. [CrossRef]
47. Gupta, J.; Mohapatra, J.; Bahadur, D. Visible Light Driven Mesoporous Ag-Embedded ZnO Nanocomposites: Reactive Oxygen Species Enhanced Photocatalysis, Bacterial Inhibition and Photodynamic Therapy. *Dalt. Trans.* **2017**, *46*, 685–696. [CrossRef] [PubMed]
48. El-Kahky, D.; Attia, M.; Easa, S.M.; Awad, N.M.; Helmy, E.A. Interactive Effects of Biosynthesized Nanocomposites and Their Antimicrobial and Cytotoxic Potentials. *Nanomaterials* **2021**, *11*, 903. [CrossRef] [PubMed]

Disclaimer/Publisher's Note: The statements, opinions and data contained in all publications are solely those of the individual author(s) and contributor(s) and not of MDPI and/or the editor(s). MDPI and/or the editor(s) disclaim responsibility for any injury to people or property resulting from any ideas, methods, instructions or products referred to in the content.

Article

Green Synthesis and Characterization of Silver Nanoparticles Using *Spondias mombin* Extract and Their Antimicrobial Activity against Biofilm-Producing Bacteria

Sumitha Samuggam [1], Suresh V. Chinni [1,*], Prasanna Mutusamy [1], Subash C. B. Gopinath [2], Periasamy Anbu [3], Vijayan Venugopal [4], Lebaka Veeranjaneya Reddy [5] and Balaji Enugutti [6]

1. Department of Biotechnology, Faculty of Applied Sciences, AIMST University, Bedong 08100, Kedah, Malaysia; sumitha@aimst.edu.my (S.S.); mutusamyprasanna@gmail.com (P.M.)
2. Institute of Nano Electronic Engineering, Faculty of Chemical Engineering Technology, Universiti Malaysia Perlis, Arau 01000, Perlis, Malaysia; subash@unimap.edu.my
3. Department of Biological Engineering, Inha University, Incheon 402-751, Korea; anbu25@yahoo.com
4. School of Pharmacy, Sri Balaji Vidyapeeth, Deemed to Be University, Puducherry 607402, India; vijayanv2@gmail.com
5. Department of Microbiology, Yogi Vemana University, Kadapa 516005, India; lvereddy@gmail.com
6. Gregor Mendel Institute (GMI), Austrian Academy of Sciences, Vienna Biocenter (VBC), Dr. Bohr-Gasse 3, 1030 Vienna, Austria; balaji.enugutti@gmi.oeaw.ac.at
* Correspondence: v_suresh@aimst.edu.my or cvsureshgupta@gmail.com; Tel.: +60-124-362-324

Abstract: Multidrug resistant bacteria create a challenging situation for society to treat infections. Multidrug resistance (MDR) is the reason for biofilm bacteria to cause chronic infection. Plant-based nanoparticles could be an alternative solution as potential drug candidates against these MDR bacteria, as many plants are well known for their antimicrobial activity against pathogenic microorganisms. *Spondias mombin* is a traditional plant which has already been used for medicinal purposes as every part of this plant has been proven to have its own medicinal values. In this research, the *S. mombin* extract was used to synthesise AgNPs. The synthesized AgNPs were characterized and further tested for their antibacterial, reactive oxygen species and cytotoxicity properties. The characterization results showed the synthesized AgNPs to be between 8 to 50 nm with -11.52 of zeta potential value. The existence of the silver element in the AgNPs was confirmed with the peaks obtained in the EDX spectrometry. Significant antibacterial activity was observed against selected biofilm-forming pathogenic bacteria. The cytotoxicity study with *A. salina* revealed the LC50 of synthesized AgNPs was at 0.81 mg/mL. Based on the ROS quantification, it was suggested that the ROS production, due to the interaction of AgNP with different bacterial cells, causes structural changes of the cell. This proves that the synthesized AgNPs could be an effective drug against multidrug resistant bacteria.

Keywords: *Spondias mombin*; AgNP; biofilm bacteria

1. Introduction

Nanotechnology is a recent new branch of science that has shown a wide range of development of novel technological advancements in environmental, biochemical, biological, and other applications [1]. Silver nanoparticles with the size of 1–100 nm are commonly applied in nanotechnology and science. In recent years, silver nanoparticles (AgNPs) have generated huge interest among scientists because of their impressive protection against numerous infective microorganisms. Several different ways of synthesizing AgNPs have been reported, including physical, biological, and chemical processes [2–5]. These approaches have their own benefits and drawbacks, based on their final applications. For instance, nanoparticles (NPs) synthesized through a chemical method can be immediately available for functionality testing [6]. However, chemically synthesized NPs exhibit many possible risks, including cytotoxicity, genotoxicity, carcinogenicity, and general toxicity [7,8]. On the other hand, physical methods are considered to take a longer time and are restricted to special requirements, including certain elevated temperatures or pressures, making the procedure expensive [8]. In contrast with these methods, biological methods (e.g., plant extracts, bacteria, and fungi) are known to be safe as they utilize very fewer toxic reactants or additives. This method is also considered to be rapid, simple, user-friendly, and inexpensive and includes the capability of synthesis in large quantities [9]. The synthesis of NPs using biological sources has gained interest in recent days. The application of plant extracts is highly recommended for the production of AgNPs [10]. Extracts from plant materials are high in secondary metabolites, including enzymes, polysaccharides, alkaloids, tannins, phenols, terpenoids and vitamins, which allow them to display excellent antimicrobial properties [11]. It is assumed that organic components from the leaf extract (flavonoids and terpenoids) help to stabilize the AgNPs [12].

Recently, the WHO published a list of antibiotic-resistant biofilm-producing bacteria including *Acinetobacter*, *Salmonella*, *Pseudomonas*, *Klebsiella*, *E. coli*, and *Proteus*. These bacteria cause deadly infection and are becoming resistant to most of the currently available antibiotics [13]. Multidrug resistance is the reason for biofilm-producing bacteria to contribute to chronic diseases [14]. The increasing occurrence of MDR bacteria against clinically important antibiotics has become the reason for the use of AgNPs to enhance the antibiotic effect, as AgNPs possess antibacterial, antiviral, antifungal, and also anti-inflammatory properties [15]. According to WHO, by 2050, MDR bacterial infection is predicted to kill more people than cancer and cost $100 trillion for healthcare. High research development costs and lack of profitability have long hindered the investment in novel antibiotic discovery. Consequently, using plant-based antimicrobials as an alternative therapeutic agent for the treatment of infections caused by MDR bacteria has gained popularity recently. Medicinal plants are rich in active compounds that have antimicrobial activity, and they are generally safer to use in terms of side effects, compared to conventional antibiotics. *Spondias mombin (S. mombin)* often gains attention among the researchers, due to its antimicrobial characteristics [16]. *Spondias mombin*, from the family of *Anacardiaceae* is also best-known as ambarella. This species has been used as a traditional medicine to treat diseases like anti-inflammatory and antithrombolytic complaints [17]. Every part of the *S. mombin* plant is reported to have its own medicinal values. For example, the bark of the tree is used as a treatment for diarrhea in countries like Cambodia and the fruit of *S. mombin* is used to cure itchiness, internal ulceration, sore throats, as well as skin inflammation. Evidence suggests that *S. mombin* leaves and fruits possess high antimicrobial, antioxidant, cytotoxic, antidiabetic, and thrombolytic ability [16].

In light of the importance of *S. mombin* and biologically synthesized AgNPs, the present research was designed to study plant-mediated AgNP synthesis and the characterization and predicted antimicrobial activity to counter different infective bacteria.

2. Results and Discussion
2.1. Silver Nanoparticle-Synthesis

The color change from colorless to yellowish brown suggested the reduction of $AgNO_3$ had occurred by using the *S. mombin* extract as a bioreducing agent. The color change proved absorption of visible light due to the excitation of the AgNP surface plasmons (Figure 1) [18]. For further confirmation of the reduction of Ag ions to AgNPs, the maximum absorbance of UV−VIS spectra of different wavelengths was obtained (Figure 1). It was observed that there was no peak obtained in the silver nitrate ($AgNO_3$) as it did not contain any reducing agent. However, the $AgNO_3$ with the plant extract showed maximum intensity between the wavelengths of 300 to 400 nm, confirming the role of *S. mombin* leaf extract as a reducing agent to form AgNPs.

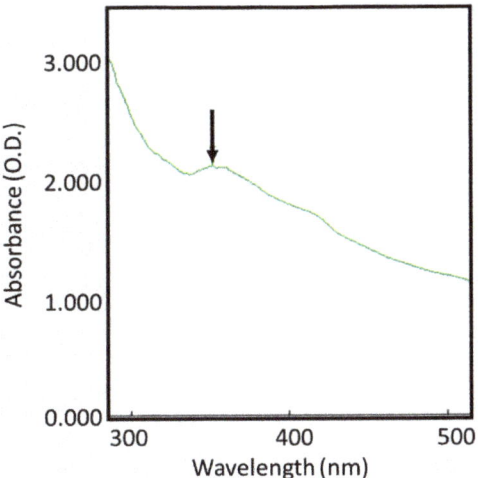

Figure 1. UV−Visible spectra of *Spondias mombin* leaf extract mediated silver nanoparticles.

A study showed that in the biosynthesis of AgNPs by the reduction of Ag ions, the terpenoids in the leaf extract play a crucial role [19]. A previous study reported this function of terpenoids from *Geranium* leaves in the biosynthesis of AgNPs [20]. A similar process might have worked in the current study, where the flavonoids and phenolic compounds from *S. mombin* extract acted as a capping and stabilizing agent in the formation of NPs.

2.2. Characterization of Synthesized Silver Nanoparticles

The particle size of the nanoparticle is an important consideration in biological applications and it strongly affects the diffusion rate via biological membranes. Previous reports showed that the smaller the size of the nanoparticle, the higher its permeability, but showed increased toxicity. Thus, a suitable size is highly recommended for specific biological functions. Hence, scanning electron microscopy (SEM) was used to study the surface morphology. Figure 2 clearly shows that the AgNPs were spherical in shape with smooth edges. The mean particle size of AgNPs was 17 nm, which is the appropriate size (8–50 nm) for biological membrane permeation, and this is the tolerable range for inducing toxicity within cells.

The elemental composition was revealed by EDX analysis of the synthesized AgNPs. EDX analysis was also used to determine the amount of each element in the formation of AgNPs. Based on Figure 2E,F, two peaks were observed in the spectrum in between 2 to 4 keV. The peak that formed at 3 keV showed the existence of an elemental Ag signal, as AgNPs have optical peaks at ~3 keV due to major emission energies. Another peak indicated the presence of elemental chlorine which could have been from the plant extract. The zeta potential was found to be -11.52 mV for the synthesized AgNPs from the *S. mombin* leaf extract (Figure 3). Based on the zeta potential value of the AgNPs, it was shown the AgNPs had moderate stability. This could be due to the existence of bioactive contents in the extract. Thus, the particles might aggregate and flocculate due to the absence of a repulsive force. It was also observed that there was the presence of some weak peaks at 28°, 54°, 57° and 86° which might have been from the organic compound in the leaf extract (Figure 4) [21,22].

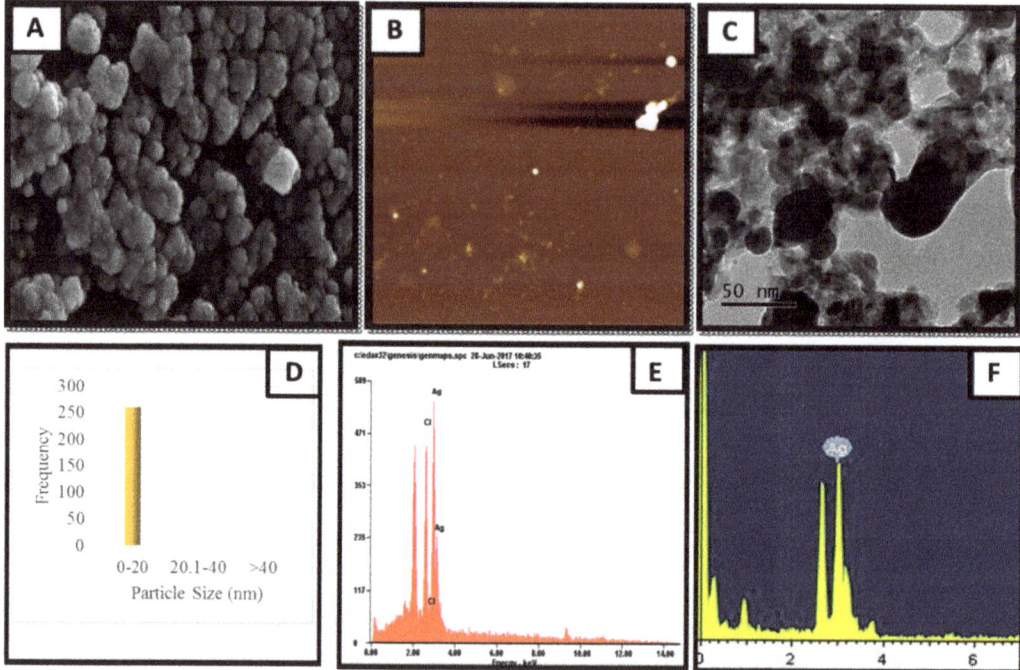

Figure 2. Structural characteristics of produced AgNPs. (**A**) SEM image with the scale of 200 nm; (**B**) spherical AgNPs observed using AFM; (**C**) TEM image with the scale of 50 nm; (**D**) histogram representing AgNP size distribution; (**E**) energy dispersive X-ray spectroscopy analysis with field emission scanning electron microscopy; (**F**) energy dispersive X-ray spectroscopy analysis with TEM.

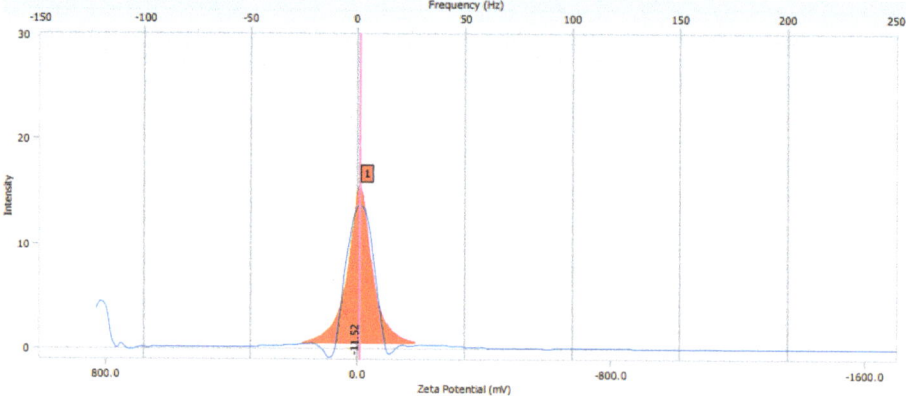

Figure 3. The zeta potential value of synthesized AgNPs.

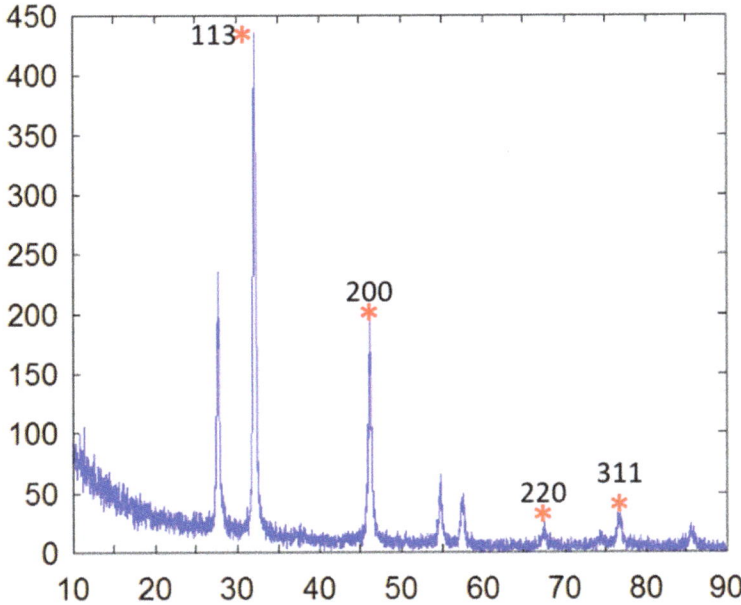

Figure 4. X-ray diffraction pattern of synthesized AgNPs. Ag peaks are marked (*) and 2θ values are given.

2.3. Antibacterial Activity of Spondias Mombin Leaf Extract

The antibacterial activity of *S. mombin* ethanolic leaf extract for selected bacteria was studied. A ciprofloxacin commercial antibiotic disc was used as positive control whereas the 10% DMSO was employed as negative control.

A clear zone of inhibition indicates a deterrent to the bacteria from growing. Results that were obtained from this study showed that *S. mombin* leaf extract had its own antimicrobial activity as they produced a clear zone of inhibition against the bacteria tested. From the results obtained, the nanoparticles showed an equal level of antimicrobial activity towards *Enterobacter cloacae*, *Escherichia coli*, *Klebsiella pneumoniae* and *Salmonella typhi*. Besides that, *Vibrio cholera* showed the lowest zone of inhibition compared to the other bacteria (Figures 5 and 6). However, there was not any zone of inhibition observed when the plant extract was tested with *Lactobacillus*. This showed that the plant extract had no antimicrobial property against *Lactobacillus*. As the bacteria have proven health benefits, this *Lactobacillus* group are classified as 'generally recognized as safe' bacteria [23]. They are categorized as nonpathogenic bacteria and the most common type of lactic acid bacteria in food and feed products [24]. The resistance of *Lactobacillus* towards *S. mombin* leaf extract could serve as an alternative treatment against human pathogenic bacteria as it does not affect the normal human flora population.

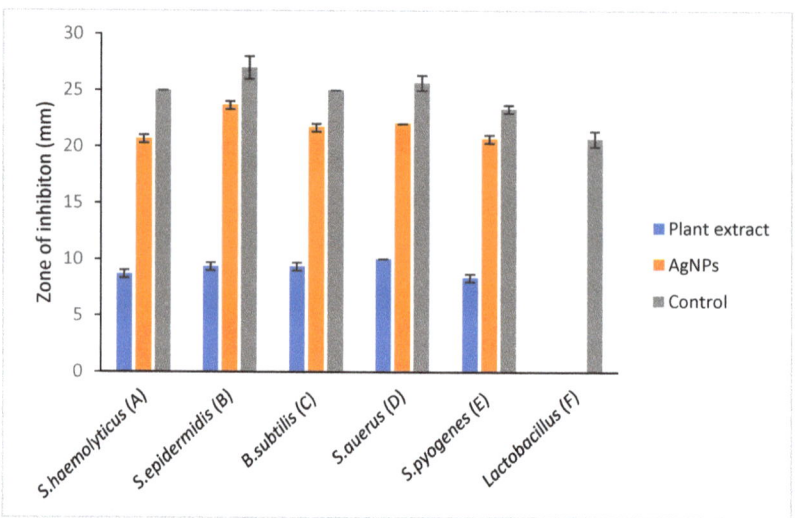

Figure 5. Antimicrobial analysis against selected Gram-positive bacteria. A: *Staphylococcus haemolyticus*, B: *Staphylococcus epidermidis*, C: *Bacillus subtilis*, D: *Staphylococcus aureus*, E: *Streptococcus pyogenes*, F: *Lactobacillus*. Comparative evaluation of selected Gram-positive bacteria vs. zone of inhibition.

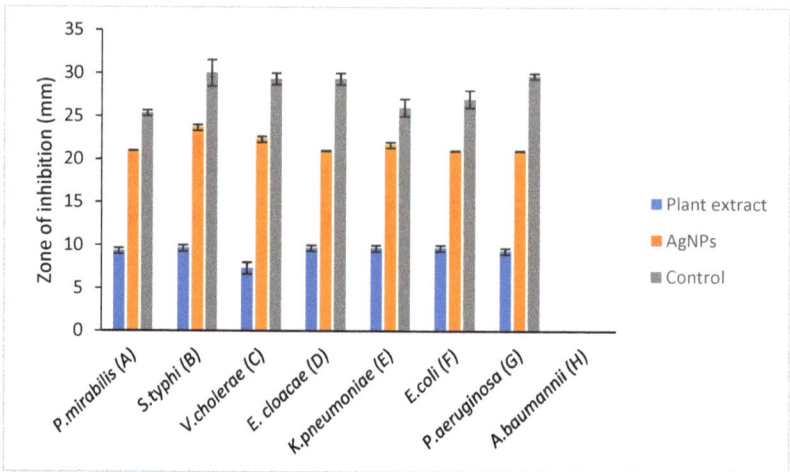

Figure 6. Antimicrobial activity against selected Gram-negative bacteria. A: Proteus mirabilis, B: Salmonella typhi, C: Vibrio cholera, D: Enterobacter cloacae, E: Klebsiella pneumoniae, F: E. coli, G: Pseudomonas aeruginosa, H: Acinetobacter baumannii. Comparative evaluation of selected Gram-negative bacteria vs. zone of inhibition.

Acinetobacter baumannii also showed no zone of inhibition for positive control which was the ciprofloxacin commercial antibiotic disc. This resistance of *A. baumannii* is mainly due to the mutation in the quinolone resistance determining region of DNA gyrase [25].

The known antimicrobial mechanism of the plant extract against various bacteria was inhibiting the cell wall synthesis, accumulating in the bacterial membrane which caused energy depletion or interference with the permeability of the cell membrane. This would eventually result in mutation, cell damage and the death of the bacteria. There is a study reporting that the phenolic and flavonoid content in the plant extract is the reason for the immune-modulator organs killing the bacteria [26].

2.4. Antibacterial Activity of Synthesized Silver Nanoparticles

The antimicrobial activity of silver nanoparticles (AgNPs) synthesis from *S. mombin* leaf extract using ethanol as solvent with disc diffusion method is tabulated in Tables 1 and 2. For the positive and negative control, ciprofloxacin commercial antibiotic disc and 10% DMSO were used, respectively.

Table 1. Antimicrobial activity of ethanolic extract of *S. mombin* and AgNP produced with ethanolic extract of *S. mombin* counter to Gram-positive bacteria. The information is presented in the table with the mean ($\pm$SE, standard error), $p < 0.05$.

Gram-Positive Bacteria	Zone of Inhibition (mm)			
	Plant Extract (Ethanolic)	Silver Nanoparticles (AgNPs)	Controls	
			Positive (Ciprofloxacin)	Negative
S. haemolyticus	8.67 ± 0.35	20.65 ± 0.35	25	-
S. epidermis	9.35 ± 0.35	23.65 ± 0.35	27 ± 1.00	-
B. subtilis	9.35 ± 0.35	21.65 ± 0.35	25	-
S. aurus	10	22	25.67 ± 0.67	-
S. pyogenes	8.33 ± 0.35	20.65 ± 0.35	23.35 ± 0.35	-
Lactobacillus	0.00	-	20.67 ± 0.67	-

Table 2. Antimicrobial activity of ethanolic extract of *S. mombin* and AgNP synthesized with ethanolic extract of *S. mombin* against Gram-negative bacteria. The information presented in the table with the mean ($\pm$SE, standard error), $p < 0.05$.

Gram-Negative Bacteria	Zone of Inhibition (mm)			
	Plant Extract (Ethanolic)	Silver Nanoparticles (AgNPs)	Controls	
			Positive (Ciprofloxacin)	Negative
P. mirabilis	9.33 ± 0.33	21	25.33 ± 0.33	-
S. typhi	9.65 ± 0.35	23.67 ± 0.33	30 ± 1.53	-
V. cholera	7.33 ± 0.67	22.33 ± 0.33	± 0.67	-
E. cloacae	9.67 ± 0.33	21	± 0.67	-
K. pneumoniae	9.65 ± 0.35	0.33	26 ± 1.00	-
E. coli	0.33	21	26 ± 1.00	-
P. aeruginosa	9.33 ± 0.33	21	29.7 ± 0.33	-
A. baumannii	0.00	-	-	-

A significant antimicrobial activity showed in the presence of *S. mombin* capped AgNPs against the selected bacteria. Silver nanoparticles synthesized from *S. mombin* leaf extract showed high antimicrobial activity for *Staphylococcus epidermidis* and *Salmonella typhi*. *Proteus mirabilis*, *Enterobacter cloacae*, *Escherichia coli*, *Pseudomonas aeruginosa* and showed a constitutive level of antimicrobial activity against the synthesized silver nanoparticle. Since there was no antibacterial activity observed in the plant extract against *Lactobacillus* and *Acinetobacter baumanii*, these strains were not tested with synthesized AgNPs.

When AgNPs contact with moisture, Ag^+ ions are released. The Ag^+ ions react with nucleic acid mainly with nucleosides forming the complex of the bacteria. AgNPs accumulate and form something called a 'pit' in the bacteria's cell wall and the nanoparticles slowly penetrate the intracellular component of the bacteria. The silver particles cause the plasma membrane to detach from the cell wall. This results in a loss of DNA replication and the protein synthesis process is also inhibited which causes the death of the bacteria. In addition to that, the hindrance of biofilm formation by AgNPs is an important mechanism, as biofilm plays a crucial part in the development of bacterial resistance against common drugs [27].

AgNPs demonstrated mediocre antibacterial action in Gram-positive bacteria compared to Gram-negative bacteria, depending on the result. This is due to the Gram-positive bacteria having a thick peptidoglycan layer. This causes difficulty in spreading AgNPs across the cell wall to disrupt the cell's activity and to inhibit its growth [28]. Gram-positive bacteria are made up of 70–100 peptidoglycans layers. Peptidoglycan consists of two polysaccharides, N-acetyl-glucosamine and N-acetyl-muramic acid, interlinked with peptide side chains and cross bridges [29]. On the other hand, compared to Gram-negative bacteria, the outer membrane of Gram-positive bacteria might cause less silver to reach the cytoplasmic membrane [30]. As a result, Gram-positive bacteria displayed a higher tolerance to synthesized silver nanoparticles relative to Gram-negative bacteria.

The oxidation of AgNP releases Ag^+ ions, and the ions are responsible for circulation in the living organism. The production of reactive oxygen species (ROS) induces oxidative stress that damages the membrane, proteins, DNA/RNA, and lipids, which enhance the cytotoxicity in prokaryotic cells. Thus, the ROS production was analyzed in these selected Gram-positive (*S. haemolyticus, S. epidermidis, B. subtilis, S. aureus, S. pyogenes*) and Gram-negative bacterial strains (*P. mirabilis, V. cholera, K. pneumoniae, E. coli, P. aeruginosa, E. cloacae, S. typhi.*) by treating with plant extract, AgNP and ciprofloxacin to quantify the amount of ROS production in contrast to the negative control (DMSO), and the findings are represented in Figures 7 and 8. However, Figure 7 shows that the ROS level in Gram-positive bacteria and plant extract showed its effect in the following order. i.e., *S. aureus, S. epidermidis, B. subtilis, S. haemolyticus,* and *S. pyogenes*. AgNP showed an excellent ROS level in all the strains as compared with ciprofloxacin. Figure 8 shows the ROS production level in the Gram-negative strains and the plant extract showed a similar level in all the strains. AgNP showed significant ROS level in *S. typhi* and *V. cholera* followed by other strains as compared with ciprofloxacin. Based on the results, it is suggested the ROS production is due to the interaction of AgNP with different bacterial cells. The interaction causes structural changes of the cell by causing toxicity by inducing oxidative stress. This affects the protein synthesis process resulting in cell death.

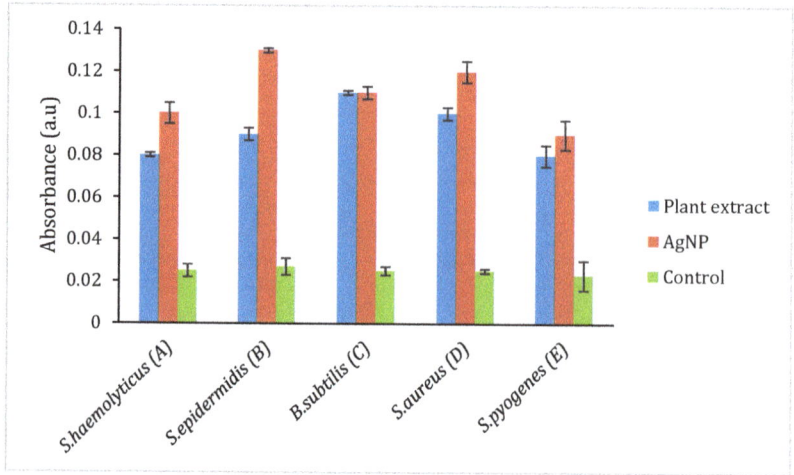

Figure 7. ROS production in selected Gram-positive bacteria. A: Staphylococcus haemolyticus, B: Staphylococcus epidermidis, C: Bacillus subtilis, D: Staphylococcus aureus, E: Streptococcus pyogenes.

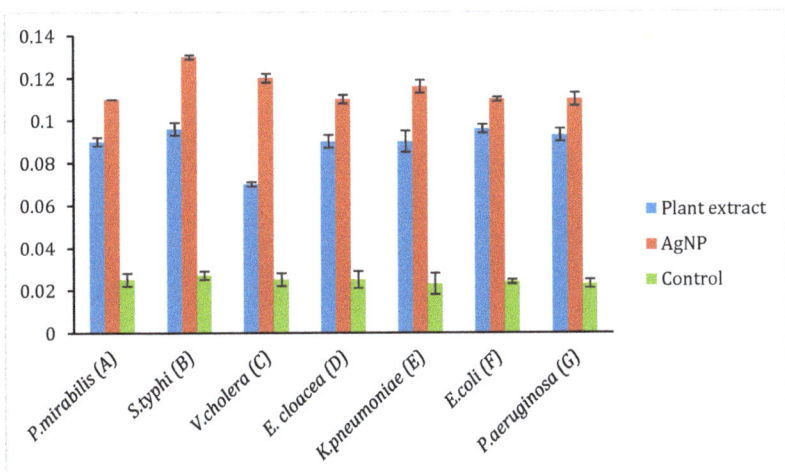

Figure 8. ROS production in selected Gram-negative bacteria A: Proteus mirabilis, B: *Salmonella typhi*, C: *Vibrio cholera*, D: *Enterobacter cloacae*, E: *Klebsiella pneumoniae*, F: *E. coli*, G: *Pseudomonas aeruginosa*.

2.5. Cytotoxic Study

The cytotoxicity study revealed that the highest mortality of 70% was obtained at 1.0 mg/mL. Figure 9 shows the plot of mortality percentage against the various concentrations of synthesized AgNPs. The graph showed a direct proportional relationship between the concentration of synthesized AgNPs and the rate of mortality. The LC_{50} of synthesized AgNPs was observed to be at 0.81 mg/mL. A study was conducted by Samuggam et al. using *Durio zibethinus* AgNPs showed the LC_{50} was 3.03 mg/mL [28]. Another study conducted by Shriniwas et al., reported the LC_{50} value of AgNPs synthesized using L. *camara* L. was 0.51 mg/mL [29]. The *A. salina* cytotoxicity depends on the AgNP size. It was stated that the cytotoxicity activity would be stronger when the size of the AgNPs was smaller [30].

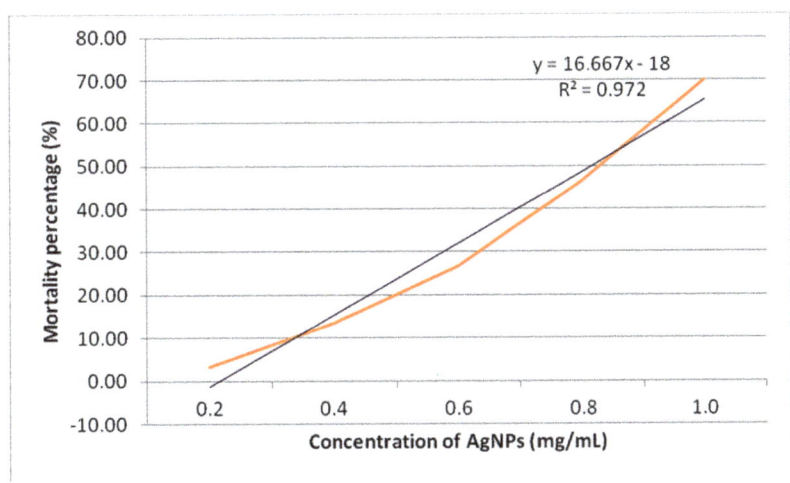

Figure 9. The cytotoxicity rate of synthesized AgNPs using *A. salina*.

3. Materials and Methods

3.1. Collecting the Plant Samples

Healthy, disease free young leaves of *S. mombin* were accumulated from Kulim, Kedah. These samples were shade dried for 2 weeks and crushed into powder form. The fine powdered leaves were stored in an airtight container at room temperature until they were required.

3.2. Preparation of S. mombin Leaf Extract

Fifty grams of powdered plant materials and 250 mL of solvent (99.98% ethanol) were added to a conical flask. This conical flask was further positioned in the incubator at 180 rpm at 37 °C for 7 days. Then, the extracted components of the plant were filtered and concentrated at temperatures around 35 °C–40 °C with the help of a rotary evaporator. The concentrated extract was air dried and the leaf extract was stored at 4 °C.

3.3. Biosynthesis of Silver Nanoparticles

AgNPs were biosynthesized according to the method mentioned previously (15–17). A millimolar solution of silver nitrate was prepared. The mixture was mixed with the magnetic mixer until it fully dissolved the silver nitrate crystals. The $AgNO_3$ solution was applied with five milliliters of plant extract slowly, until the hue shifted from pale yellow to brown. For 19 h in the dark room, the solution was incubated. After 19 h, the solution was centrifuged, for 15 min at 4000 rpm and the supernatant discarded. The pellet was then cleaned, scattered, and poured out into the glass of the clock with purified water. The pellet was air dried and stored for further use at 4 °C.

3.4. Characterization of Synthesized Silver Nanoparticles

To detect the reduction of the aqueous silver ion by scanning from 300 to 900 nm to obtain the maximum absorption strength of AgNPs, the UV−visible spectrophotometer (Beckman Coulter DU 800 Spectrophotometer, Williamston, SC, USA) was used. Scanning electron microscopy (SEM) and transmission electron microscopy (TEM) were used to analyze the morphological and structural features of the synthesized AgNPs. SEM and TEM analysis were performed on a Hitachi, S-4300 SE, Japan and a JEM-2100F, JEOL, Japan, respectively. The EDX analysis was performed to study the elemental composition of the AgNPs. Both SEM and TEM were equipped with energy-dispersive X-ray analysis. AgNP FESEM photographs were taken under a high-energy electron beam with a working distance of 15 kV and 4.5 mm. Surface texture analysis was carried using atomic force microscopy (AFM) using Nano Scope, Ica, Vecco, Plainsview, NY, USA. The sample for AFM analysis was conducted by preparing a thin pellet of AgNPs on a glass slide which could dry for 5 min. Crystalline nature of AgNP was determined using an X-ray Diffractometer (DMAX-2500, Rigaku, Tokyo, Japan). The diffraction angle was varied from 10° to 90° at 40 kV and 100 mA with Cu Kα radiation source. The size distribution and stability of AgNPs were studied using particle size analyzer (PHOTAL OTSUKA ELECTRONICS, ELC-Z model, Osaka, Japan).

3.5. Verification of the Antibacterial Activity of Synthesized Silver Nanoparticles

Kirby−Bauer antibiotic test (disc diffusion test) was used in this project. In this test, 10 bacteria were used as the testing microorganisms which consisted of Gram positive and Gram negative. They were *Bacillus subtilis*, *Escherichia coli*, *Staphylococcus aureus*, *Enterobacter cloacae*, *Staphylococcus epidermidis*, *Klebsiella pneumoniae*, *Vibrio cholera*, *Salmonella typhi*, *Staphylococcus haemolyticus*, *Proteus mirabilis*, *Streptococcus pyogenes*, *Pseudomonas aeruginosa*, *Lactobacillus* and *Acinetobacter baumannii*. The antimicrobial activity between the AgNPs and leaf extract was analyzed by the zone of inhibition formed on the agar plates.

3.6. Reactive Oxygen Species (ROS) Quantification

The Choi et al., 2006 method was used to determine the amount of reactive oxidative species (ROS) released by the microbes. To conclude, a total of 200 mL of bacterial strain was applied with 1 mL of plant extract, AgNP, ciprofloxacin (positive control) and DMSO (negative control) and kept in 37 °C incubator shaker. Once 6 h of incubation had been achieved, the bacteria suspension was centrifuged at 11,000× g for 11 min at low temperature to obtain the pellet. The pellet was applied with 2% Nitro Blue Tetrazolium (NBT) mixture. This pellet was kept at room temperature for 60 min in dark conditions. After centrifugation of the solution, the supernatant was removed, and the pellet was rinsed twice using PBS before another centrifugation at 9000× g for 3 min. The obtained pellet containing cells membrane was disrupted by treating with 2 M KOH solution. A sample of 50% DMSO was combined with the solution and followed by 10 min incubation at room temperature to dissolve formazan crystals. The solution was again centrifuged and 100 µL of the supernatant was distributed to 96 well plates. The absorbance was calculated at 620 nm using ELISA reader.

3.7. Cytotoxicity Study

The cytotoxicity study was done using *Artemia salina* (*A. salina*) according to Samuggam et al. [28]. Total of 10 larvae of *A. salina* were incubated at different concentrations of AgNPs in range of 0.2 to 1.0 mg/mL in 1 milliliter of sterilized seawater. This *A. salina* was incubated for 16 h and 8 h of light and dark, respectively, at 25 °C for 24 h. The assay was carried out in triplicate. Based on the larval mortality percentage, the LC_{50} values were determined.

4. Conclusions

In conclusion, *Spondias mombin* mediated silver nanoparticles proved their antibacterial ability against biofilm-producing bacteria, *S. haemolyticus*, *S. epidermidis*, *B. subtilis*, *S. aureus*, *S. pyogenes*, *Enterobacter cloacae*, *Escherichia coli*, *Klebsiella pneumoniae* and *Salmonella typhi*. The property of antibacterial activity of this plant extract was improved by synthesizing *Spondias mombin* leaf extract with capped silver nanoparticles. The production of ROS and cytotoxicity studies suggested the interaction of AgNPs with the bacterial cells caused structural changes which led to cell death and less cytotoxicity, respectively. Thus, these synthesized silver nanoparticles have the potential to be an effective drug against biofilm-producing bacteria.

Author Contributions: Conceptualization, S.V.C., S.S. and S.C.B.G.; methodology, S.V.C., S.S. and S.C.B.G., P.A. and P.M.; software, S.S., B.E. and S.V.C.; validation, V.V., L.V.R., S.S. and S.V.C.; formal analysis, S.V.C., S.S. and S.C.B.G., P.A. and P.M.; investigation, S.V.C., S.S. and S.C.B.G., P.A., V.V., L.V.R. and P.M.; resources, S.V.C., S.S. and S.C.B.G. and P.A.; data curation, S.V.C., S.S. and S.C.B.G., B.E. and P.A.; writing—original draft preparation, S.V.C., S.S. and P.M.; writing—review and editing, S.V.C., S.S., B.E., V.V., L.V.R. and P.M.; visualization, S.V.C., S.S. and S.C.B.G., P.A., V.V., L.V.R. and Prasanna M; supervision, S.V.C., S.S. and S.C.B.G.; project administration, S.V.C., S.S. and S.C.B.G.; funding acquisition, S.V.C.; All authors have read and agreed to the published version of the manuscript.

Funding: This work was supported by FRGS, Malaysia, FRGS/1/2018/STG03/AIMST/02/1.

Institutional Review Board Statement: Not applicable.

Informed Consent Statement: Not applicable.

Data Availability Statement: Data is contained within the article.

Acknowledgments: The authors are grateful to AIMST University, Malaysia for the support to successfully accomplish this research.

Conflicts of Interest: The authors declare no conflict of interest.

Sample Availability: Samples of the compounds or not available from the authors.

References

1. Garibo, D.; Borbón-Nuñez, H.A.; De León, J.N.D.; Mendoza, E.G.; Estrada, I.; Toledano-Magaña, Y.; Tiznado, H.; Ovalle-Marroquin, M.; Soto-Ramos, A.G.; Blanco, A.; et al. Green synthesis of silver nanoparticles using Lysiloma acapulcensis exhibit high-antimicrobial activity. *Sci. Rep.* **2020**, *10*, 1–11. [CrossRef]
2. Chen, H.; Roco, M.C.; Li, X.; Lin, Y.-L. Trends in nanotechnology patents. *Nat. Nanotechnol.* **2008**, *3*, 123–125. [CrossRef]
3. Chinni, S.; Gopinath, S.; Anbu, P.; Fuloria, N.; Fuloria, S.; Mariappan, P.; Krusnamurthy, K.; Reddy, L.V.; Ramachawolran, G.; Sreeramanan, S.; et al. Characterization and Antibacterial Response of Silver Nanoparticles Biosynthesized Using an Ethanolic Extract of Coccinia indica Leaves. *Crystals* **2021**, *11*, 97. [CrossRef]
4. Lashin, I.; Fouda, A.; Gobouri, A.; Azab, E.; Mohammedsaleh, Z.; Makharita, R. Antimicrobial and In Vitro Cytotoxic Efficacy of Biogenic Silver Nanoparticles (Ag-NPs) Fabricated by Callus Extract of Solanum incanum L. *Biomolecules* **2021**, *11*, 341. [CrossRef] [PubMed]
5. Das, C.G.; Kumar, G.; Dhas, S.; Velu, K.; Govindaraju, K.; Joselin, J.; Baalamurugan, J. Antibacterial activity of silver nanoparticles (biosynthe-sis): A short review on recent advances. *Biocatal. Agric. Biotechnol.* **2020**, *27*, 101593. [CrossRef]
6. Natsuki, J.; Natsuki, T.; Hashimoto, Y. A Review of Silver Nanoparticles: Synthesis Methods, Properties and Applications. *Int. J. Mater. Sci. Appl.* **2015**, *4*, 325. [CrossRef]
7. Ahmed, S.; Ahmad, M.; Swami, B.L.; Ikram, S. A review on plants extract mediated synthesis of silver nanoparticles for antimi-crobial applications: A green expertise. *J. Adv. Res.* **2016**, *7*, 17–28. [CrossRef]
8. Iravani, S.; Korbekandi, H.; Mirmohammadi, S.V.; Zolfaghari, B. Synthesis of silver nanoparticles: Chemical, physical and bio-logical methods. *Res. Pharm. Sci.* **2014**, *9*, 385–406.
9. Rajeshkumar, S.; Bharath, L.V. Mechanism of plant-mediated synthesis of silver nanoparticles—A review on biomolecules involved, characterisation and antibacterial activity. *Chem. Biol. Interact.* **2017**, *273*, 219–227. [CrossRef] [PubMed]
10. Pantidos, N.; Horsfall, L.E. Biological Synthesis of Metallic Nanoparticles by Bacteria, Fungi and Plants. *J. Nanomed. Nanotechnol.* **2000**, *5*. Available online: https://www.omicsonline.org/open-access/biological-synthesis-of-metallic-nanoparticles-by-bacteria-fungi-and-plants-2157-7439.1000233.php?aid=31363 (accessed on 18 June 2018). [CrossRef]
11. Cowan, M.M. Plant Products as Antimicrobial Agents. *Clin. Microbiol. Rev.* **1999**, *12*, 564–582. [CrossRef]
12. Siddiqi, K.S.; Husen, A.; Rao, R.A.K. A review on biosynthesis of silver nanoparticles and their biocidal properties. *J. Nanobiotechnol.* **2018**, *16*, 14. [CrossRef] [PubMed]
13. WHO Publishes List of Bacteria for Which New Antibiotics Are Urgently Needed. Available online: https://www.who.int/news/item/27-02-2017-who-publishes-list-of-bacteria-for-which-new-antibiotics-are-urgently-needed (accessed on 27 March 2021).
14. Sharma, D.; Misba, L.; Khan, A.U. Antibiotics versus biofilm: An emerging battleground in microbial communities. *Antimicrob. Resist. Infect. Control.* **2019**, *8*, 1–10. [CrossRef]
15. Kuppusamy, P.; Yusoff, M.M.; Maniam, G.P.; Govindan, N. Biosynthesis of metallic nanoparticles using plant derivatives and their new avenues in pharmacological applications–An updated report. *Saudi Pharm. J.* **2016**, *24*, 473–484. [CrossRef] [PubMed]
16. Islam, S.M.d.A.; Ahmed, K.T.; Manik, M.K.; Wahid, M.d.A.; Kamal, C.S.I. A comparative study of the antioxidant, antimicrobial, cyto-toxic and thrombolytic potential of the fruits and leaves of Spondias dulcis. *Asian Pac J Trop Biomed.* **2013**, *3*, 682–691. [CrossRef]
17. Fernandes, F.A.; Salgado, H. Antimicrobial Activity of Spondias dulcis Parkinson Extract Leaves Using Microdilution and Agar Diffusion: A Comparative Study. *EC Microbiol.* **2018**, *14*, 9.
18. Zhang, X.-F.; Liu, Z.-G.; Shen, W.; Gurunathan, S. Silver Nanoparticles: Synthesis, Characterization, Properties, Applications, and Therapeutic Approaches. *Int. J. Mol. Sci.* **2016**, *17*, 1534. [CrossRef] [PubMed]
19. Mohanta, Y.K.; Panda, S.K.; Bastia, A.K.; Mohanta, T.K. Biosynthesis of Silver Nanoparticles from Protium serratum and Investi-gation of their Potential Impacts on Food Safety and Control. *Front. Microbiol.* **2017**, *8*. Available online: https://www.frontiersin.org/articles/10.3389/fmicb.2017.00626/full (accessed on 24 January 2021). [CrossRef]
20. Shankar, S.S.; Ahmad, A.; Sastry, M. Geranium Leaf Assisted Biosynthesis of Silver Nanoparticles. *Biotechnol. Prog.* **2003**, *19*, 1627–1631. [CrossRef]
21. Duraisamy, S.; Kasi, M.; Balakrishnan, S.; Al-Sohaibani, S.; Murugan, K.; Senthilkumar, B.; Senbagam, D. Biosynthesis of silver nanoparticles using Acacia leucophloea extract and their antibacterial activity. *Int. J. Nanomed.* **2014**, *9*, 2431. [CrossRef]
22. Suvith, V.; Philip, D. Catalytic degradation of methylene blue using biosynthesized gold and silver nanoparticles. *Spectrochim. Acta Part A Mol. Biomol. Spectrosc.* **2014**, *118*, 526–532. [CrossRef] [PubMed]
23. Georgieva, R.; Yocheva, L.; Tserovska, L.; Zhelezova, G.; Stefanova, N.; Atanasova, A.; Danguleva, A.; Ivanova, G.; Karapetkov, N.; Rumyan, N.; et al. Antimicrobial activity and antibiotic susceptibility of Lactobacillus and Bifidobacterium spp. intended for use as starter and probiotic cultures. *Biotechnol. Biotechnol. Equip.* **2015**, *29*, 84. [CrossRef] [PubMed]
24. Kang, C.-G.; Hah, D.-S.; Kim, C.-H.; Kim, Y.-H.; Kim, E.-K.; Kim, J.-S. Evaluation of Antimicrobial Activity of the Methanol Extracts from 8 Traditional Medicinal Plants. *Toxicol. Res.* **2011**, *27*, 31–36. [CrossRef]
25. Jeong, S.H.; Yeo, S.Y.; Yi, S.C. The effect of filler particle size on the antibacterial properties of compounded polymer/silver fi-bers. *J. Mater Sci.* **2005**, *40*, 5407–5411. [CrossRef]
26. Mohamed, D.S.; El-Baky, R.M.A.; Sandle, T.; Mandour, S.A.; Ahmed, E.F. Antimicrobial Activity of Silver-Treated Bacteria against other Multi-Drug Resistant Pathogens in Their Environment. *Antibiotics* **2020**, *9*, 181. [CrossRef] [PubMed]

27. Wang, L.; Hu, C.; Shao, L. The antimicrobial activity of nanoparticles: Present situation and prospects for the future. *Int. J. Nanomed.* **2017**, *12*, 1227–1249. [CrossRef] [PubMed]
28. Sumitha, S.; Vasanthi, S.; Shalini, S.; Chinni, S.V.; Gopinath, S.C.B.; Anbu, P.; Bahari, M.B.B.; Harish, R.; Kathiresan, S.; Ravichandran, V. Phyto-Mediated Photo Catalysed Green Synthesis of Silver Nanoparticles Using Durio Zibethinus Seed Extract: Antimicrobial and Cytotoxic Activity and Photocatalytic Applications. *Molecules* **2018**, *23*, 3311. [CrossRef]
29. Patil, P.S.; Kumbhar, S.T. Antioxidant, antibacterial and cytotoxic potential of silver nanoparticles synthesized using terpenes rich extract of Lantana camara L. leaves. *Biochem. Biophys. Rep.* **2017**, *10*, 76–81.
30. Kittler, S.; Greulich, C.; Diendorf, J.; Koller, M.; Epple, M. Toxicity of Silver Nanoparticles Increases during Storage Because of Slow Dissolution under Release of Silver Ions. *Chem. Mater.* **2010**, *22*, 4548–4554. [CrossRef]

Article

Biogenic Synthesis of NiO Nanoparticles Using *Areca catechu* Leaf Extract and Their Antidiabetic and Cytotoxic Effects

Shwetha U R [1], Rajith Kumar C R [1], Kiran M S [1], Virupaxappa S. Betageri [1,*], Latha M S [2], Ravindra Veerapur [3], Ghada Lamraoui [4], Abdulaziz A. Al-Kheraif [5], Abdallah M. Elgorban [6], Asad Syed [6], Chandan Shivamallu [7] and Shiva Prasad Kollur [8,*]

[1] Research Centre, Department of Chemistry, GM Institute of Technology, Davangere 577 006, Karnataka, India; shwethaur@gmail.com (S.U.R.); rajithcr91@gmail.com (R.K.C.R.); kiranmsafare@gmail.com (K.M.S.)
[2] Department of Chemistry, R L Science Institute, Belagavi 590 001, Karnataka, India; lathamschem97@gmail.com
[3] Department of Metallurgy and Materials Engineering, Malawi Institute of Technology, Malawi University of Science and Technology, Limbe P.O. Box 5916, Malawi; rveerapur@must.ac.mw
[4] Faculty of Nature and Life Sciences, Earth and Universe Sciences, University of Tlemcen, Tlemcen 13000, Algeria; lamraouig@gmail.com
[5] Dental Biomaterials Research Chair, Dental Health Department, College of Applied Medical Sciences, King Saud University, P.O. Box 10219, Riyadh 11433, Saudi Arabia; aalkhuraif@ksu.edu.sa
[6] Department of Botany and Microbiology, College of Science, King Saud University, P.O. Box 2455, Riyadh 11451, Saudi Arabia; aelgorban@ksu.edu.sa (A.M.E.); assyed@ksu.edu.sa (A.S.)
[7] Department of Biotechnology and Bioinformatics, School of Life Sciences, JSS Academy of Higher Education and Research, Mysuru 570 015, Karnataka, India; chandans@jssuni.edu.in
[8] Department of Sciences, Amrita School of Arts and Sciences, Amrita Vishwa Vidyapeetham, Mysuru Campus, Mysuru 570 026, Karnataka, India
* Correspondence: virupaxb@gmail.com (V.S.B.); shivachemist@gmail.com (S.P.K.)

Citation: U R, S.; C R, R.K.; M S, K.; Betageri, V.S.; M S, L.; Veerapur, R.; Lamraoui, G.; Al-Kheraif, A.A.; Elgorban, A.M.; Syed, A.; et al. Biogenic Synthesis of NiO Nanoparticles Using *Areca catechu* Leaf Extract and Their Antidiabetic and Cytotoxic Effects. *Molecules* **2021**, *26*, 2448. https://doi.org/10.3390/molecules26092448

Academic Editor: Nagaraj Basavegowda

Received: 18 December 2020
Accepted: 10 March 2021
Published: 22 April 2021

Publisher's Note: MDPI stays neutral with regard to jurisdictional claims in published maps and institutional affiliations.

Copyright: © 2021 by the authors. Licensee MDPI, Basel, Switzerland. This article is an open access article distributed under the terms and conditions of the Creative Commons Attribution (CC BY) license (https://creativecommons.org/licenses/by/4.0/).

Abstract: Nanoworld is an attractive sphere with the potential to explore novel nanomaterials with valuable applications in medicinal science. Herein, we report an efficient and ecofriendly approach for the synthesis of Nickel oxide nanoparticles (NiO NPs) via a solution combustion method using *Areca catechu* leaf extract. As-prepared NiO NPs were characterized using various analytical tools such as powder X-ray diffraction (XRD), scanning electron microscopy (SEM), transmission electron microscopy (TEM), and UV-Visible spectroscopy (UV-Vis). XRD analysis illustrates that synthesized NiO NPs are hexagonal structured crystallites with an average size of 5.46 nm and a hexagonal-shaped morphology with slight agglomeration. The morphology, size, and shape of the obtained material was further confirmed using SEM and TEM analysis. In addition, as-prepared NiO NPs have shown potential antidiabetic and anticancer properties. Our results suggest that the inhibition of α-amylase enzyme with IC 50 value 268.13 µg/mL may be one of the feasible ways through which the NiO NPs exert their hypoglycemic effect. Furthermore, cytotoxic activity performed using NiO NPs exhibited against human lung cancer cell line (A549) proved that the prepared NiO NPs have significant anticancer activity with 93.349 µg/mL at 50% inhibition concentration. The biological assay results revealed that NiO NPs exhibited significant cytotoxicity against human lung cancer cell line (A549) in a dose-dependent manner from 0–100 µg/mL, showing considerable cell viability. Further, the systematic approach deliberates the NiO NPs as a function of phenolic extracts of *A. catechu* with vast potential for many biological and biomedical applications.

Keywords: *Areca catechu*; NiO NPs; TEM; antidiabetic activity; anticancer potential

1. Introduction

For the past few years, nanotechnology has acquired marvelous impetus by creating new scientific ideas in this rapidly growing technological era [1,2]. Nanomaterials have revealed many technological insights with their tremendous applications and specific properties [3,4]. Surface morphology, characteristic size, and shape are key features for nanomaterials, which make them highly attractive and more reactive for researchers [5,6]. Biologically fabricated nanoparticles with their immense applications in various fields are growing continuously through the collaboration of different natural science sectors. The world of nanotechnology may furnish a novel resource for the evaluation and development of safer, newer, and effective drug formulations in the treatment of infectious diseases [7].

Recently, the interest in synthesizing metal oxide nanoparticles has increasingly been employed in various fields due to their potential applications in memory storage devices, photocatalytic sensors, magnetic resonance imaging, drug delivery, catalysis, and biomedicine [8]. Nanoparticles exhibits cytotoxic activity due to their higher adsorption ability over bulk materials [9]. Hence, they are used to treat various tumor and cancer cells [10]. Nickel oxide is a p-type semiconductor metal oxide possessing a band gap from 3.6 to 4.0 eV that has great importance and has received enormous consideration in research owing to its peculiar properties like large surface area, high chemical stability, good electronic conductivity, and super conductance characteristics [11,12]. Its ecofriendly nature and high reactivity makes it a potential candidate for applications in the field of magnetism, electronics, energy technology gas sensors, electrochemical super capacitors, catalysis, battery cathodes, magnetic materials, fuel cells, optical fibers, and biomedicines [13,14]. Moreover, NiO nanostructures have motivated young researchers due to their easy availability with low cost, quantum size confinement, and surface-to-volume effect [15,16]. NiO NPs are synthesized by different physical and chemical methods, namely, Sol-gel, hydrothermal, precipitation, solvothermal, etc. However, the biogenic synthesis approach has drawn the attention of researchers due to its biocompatibility and ecofriendly process, which involves green synthetic routes that are less toxic. Exploiting the potential of medicinal plants is one of the green synthesis routes, which includes algae, microorganisms, plants, etc., and is significant because the current therapeutic approaches have toxicity problems and microbial multidrug resistance issues. Metal nanoparticles have received great attention across the globe, so, in this study, we discuss and focus on metallic nanoparticles obtained by green synthesis using medicinal plants. We also discuss medicinal properties like antidiabetic and anticancer activities of synthesized nanoparticles. The biomolecules, secondary metabolites, and coenzymes present in the plants help with the easy reduction of metal ions to nanoparticles. Such nanoparticles are considered as potential antioxidants and promising candidates in cancer treatment. Thus, the synthesis of ecofriendly nanoparticles from combustion solutions is one of the simplest and easiest synthetic approaches towards uniform mixing of plant extract with precursor/oxidizing agents [17].

Plants are known for their medicinal values in terms easier availability and large number of biologically active components. *A. catechu* is one of the known fruit plants belonging to the Palmaceae family and is cultivated in most Asian countries [18]. Medicinal properties of this plant's extracts are due to the presence of various phytochemicals that are present in the different parts of the plant [19]. Perusal of the literature shows that *Areca* leaves possess more bioactive molecules, namely, arecoline, arecolidine, arecaidine, guvacoline, guvacine, and isoguvacine. Use of plant extracts for the synthesis of nanoparticles is desirable due to the various plant metabolites like polyphenols, alkaloids, phenolic acids, and terpenoids, which play a major role in the bioreduction of metal ions, yielding nanoparticles. Plant act as bioreactors in the binding and reduction of metal ions, thereby influencing the formation of nanoparticles.

In recent years, solution combustion synthesis is emerging as one of the efficient methods to produce nanomaterials with a controlled size and shape. It is also used as a rapid heating method for metal oxides synthesis. Beyond rapid heating, this green synthesis method gives good product yield in less time when compared to other conventional methods. The present study sheds light on the synthesis of highly efficient, cost effective, nanosized NiO nanoparticles by using the solution combustion synthesis method. Solution combustion synthesis is a green, efficient, simple, fast, and high-yield method. The novelty of the study is the use of Areca catechu leaf extract as a reducing and stabilizing agent for NiO nanoparticles synthesis. Temperature plays a pivotal role here. The solution combustion reaction depends on various process parameters, and it plays a significant role in phase formation, phase stability, and physical characteristics. The reaction temperature is a crucial parameter in the synthesis of materials. The released heat of the combustion reaction fulfils the energy requirement for the formation of oxides. The presence of phytochemical constituents in the plant extract; concentration of plant extract; and reaction conditions like temperature, reducing agent concentration, reaction time, and size of nanoparticles all influence the stability of NiO NPs [20].

The size and morphology of the nanoparticles play a significant role in developing the chemical and physical properties and largely influence their existing applications. Therefore, much effort was dedicated to the fabrication of NiO NPs with different sizes and morphologies. The decrease in dimension leads to an increase in the surface area and this enhances the biological properties.

In the current study, *A. catechu* leaf extract is used as a reducing and stabilizing agent to synthesize NiO NPs. Prepared nanoparticles were characterized using XRD, SEM with EDAX, and HR-TEM. Furthermore, we investigated the cytotoxicity of NiO NPs by examining cell viability and antidiabetic activity. This study provides detailed information about the cytotoxic effects of as-prepared NiO NPs against human lung cancer cells and offers a sound basis for the clarification of its toxicity mechanisms.

2. Materials and Methods

All the chemicals were analytical grade, procured from SD Fine and Himedia Laboratory Pvt. Ltd., India, and used without further purification. The morphology of as-prepared NiO NPs was observed by Transmission Electron Microscopy (TEM-1011, JEOL, Tokyo, Japan). SEM with Energy dispersive X-ray Analysis was utilized to evaluate the elemental study (Hitachi S3400n, Tokyo, Japan). X-ray diffraction examination of NiO NPs was done on a PANalytical X'Pert-PRO (Rigaku Smart Lab). UV-Visible spectrophotometer (Shimadzu UV-2450, Kyoto, Japan) was used to record electronic absorption spectra.

2.1. Preparation of Areca Catechu Leaf Extract and Synthesis of NiO NPs

Areca Catechu leaves were collected from the local areas near Davanagere. Freshly collected leaves were washed with double distilled water, dried, and grinded well to get fine powder. To prepare the leaf extract, 10 g of *A. catechu* leaf powder was boiled in 100 mL distilled water for 30 min at 60 °C. Further, the extract was filtered and dried under vacuum using a rotary evaporator.

The solution combustion method was used to synthesis NiO NPs. In a typical experiment, 10 mL of *A. catechu* leaf extract and 1 g of nickel nitrate hexahydrate $Ni(NO_3)_2 6H_2O$) were taken in a silica crucible and placed in a preheated muffle furnace maintained at 500 °C. An exothermic, vigorous reaction leads to the formation of fine, black colored NiO NPs. The obtained product was kept in an airtight container for further analysis [21].

2.2. Antidiabetic Activity: Inhibition of Alpha Amylase Enzyme Assay

Pancreatic α-amylase belongs to the class of α-1,4-gluconohydrolases and is one of the important target enzymes for the conventional treatment of diabetes. It catalyzes the initial step in hydrolysis of starch to maltose and maltotriose, which are then acted upon by α-glucosidases, broken down into glucose, and enter the blood stream. Naturally available α-amylase inhibitors from medicinally important plants are shown to be very effective in managing postprandial hyperglycemia, which is a major concern in type 2 diabetes [22].

In a fresh tube, 1 mL of phosphate buffered saline (PBS) solution was mixed with 0.5 mL of different concentrations (100, 200, 300, 400, and 500 μg/mL) of samples or the standard solution, then 200 μL of 0.5 mg/mL α-amylase was added followed by 200 μL of 5 mg/mL starch solution and incubated for 10 min at room temperature. Control was taken as starch with amylase and without α-amylase. Then, the reaction mixture was stopped by adding 400 μL of Dextrose normal saline (DNS) solution, followed by heating the mixture in a boiling water bath for 5 min, then cooling. The reaction without *A. catechu* leaf extract was used as a control. Metformin was used as a standard drug [23]. Inhibition of enzyme activity was calculated using the following formula:

% Inhibition of enzyme activity = Abs sample − Abs control / Abs sample × 100. (1)

2.3. Anticancer Activity: Cytotoxicity Assay of NiO NPs

The cytotoxicity assay of biosynthesized NiO NPs was performed against human lung cancer cell line (A549). The cell lines were cultivated in Dulbecco's Modified Eagle's Medium (DMEM) with fetal bovine serum, with antibiotics as supplements. Temperature was maintained around 37 °C with humidified 5% CO_2 atmosphere for about 24 h. The cells were seeded in 96-well plates at a density 25×10^3 cells/well. Cytotoxicity of biosynthesized NiO NPs was studied using 3-(4,5-dimethylthiazol-2-yl)-2,5-diphenyl tetrazolium bromide (MTT) assay. Here, human cancer cell lines were treated with different concentrations of NiO NPs (20 to 100 mg/mL from stock). The plate was removed from the incubator and the drug-containing media was aspirated. A total of 100 µL of medium containing 10% MTT reagent was then added to each well to get a final concentration of 0.5 mg/mL, and the plate was incubated at 37 °C and 5% CO_2 atmosphere for 3 h. The culture medium was removed completely without disturbing the crystals formed. Then, 100 µL of solubilization solution (DMSO) was added and the plate was gently shaken in a gyratory shaker to solubilize the formed formazan [22].

The absorbance was measured using a microplate reader at a wavelength of 570 nm and also at 630 nm. The percentage growth inhibition was calculated, after subtracting the background, the blank, and the concentration of test drug needed to inhibit cell growth by 50% (IC_{50}). Yellow color MTT dye turning to purple color due to the reduction of formazon crystals in the presence of cytotoxic activity shows in the mitochondrial succinate dehydrogenase enzyme in viable cells. The amount of 50% inhibition concentration was obtained by plotting the dose-dependent curve [24].

3. Results and Discussion

3.1. XRD Analysis

The XRD pattern of green synthesized NiO NPs from *Areca catechu* leaf extract show strong diffraction peaks at 37.23°, 43.29°, 62.88°, and 75.45°, which are assigned to the crystal planes (111), (200), (220), and (311), respectively, as shown in Figure 1, and are further well matched with JCPDS card no. 4-835. These planes indicate the formation of FCC cubical structure for NiO NPs. Further, no impurities were observed, which suggests a high purity of monophasic NiO NPs. The average crystalline size found to be 5.63 nm, calculated by the Debye–Scherer formula [25]. Moreover, the EDAX spectra of nanoparticles displayed the peaks of Ni and O, as seen in Figure 2, suggesting the chemical nature of the prepared material. The obtained profile of the synthesized nanoparticles confirmed the presence of nickel and oxygen in the nanoparticles.

3.2. UV-Visible Spectral Analysis

It is clear from the UV-Visible spectrum of as-prepared NiO NPs (Figure 3) that the maximum absorption band observed at 380 nm reveals the formation of pure NiO NPs. This absorption in the UV region can be attributed to the electronic transition from the valence band to the conduction band in the NiO semiconducting nanocrystals.

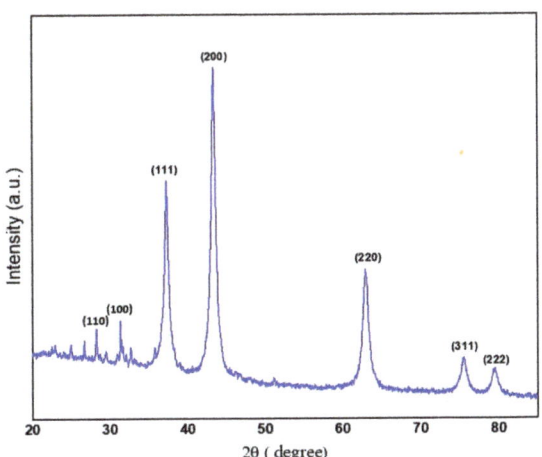

Figure 1. X-ray diffraction patterns revealing the crystal planes of as-prepared NiO NPs.

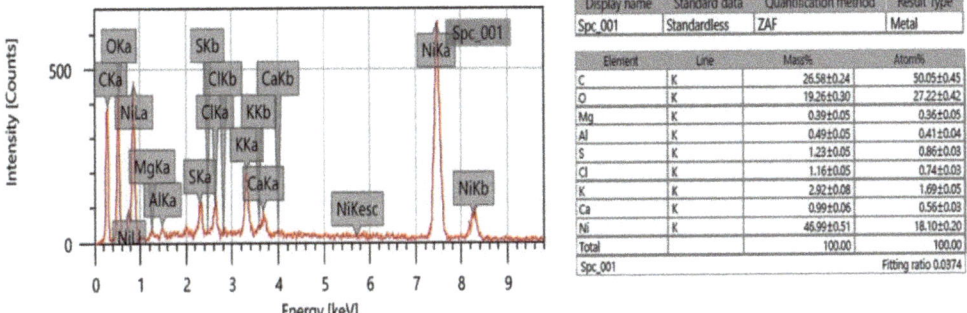

Figure 2. Energy-dispersive X-ray (EDAX) spectra depicting the chemical composition of the synthesized NiO NPs.

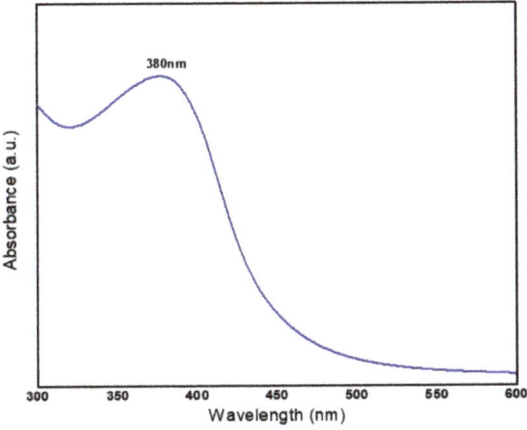

Figure 3. UV-Visible Spectrum of as-prepared NiO NPs.

3.3. SEM Analysis

The surface morphological features of synthesized NiO NPs was studied using scanning electron microscope (SEM). In Figure 4, the SEM micrographs show the agglomeration with irregularly shaped nanoparticles. It can also be seen that the particles have a hexagonal shape with some degree of agglomerations, which may be attributed to the fact that NiO nanoparticles have high surface energy and high surface tension.

Figure 4. SEM images showing the morphology of as-prepared NiO NPs with different magnifications.

3.4. TEM Analysis

The formation of NiO NPs was perceived in the TEM images (Figure 5), which specifies the particle size within the range of 5 to 15 nm. Further, this supports the average crystal size from the XRD pattern. Figure 5b,c represent the HR-TEM micrographs that show particles in the hexagonal and rhombohedral shapes with an interplanar spacing of 0.21 nm. The selected area electron diffraction (SAED) pattern depicted in Figure 5d indicates the presence of the (111), (200), and (220) planes of the synthesized rhombohedral NiO NPs.

3.5. Antidiabetic Studies

In Vitro Alpha Amylase Inhibition Method

In our digestive system, pancreatic α-amylase is a key enzyme that catalyzes the initial step in the hydrolysis of starch. It is the main source of glucose in the diet. α-amylase inhibitors are those that inhibit the amylase activity that results in the delay of carbohydrate digestion and prolongs overall carbohydrate digestion time, causing a reduction in the rate of glucose absorption and consequently reducing the postprandial plasma glucose rise.

The α-amylase inhibitor effectiveness of NiO NPs was compared with standard drug Metformin. The values were presented with graphical representation of the same in Figure 6. Alpha amylase is an enzyme that hydrolyses α-bonds of large α-linked polysaccharides such as glycogen and starch to yield glucose and maltose. α-amylase inhibitors bind to α-bond of polysaccharide and prevent the breakdown of polysaccharides in mono- and disaccharide. Standard drug Metformin showed inhibitory effects on the α-amylase activity with an IC_{50} value of 232.12 µg/mL. Prepared NiO NPs from Areca leaves exhibited α-amylase inhibitory activity with an IC_{50} value of 268.13 µg/mL. As a result, as-synthesized NiO NPs showed significant antidiabetic activity compared to Metformin. Moreover, drugs that inhibit carbohydrate hydrolyzing enzymes have been demonstrated to decrease postprandial hyperglycemia and improve impaired glucose metabolism without promoting insulin secretion of noninsulin-dependent diabetic patients. The results of in vitro studies showed that NiO NPs inhibits α-amylase activity [26].

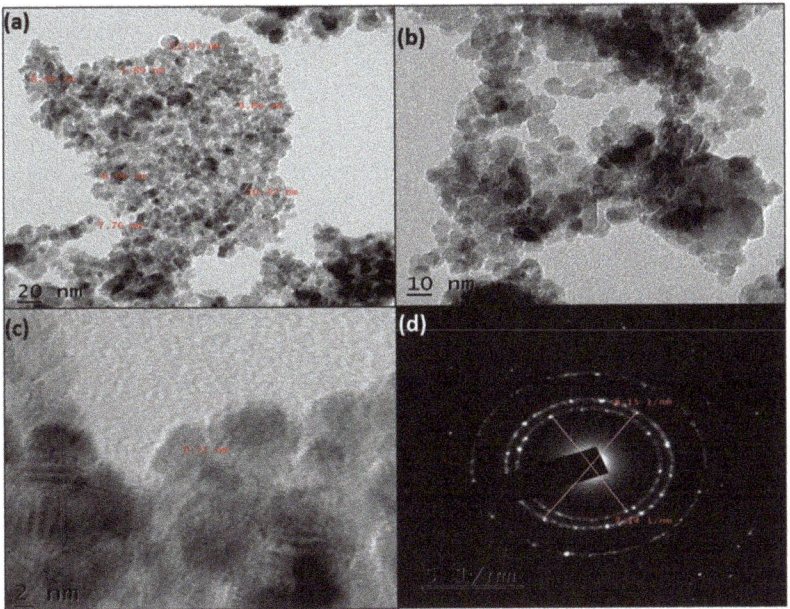

Figure 5. (**a**,**b**) TEM images, (**c**) HR-TEM image, and (**d**) SAED of as-prepared NiO NPs.

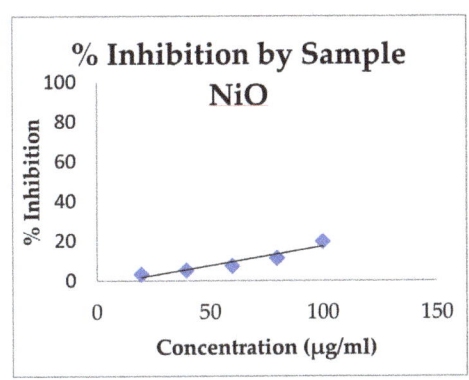 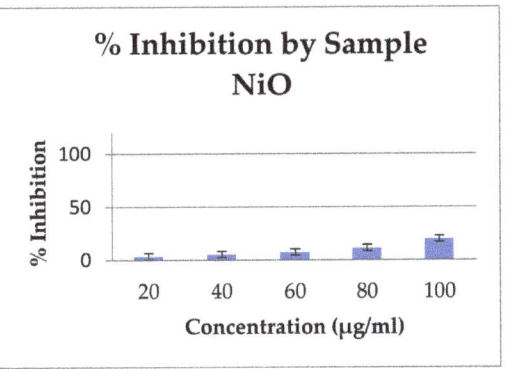

Figure 6. Antidiabetic potential of as-prepared NiO NPs showing inhibition of α-amylase activity at different concentrations.

As-prepared NiO NPs showed a percentage inhibition of 3.35 and 19.77 at 20 µg/mL and 100 µg/mL, respectively. The IC_{50} value of the extract was found to be 268.13 µg/mL, whereas the IC_{50} value of metformin was observed to be 232.12 µg/mL (Table 1). The concentration-based inhibition was noticed and the same has been depicted in Figure 6. Metformin is a standard antidiabetic drug and is competitively and reversibly inhibiting the pancreatic α-amylase. The retardation of glucose diffusion is also due to the inhibition of α-amylase, thereby limiting the release of glucose from the starch. The inhibition of α-amylase activity by medicinal plants might be attributed to several possible factors such as fiber concentration; the presence of inhibitors on fibers; and the encapsulation of starch and enzymes by the fibers present in the sample, thereby reducing accessibility of starch to the enzyme and direct adsorption of the enzyme on fibers, leading to decreased amylase activity. Thus, the inhibition of α-amylase activity is important to control postprandial hyperglycemia in the treatment of diabetes [27].

Table 1. Antidiabetic activity of NiO NPs by α-amylase (pancreatic) inhibition assay by DNS method.

Sl. No	Concentrations µg/mL	% Inhibition by Sample NiO NPs	% Inhibition by Standard Drug Metformin
1	20	3.35088	5.08616
2	40	5.39944	8.87984
3	60	7.45572	13.34370
4	80	11.44088	16.51507
5	100	19.77022	22.59454

3.6. Cytotoxicity Studies

The evaluation of cytotoxicity of biosynthesized NiO NPs against A549 cell line cancer cells was measured based on cellular reduction of MTT during in vitro analysis. The as-prepared NiO NPs was screened against cell lines with the respective positive control Cisplatin, as shown in the Figure 7. NiO NPs treatment enhanced the cell death and also inhibited A549 cell population in a concentration-dependent manner. After treatment with different concentrations (20, 40, 60, 80, and 100 µg/mL), the plating efficiency of A549 cells declines, as proved by the reduction in the number of cancer cells formed. Exposure of various concentrations NiO NPs shows a decline in cell survival and plating efficiency. When compared with regular cisplatin, minimum inhibition was observed at 20 µg/mL and maximum at 100 µg/mL. The viability assay of cytotoxicity of NiO NPs against the cancer cell line is shown in Figure 8. Further, the IC_{50} density was found to be 93.349 µg/mL. The healthy and rapidly growing cells exhibit high rates of MTT reduction to formazan while the dead or inactive cells fail to do so. Viability in the MTT assay is connected linearly with enzyme activity and indirectly to the number of viable cells. The decrease in cell viability with the increasing concentration of NiO NPs shows significant cytotoxicity to accumulate in the internal cells and higher stress, ultimately leading to apoptosis [28,29].

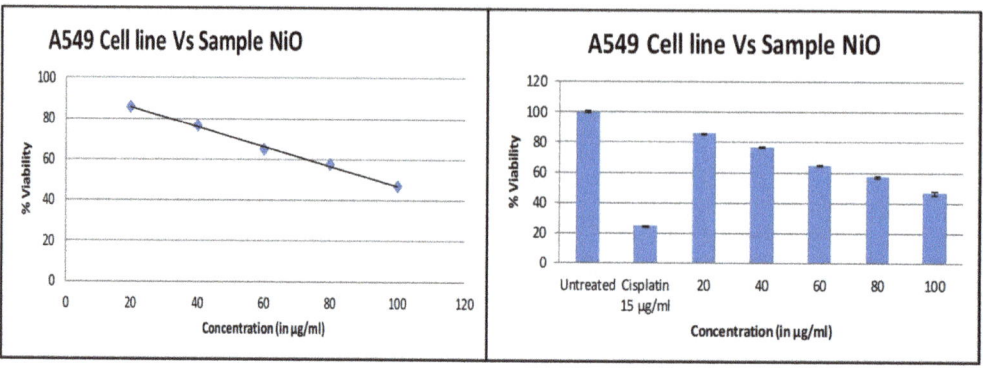

Figure 7. Graph representing the screening of anticancer activity with respect to the standard control for different concentrations of synthesized NiO NPs.

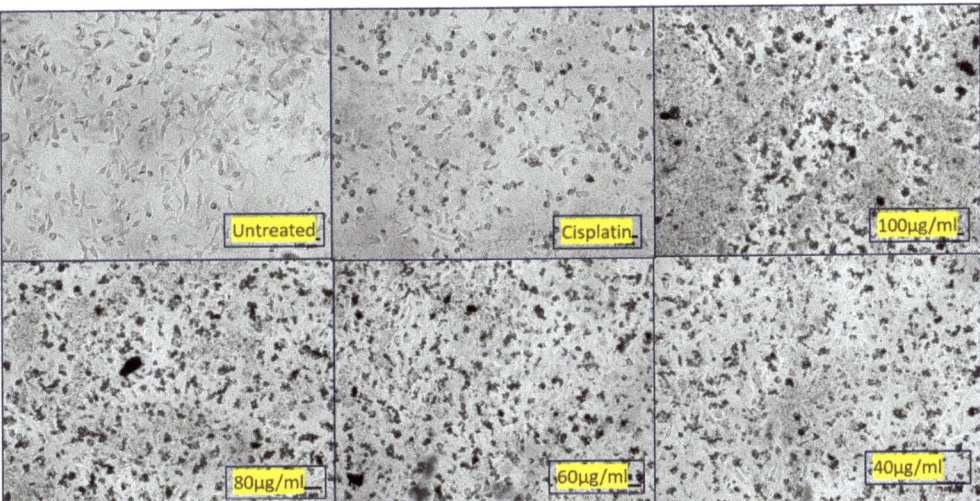

Figure 8. The viability assay of cytotoxicity of NiO NPs against cancer cell line (A549) treated with different concentrations of NiO NPs.

4. Conclusions

In summary, we have reported the synthesis of NiO NPs by an ecofriendly approach via solution combustion method using the *Areca catechu* leaf extract. Areca is the important plant in Asia both in an agricultural role and as a traditional medicine. Preliminary phytochemicals like phenolic compounds, alkaloids, glycosides, and tannins are well-reported in literature. The X-ray diffractogram revealed the formation of hexagonal NiO NPs with a well crystalline nature and a very fine crystallite size of 5.63 nm. Further, the morphological characteristics determined by SEM and TEM analysis disclosed a size and shape of as-prepared nanostructures. Further, the antidiabetic activity of as-prepared NiO NPs was carried out using glucose uptake by yeast cell and α-amylase inhibition, which demonstrated significant antidiabetic activity. In addition, the prepared material showed potential anticancer activity against human lung cancer cell lines. The chemical constituents of areca plant had proven diverse pharmacological actions and were used as antidiabetic and anticancer agents. Overall, the present study clearly indicated that biosynthesized NiO NPs from *Areca catechu* leaves are a promising avenue for the prevention of diabetes and cancer diseases.

Author Contributions: Conceptualization, S.U.R., R.K.C.R. and K.M.S.; methodology, S.U.R. and S.P.K.; software, R.V. and G.L.; validation, V.S.B., C.S. and L.M.S.; formal analysis, S.U.R., S.P.K., C.S. and A.S.; investigation, S.P.K., R.V., A.A.A.-K. and L.M.S.; resources, A.A.A.-K., A.M.E. and V.S.B.; data curation, S.U.R., R.K.C.R. and R.V.; writing—original draft preparation, S.U.R., S.P.K., and V.S.B.; writing—review and editing, V.S.B. and S.P.K.; visualization, R.V. and C.S.; supervision, V.S.B.; project administration, L.M.S., C.S. and A.S.; funding acquisition, A.A.A.-K., A.M.E. and A.S. All authors have read and agreed to the published version of the manuscript.

Funding: This research was funded by the Deanship of Scientific Research, King Saud University through the Vice Deanship of Scientific Research Chairs.

Institutional Review Board Statement: Not applicable.

Informed Consent Statement: Not applicable.

Data Availability Statement: Data is contained within the article.

Acknowledgments: Authors thank the Director, Indian Institute of Science, Bengaluru, India for analytical facilities. KSP is grateful to the Director, Amrita Vishwa Vidyapeetham, Mysuru campus for infrastructure support. CS acknowledge the support and infrastructure provided by the JSS Academy of Higher Education and Research (JSSAHER), Mysuru, India. The authors are grateful to the Deanship of Scientific Research, King Saud University for funding through Vice Deanship of Scientific Research Chairs.

Conflicts of Interest: The authors declare that there are no conflicts of interest.

References

1. Mnyusiwalla, A.; Daar, A.S.; Singer, A.P. Mind the gap: Science and ethics in nanotechnology. *Nanotechnol.* **2003**, *14*, R9–R13. [CrossRef]
2. Antonietti, M. Small is beautiful: Challenges and perspectives of Nano/Meso/Microscience. *Small* **2016**, *12*, 2107–2114. [CrossRef] [PubMed]
3. Na, Y.; Yang, S.; Lee, S. Evaluation of citrate-coated magnetic nanoparticles as draw solute for forward osmosis. *Desalination* **2014**, *347*, 34–42. [CrossRef]
4. Davar, F.; Fereshteh, Z.; Salavati-Niasari, M. Nanoparticles Ni and NiO: Synthesis, characterization and magnetic properties. *J. Alloy. Compd.* **2009**, *476*, 797–801. [CrossRef]
5. Kreyling, W.G.; Semmler-Behnke, M.; Chaudhry, Q. A complementary definition of nanomaterial. *Nano Today* **2010**, *5*, 165–168. [CrossRef]
6. Krishnamurthy, N.; Vallinayagam, P.; Madhavan, D. *Engineering Chemistry*; PHI Learning Pvt Ltd.: New Dehli, India, 2014; pp. 154–160.
7. Dipankar, C.; Murugan, S. The green synthesis, characterization and evaluation of the biological activities of silver nanopar-ticles synthesized from Iresine herbstii leaf aqueous extracts. *Colloids Surf. B Biointerfaces.* **2012**, *98*, 112–119. [CrossRef]
8. Din, M.; Amna, G.N.; Aneela, R.; Aihetasham, A.; Maria, M. Single step green synthesis of stable nickel and nickel oxide na-noparticles from Calotropis gigantea: Catalytic and antimicrobial potentials. *Environ. Nanotechnol. Monitor. Manage.* **2018**, *9*, 29–36. [CrossRef]
9. Costa, M.; Heck, J.D. Perspectives on the mechanism of nickel carcinogenesis. *Adv. Inorg. Biochem.* **1984**, *6*, 285–309.
10. Saleem, S.; Ahmed, B.; Khan, M.S.; Al-Shaeri, M.; Musarrat, J. Inhibition of growth and biofilm formation of clinical bacterial isolates by NiO nanoparticles synthesized from Eucalyptus globulus plants. *Microb. Pathog.* **2017**, *111*, 375–387. [CrossRef]
11. Wu, L.; Wu, Y.; Wei, H.; Shi, Y.; Hu, C. Synthesis and characteristics of NiO nanowire by a solution method. *Mater. Lett.* **2004**, *58*, 2700–2703. [CrossRef]
12. Sasi, B.; Gopchandran, K.; Manoj, P.; Koshy, P.; Rao, P.P.; Vaidyan, V. Preparation of transparent and semiconducting NiO films. *Vac.* **2002**, *68*, 149–154. [CrossRef]
13. Mariam, A.A.; Kashif, M.; Arokiyaraj, S.; Bououdina, M.; Sankaracharyulu, M.; Jayachandran, M.; Hashim, U. Bio-Synthesis of NiO and Ni nanoparticles and their characterization. *Dig. J. Nanomater. Biostruct.* **2014**, *9*, 1007–1019.
14. Pandian, C.J.; Palanivel, R.; Dhananasekaran, S. Green synthesis of nickel nanoparticles using Ocimum sanctum and their application in dye and pollutant adsorption. *Chin. J. Chem. Eng.* **2015**, *23*, 1307–1315. [CrossRef]
15. A review on nickel nanoparticles as effective therapeutic agents for inflammation. *Inflamm. Cell Signal.* **2014**, *1*. [CrossRef]
16. Kumar, C.R.; Betageri, V.S.; Nagaraju, G.; Pujar, G.; Suma, B.; Latha, M. Photocatalytic, nitrite sensing and antibacterial studies of facile bio-synthesized nickel oxide nanoparticles. *J. Sci. Adv. Mater. Devices* **2020**, *5*, 48–55. [CrossRef]
17. Lingaraju, K.; Naika, H.R.; Nagabhushana, H.; Jayanna, K.; Devaraja, S.; Nagaraju, G. Biosynthesis of Nickel oxide Nanoparticles from *Euphorbia heterophylla* (L.) and their biological application. *Arab. J. Chem.* **2020**, *13*, 4712–4719. [CrossRef]
18. Raghavendra, N.; Bhat, J.I. Natural Products for Material Protection: An Interesting and Efficacious Anticorrosive Property of Dry Arecanut Seed Extract at Electrode (Aluminum)–Electrolyte (Hydrochloric Acid) Interface. *J. Bio. Tribo Corrosion* **2016**, *2*, 21. [CrossRef]
19. Khan, S.; Mehmood, M.H.; Ali, A.N.A.; Ahmed, F.S.; Dar, A.; Gilani, A.-H. Studies on anti-inflammatory and analgesic activities of betel nut in rodents. *J. Ethnopharmacol.* **2011**, *135*, 654–661. [CrossRef]
20. Udayabhanu Nethravathi, P.C.; PavanKumar, M.A.; Suresh, D.; Lingaraju, K.; Rajanaika, H.; Nagabhushana, H.; Sharma, S.C. Tinospora cordifolia mediated facile green synthesis of cupric oxide nanoparticles and their photocatalytic, antioxidant and antibacterial properties. *Mater. Sci. Semiconductor Process.* **2015**, *33*, 81–88. [CrossRef]
21. Cirillo, V.P. *The Transport of Non-Fermentable Sugars Across the Yeast Cell Membrane, Membrane Transport and Metabolism*; Academic Press: New York, NY, USA, 1961; pp. 343–351.
22. Kiran, M.S.; Betageri, V.S.; Kumar, C.R.R.; Vinay, S.P.; Latha, M.S. In-Vitro Antibacterial, Antioxidant and Cytotoxic Potential of Silver Nanoparticles Synthesized Using Novel Eucalyptus tereticornis Leaves Extract. *J. Inorg. Organomet. Polym. Mater.* **2020**, *30*, 2916–2925. [CrossRef]
23. Rijuta, G.; Saratale; Han Seung, S.; Gopalakrishnan, K.; Giovanni, B.; Dong-Su, K.; Ganesh, D.; Saratale. Exploiting antidiabetic activity of silver nanoparticles synthesized using Punicagranatum leaves and anticancer potential against human liver cancer cells (HepG2), Artificial Cells. *Nanomed. Biotechnol.* **2018**, *46*, 211–222. [CrossRef]
24. Miller, G.L. Use of Dinitrosalicylic Acid Reagent for Determination of Reducing Sugar. *Anal. Chem.* **1959**, *31*, 426–428. [CrossRef]

25. Suresh, D.; Nethravathi, P.; Udayabhanu; Rajanaika, H.; Nagabhushana, H.; Sharma, S. Green synthesis of multifunctional zinc oxide (ZnO) nanoparticles using Cassia fistula plant extract and their photodegradative, antioxidant and antibacterial activities. *Mater. Sci. Semicond. Process.* **2015**, *31*, 446–454. [CrossRef]
26. Shwetha, U.R.; Latha, M.S.; Kumar, C.R.R.; Kiran, M.S.; Betageri, V.S. Facile Synthesis of Zinc Oxide Nanoparticles Using Novel Areca catechu Leaves Extract and Their In Vitro Antidiabetic and Anticancer Studies. *J. Inorg. Organomet. Polym. Mater.* **2020**, *30*, 4876–4883. [CrossRef]
27. Sun, C.; Li, H.; Chen, L. Nanostructured ceria-based materials: Synthesis, properties, and applications. *Energy Environ. Sci.* **2012**, *5*, 8475–8505. [CrossRef]
28. Vani, M.; Vasavi, T.; Uma Maheswari Devi, P. Evaluation of in vitro antidiabetic activity of methanolic extract of seagrass *Halophila beccarii*. *Asian J. Pharm. Clin. Res.* **2018**, *11*, 150–153. [CrossRef]
29. Ahmad, J.; Alhadlaq, H.A.; Siddiqui, M.A.; Saquib, Q.; Al-Khedhairy, A.A.; Musarrat, J.; Ahamed, M. Concentration-dependent induction of reactive oxygen species, cell cycle arrest and apoptosis in human liver cells after nickel nanoparticles exposure. *Environ. Toxicol.* **2013**, *30*, 137–148. [CrossRef]

Article

Algal-Derived Synthesis of Silver Nanoparticles Using the Unicellular *ulvophyte* sp. MBIC10591: Optimisation, Characterisation, and Biological Activities

Reham Samir Hamida [1], Mohamed Abdelaal Ali [2], Mariam Abdulaziz Alkhateeb [3], Haifa Essa Alfassam [3], Maha Abdullah Momenah [3,*] and Mashael Mohammed Bin-Meferij [3,4]

1 Nanobiology Lab, Institute for Protein Research, Osaka University, Osaka 565-0871, Japan
2 Plant Production Department, Arid Lands Cultivation Research Institute, City of Scientific Research and Technological Applications (SRTA-CITY) New Borg El-Arab, Alexandria 21934, Egypt
3 Department of Biology, College of Science, Princess Nourah bint Abdulrahman University, Riyadh 11671, Saudi Arabia
4 Histopathology Unit, Research Department, Health Sciences Research Center (HSRC), Princess Nourah bint Abdulrahman University, Riyadh 11671, Saudi Arabia
* Correspondence: mamomenah@pnu.edu.sa

Abstract: Algal-mediated synthesis of nanoparticles (NPs) is an eco-friendly alternative for producing NPs with potent physicochemical and biological properties. Microalgae represent an ideal bio-nanofactory because they contain several biomolecules acting as passivation and stabilising agents during the biogenesis of NPs. Herein, a novel microalgae sp. was isolated, purified, and identified using light and electron microscopy and 18s rRNA sequencing. The chemical components of their watery extract were assessed using GC-MS. Their dried biomass was used to synthesise silver (Ag) NPs with different optimisation parameters. Ag-NPs were physiochemically characterised, and their anticancer and antibacterial effects were examined. The data showed that the isolated strain was 99% similar to the unicellular *ulvophyte* sp. MBIC10591; it was ellipsoidal to spherical and had a large cup-shaped spongiomorph chloroplast. The optimum parameters for synthesising Ag-NPs by unicellular *ulvophyte* sp. MBIC10591 (Uv@Ag-NPs) were as follows: mixture of 1 mM of $AgNO_3$ with an equal volume of algal extract, 100 °C for 1 h, and pH of 7 under illumination for 24 h. TEM, HRTEM, and SEM revealed that Uv@Ag-NPs are cubic to spherical, with an average nanosize of 12.1 ± 1.2 nm. EDx and mapping analysis showed that the sample had 79% of Ag, while FTIR revealed the existence of several functional groups on the NP surface derivatives from the algal extract. The Uv@Ag-NPs had a hydrodynamic diameter of 178.1 nm and a potential charge of −26.7 mV and showed marked antiproliferative activity against PC3, MDA-MB-231, T47D, and MCF-7, with IC_{50} values of 27.4, 20.3, 23.8, and 40 µg/mL, respectively, and moderate toxicity against HFs (IC_{50} of 13.3 µg/mL). Uv@Ag-NPs also showed marked biocidal activity against Gram-negative bacteria. *Escherichia coli* was the most sensitive bacteria to the NPs with an inhibition zone of 18.9 ± 0.03 mm. The current study reports, for the first time, the morphological appearance of the novel unicellular *ulvophyte* sp., MBIC10591, and its chemical composition and potential to synthesise Uv@Ag-NPs with smaller sizes and high stability to act as anti-tumour and microbial agents.

Keywords: green synthesis; microalgae; anticancer; antibacterial; optimisation parameter

1. Introduction

Green synthesis has become a reliable and sustainable method for the biogenesis of several nanomaterials such as metallic nanoparticles (NPs) (M-NPs), metal oxide NPs, bimetallic NPs, and quantum dots [1]. NPs represent potent alternative drugs for several diseases, such as infectious diseases [2], cancers [3], diabetes [4], and wound healing [5], due to their unique features including their smaller size to larger surface area, various shapes, and sufficient reactivity facilitating their use for drug delivery, sensing, and catalysis, among others [6–8]. Generally, NPs are synthesised by three main methods: physical, chemical, and biological (green) syntheses [9]. Phycosynthesis is a green synthesis (bottom-up) route that uses algal cells and their biocomponents to produce NPs with various shapes and sizes [10]. Microalgae are considered model microorganisms for the biogenesis of NPs due to their potential to hyper-accumulate heavy metals and be redesigned to more malleable shapes [11]. Moreover, microalgae contain diverse biomolecules such as lipids, proteins, carbohydrates, vitamins, pigments (such as phycocyanin, chlorophyll, and carotenoids), antioxidants, and others that precipitate during the biogenesis of NPs as reducing and stabilising agents [12]. The *Chlorophyta* phylum includes several species that are sources of several secondary metabolites that act as new drugs in the nutraceutical and pharmaceutical industries [13]. The unicellular *ulvophyte* sp. MBIC10591 is a strain belonging to the *Chlorophyta* phylum. Unfortunately, there are no publications on its morphology and applications. This strain was isolated and deposited by Japanese scientists Suda et al. in GenBank for the first time in 2001 with the accession number AB058370. Several studies used microalgae and cyanobacteria to produce several types of metallic and metallic oxide NPs such as Au- [14], Ag- [15], ZnO- [16], and TiO_2-NPs [17]. Hamida et al. synthesised hexagonal Ag-NPs using the novel microalgae strain *Coelastrella aeroterrestrica* BA_Chlo4 with a smaller diameter of 14.5 ± 0.5 nm [12]. The NPs showed marked activity against different tumour cells, including MCF-7 and MDA, HCT-116, and HepG2 cells, with low toxicity against the normal cells (HFs and Vero). They also demonstrated moderate antioxidant activity and marked biocidal activity against both Gram-positive and -negative bacteria. The biological synthesis method requires the adjustment of physicochemical and biological parameters to obtain M-NPs with controlled sizes, shapes, and dispersity. Several studies have reported that the precursor concentrations, reactant ratios, temperature, pH, reaction time, time of exposure, and illumination conditions are important factors that influence the physicochemical and biological properties [18–20]. It was found that the increase in precursor concentration caused an increase in NP intensity, suggesting polydispersity and agglomeration of NPs at higher concentrations [21]. The change in the temperatures of the NP synthesis process may result in the formation of smaller or larger NPs. It was found that the reduction in NP size at higher temperatures could be attributed to an increase in the nucleation kinetics constant instead of the decreased growth kinetics constant, considering the concentrations of the precursors [22]. Among M-NPs, Ag-NPs have more effectiveness against microbes and cancerous cells. Aziz et al. used *Chlorella pyrenoidosa* as a source of reducing and stabilising agents to fabricate Ag-NPs and found that the resultant biogenic NPs exhibited a marked antibacterial activity against *Klebsiella pneumoniae, Aeromonas hydrophila, Acenetobacter* sp., and *Staphylococcus aureus* [23]. Ag-NPs are also used in sunscreen lotions, burn treatments, wound dressings, textiles, dental materials, and bone implants [24–26]. The Ag-NP mechanisms inside living cells have been reported to depend on their potential to facilitate oxidative stress by promoting the formation of reactive oxygen species. Moreover, their small sizes, potential charge, and surface chemistry enable their interactions with cellular proteins and DNA, resulting in cellular growth inhibition and death [3,27,28]. The current study revealed, for the first time, the morphology and chemical components of the novel unicellular *ulvophyte* sp. MBIC10591 and its potential for the biogenesis of Ag-NPs under optimum conditions and anticancer and antibacterial activities.

2. Results and Discussions
2.1. Algal Identification
2.1.1. Morphological Appearance

Light and inverted light micrographs revealed that the unicellular *ulvophyte* sp. MBIC10591 was spherical. Several unicellular vegetative cells were detected with cup-shaped spongiomorph chloroplasts with pyrenoids surrounded by several starch grains. Single cells were the most dominant; however, package cells with parenchyma-like structures containing more daughter cells were detected. All cells were surrounded by thick cell walls (Figure 1A–D). SEM micrographs showed cells with widely ellipsoidal to spherical shapes with sizes of 13.8 × 12.7 µm. Several irregular ribs existed on algal surfaces, and parenchyma-like structure package cells with more than nine daughter cells were observed (Figure 2A–D). The daughter cells were surrounded by thick cell walls. Unfortunately, no publications have demonstrated the morphological appearance of this isolate. The strain was deposited in 2001 for the first time by Japanese scientists Suda et al. in GenBank with the accession number AB058370. However, the current isolate shared several features with *Desmochloris* sp. in that its cells are distinguished by their spherical to ellipsoidal shapes and cup-shaped spongiomorph chloroplasts [29,30].

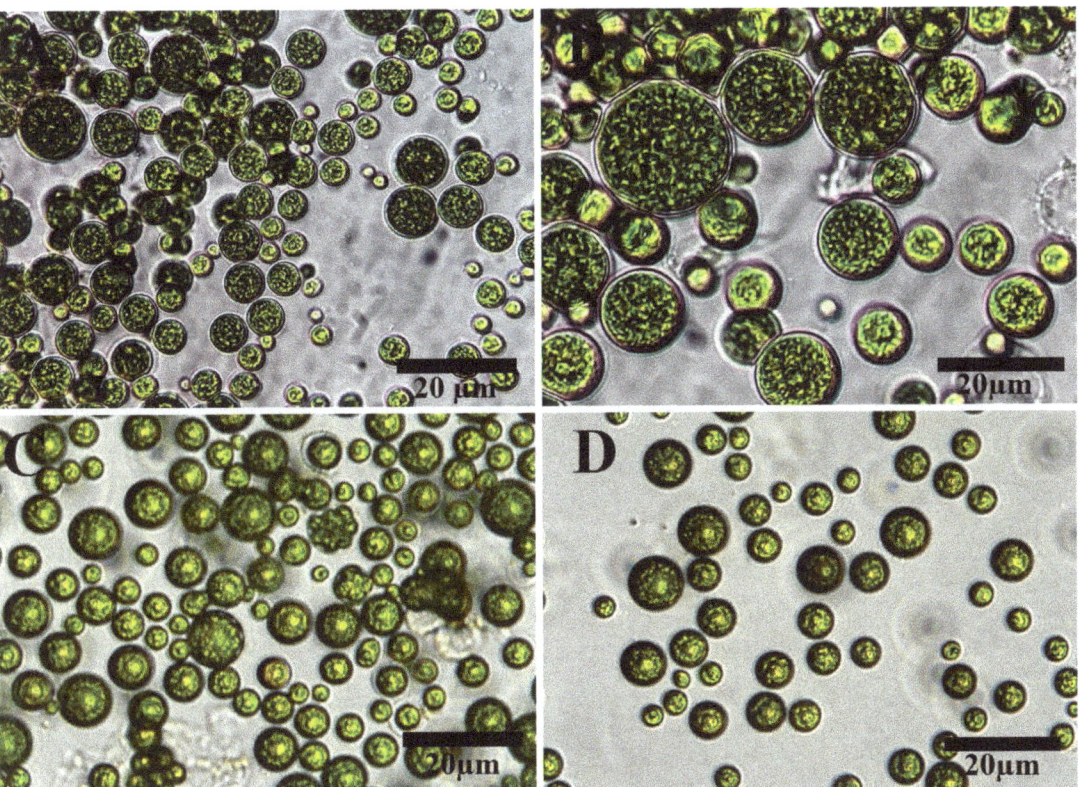

Figure 1. Light (**A**,**B**) and inverted light (**C**,**D**) microscopy of unicellular *ulvophyte* sp. MBIC10591. Scale bar = 20 µm.

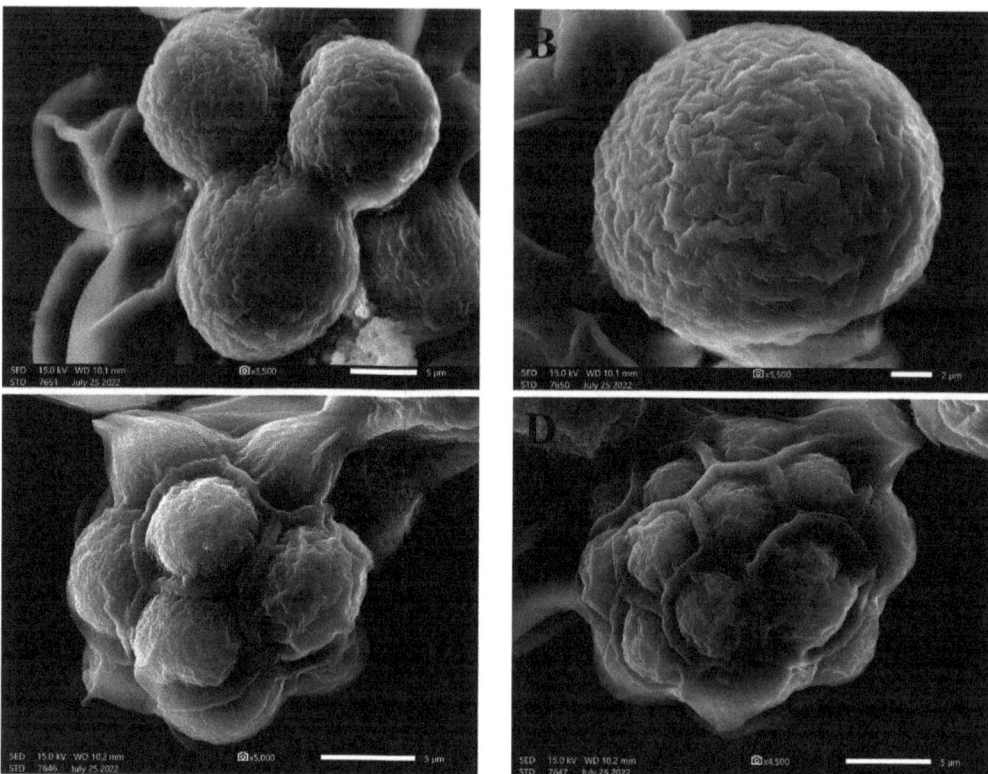

Figure 2. SEM micrographs of unicellular *ulvophyte* sp. MBIC10591 showing the morphology of single cells of *ulvophyte* sp. MBIC10591 (**A,B**) and package cells with a parenchyma-like structure containing more daughter cells (**C,D**). Scale bar = 5 μm (**A,C,D**) and 2 μm (**B**).

2.1.2. Molecular Identification

The 18s rRNA analysis revealed that the current strains were 99% similar to the unclassified unicellular *ulvophyte* sp. MBIC10591 with a query covering of 89%. The sequence was deposited in GenBank, NCBI, with accession number OP605382. The phylogenetic tree demonstrated that the unicellular *ulvophyte* sp. MBIC10591 may be clustered within *Desmochloris* sp. (Figure 3).

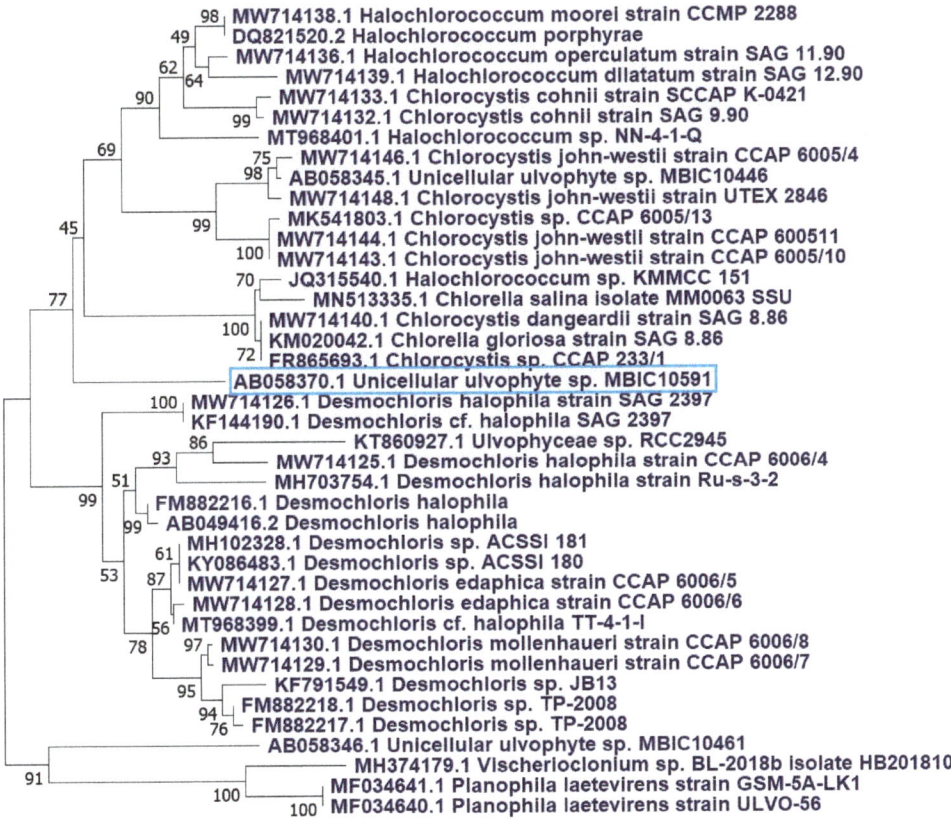

Figure 3. Phylogenetic tree of the unicellular *ulvophyte* sp. MBIC10591 (blue frame) inferred from 18S r RNA. Tree was constructed by cluster method using MEGA4 software version 10.2.6. Number at each branch refers to the bootstrap values for % of 1000 replicate trees calculated by neighbour joining statistical method.

2.1.3. GC-MS Analysis

The GC-MS chromatograph demonstrated, for the first time, the volatile organic molecules of the watery unicellular *ulvophyte* sp. MBIC10591 extract with a retention time of 4–44 min. The data showed 34 peaks corresponding to 24 algal bio-compounds. These 24 biomolecules included fatty acids (FA), FA esters, vitamins, alcohols, phenols, hydrocarbons, organosulphur compounds, amino-acid-like compounds, and polysaccharides (Table 1 and Figure 4). It was found that the unicellular *ulvophyte* sp. MBIC10591 was enriched with various molecules that act as antioxidant, antimicrobial, anticancer, and anti-hypercholesterolemic agents such as D-fructose, diethyl mercaptal, pentaacetate, 25,26,27-trinorcholecalcifer-24-al, trisulphide, and di-2-propenyl, among others [30–32]. Based on the GC-MS spectra, the main organic molecules were speculated to be lipids and hydrocarbons that could precipitate while stabilising NPs. However, the existence of alcohols and phenols may indicate that hydroxyl groups have significant roles in the biogenesis of Uv@Ag-NPs. Olasehinde et al. analysed the ethanolic and dichloromethane extracts of *Chlorella sorokiniana* and *Chlorella minutissima* and found that the microalgae were enriched with phenols, sterols, steroids, fatty acids, and terpenes that have modulatory activities for some mediators of Alzheimer's disease [33]. GC-MS analysis of the aqueous extract of *Coelastrella aeroterrestrica* BA_Chlo4 showed that the dominant biomolecules of the algal extract were fatty acids and hydrocarbons [12].

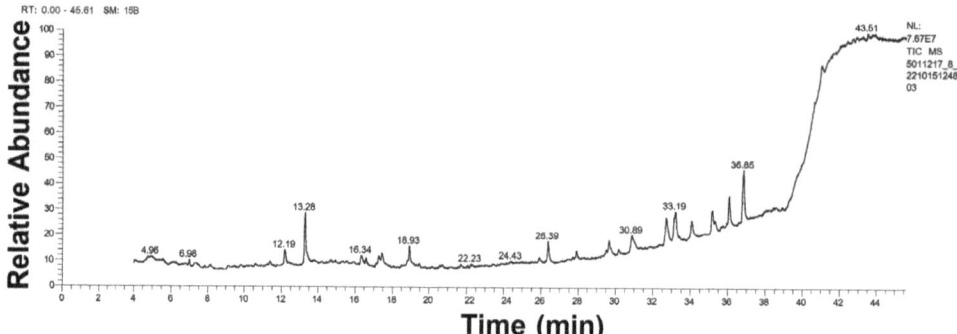

Figure 4. GC-MS chromatogram of unicellular *ulvophyte* sp. MBIC10591 watery extract.

Table 1. Chemical composition analysis of the unicellular *ulvophyte* sp. MBIC10591 using GC-MS.

No.	Biomolecule Name	Retention Time	Area %	Mentioned Factor	Molecular Formula	Molecular Weight
1	D-fructose, diethyl mercaptal, pentaacetate	4.04, 4.09, 16.34	0.49, 0.63, 3.25	659, 655, 659	$C_{20}H_{32}O_{10}S_2$	496
2	Methyl 4,7,10,13-hexadecatetraenoate	6.15	0.89	708	$C_{17}H_{26}O_2$	262
3	2-Aminoethanethiolsulphuric acid	6.99, 11.39, 21.7	1.10, 0.96, 0.71	699, 727, 760	$C_2H_7NO_3S_2$	157
4	25,26,27-Trinorcholecalcifer-24-al	7.34	2.05	766	$C_{24}H_{36}O_2$	356
5	Trisulfide, di-2-propenyl	12.19	3.34	760	$C_6H_{10}S_3$	178
6	1,2-Diacetin	13.28	9.63	927	$C_7H_{12}O_5$	176
7	[5,5-dimethyl-6-(3-methyl-buta-1,3-dienyl)-7-oxa-bicyclo [4.1.0]hept-1-yl]-methanol	16.59	1.22	748	$C_{14}H_{22}O_2$	222
8	3a,4,7,7a-Tetrahydrodimethyl-4,7-methanoindene	17.26	1.58	775	$C_{12}H_{16}$	160
9	Phenol, 2,6-bis(1,1-dimethylethyl)-	17.45	2.06	749	$C_{14}H_{22}O$	206
10	Methyl 6,9-octadecadiynoate	18.83	0.94	739	$C_{19}H_{30}O_2$	290
11	1H-Cycloprop[e]azulen-7-ol, decahydro-1,1,7-trimethyl-4-methylene-, [1ar-(1aalpha,4aalpha,7beta,7abeta,7balpha)]-	18.93	3.61	899	$C_{15}H_{24}O$	220
12	3-Oxo-20-methyl-11-à-hydroxyconanine-1,4-diene	19.47	0.73	757	$C_{22}H_{31}NO_2$	341
13	2,5-Octadecadiynoic acid, methyl Ester	20.72	0.76	786	$C_{19}H_{30}O_2$	290
14	(5e,7e)-9,10-Seccocholesta-5,7,10-triene-3,25,26-triol #	22.23	0.64	753	$C_{27}H_{44}O_3$	416
15	9-Oximino-2,7-diethoxyfluorene	25.92, 27.92	1.43, 1.75	745, 747	$C_{17}H_{17}NO_3$	283
16	Methyl 14-methylpentadecanoate	26.39	4.38	742	$C_{17}H_{34}O_2$	270
17	1-Heptatriacotanol	29.52	0.85	780	$C_{37}H_{76}O$	536
18	9-Octadecenoic acid (z)-, 2-Hydroxy-1-(hydroxymethyl)ethyl Ester	29.67	3.40	792	$C_{21}H_{40}O_4$	356
19	Cyclopropanebutanoic acid, 2-[[2-[[2-[(2 pentylcyclopropyl)meth yl]cyclopropyl]methyl]cyclopropyl] methyl]-, methyl ester	30.18	1.23	813	$C_{25}H_{42}O_2$	374

Table 1. *Cont.*

No.	Biomolecule Name	Retention Time	Area %	Mentioned Factor	Molecular Formula	Molecular Weight
20	9-(2′,2′-Dimethylpropanoilhydrazono)-3,6-dichloro-2,7-bis-[2-(diethylamino)-ethoxy]fluorene	30.89	5.84	761	$C_{30}H_{42}Cl_2N_4O_3$	576
21	1,2-Benzenedicarboxylic acid	32.70, 34.07, 35.15, 36.85, 36.07	7.78, 3.83, 4.94, 6.02, 12.89	789, 783, 776, 795, 779	$C_{24}H_{38}O_4$	390
22	Widdrol hydroxyether	33.12, 33.19	3.17, 5.39	770, 752	$C_{15}H_{26}O_2$	238
23	9,12,15-octadecatrienoic acid, 2,3-bis [(trimethylsilyl)oxy]propyl ester, (z,z,z)	35.31	1.25	734	$C_{27}H_{52}O_4Si_2$	496
24	Tetraneurin-a-diol	35.73	0.71	805	$C_{15}H_{20}O_5$	280

2.2. Uv@Ag-NPs Synthesis

2.2.1. Optimisation Parameters of Uv@Ag-NPs Synthesis

To obtain smaller nanoparticles with high stability, various parameters were studied, including precursor concentrations, the ratio between algal extract and precursor, temperature, pH, illumination, and time of incubation (Figure 5A–G).

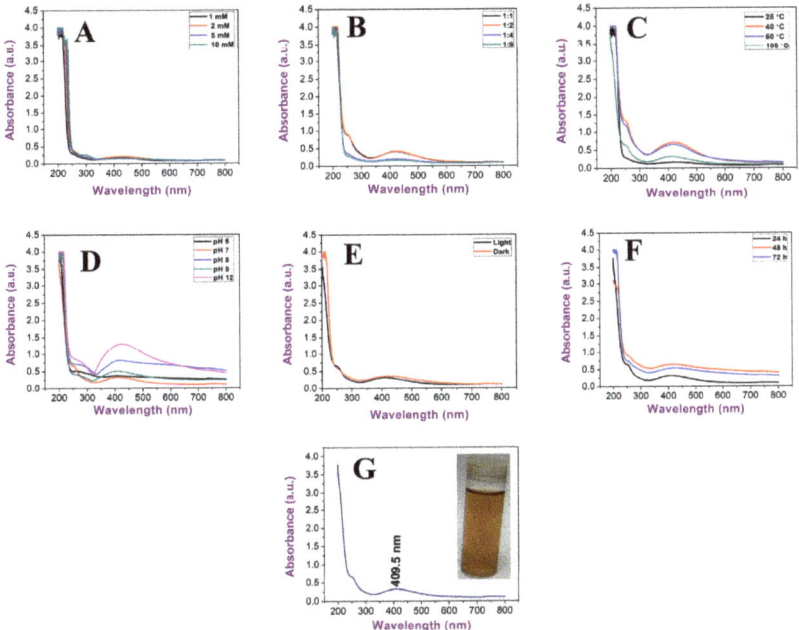

Figure 5. UV–Vis spectroscopy graphs illustrating the influence of (**A**) AgNO$_3$ concentration, (**B**) ratio between algal extract and AgNO$_3$, (**C**) temperature, (**D**) pH, (**E**) illumination conditions, and (**F**) incubation duration and (**G**) Uv@Ag-NPs under optimum conditions.

The data revealed an increment in wavelengths of Uv@Ag-NPs from 1 mM (425 nm) and 2 mM (425.5 nm) to 5 mM (428 nm) at a constant ratio of 1:9 of algal extract to AgNO$_3$, temperature of 25 °C, pH of 7, and light illumination for 24 h. On the other hand, with 10 mM of AgNO$_3$, no NPs were produced, suggesting that the higher concentrations above 5 mM significantly slowed the generation of nuclei and growth down. Therefore, it took a longer time to complete the reduction in precursors. The concentration of the NP in their suspension at 1, 2, and 5 mM was low, and the suspension had a faint golden-yellow colour. The data revealed that 1 mM of AgNO$_3$ was the optimum for Uv@Ag-NP synthesis. Khan et al. showed that the intensity of Ag-NPs synthesised from the *Piper betle* leaf extract increased at higher concentrations of 3 and 4 mM with high wavelength values relative to the other lower concentrations of 1 and 2 mM of AgNO$_3$; this suggests polydispersity and agglomeration of Ag-NPs at higher concentrations [21]. Changing the ratio of the algal extract to AgNO$_3$ from 1:9 to 1:1, 1:2, and 1:4 at a constant 1 mM AgNO$_3$ caused a reduction in wavelength from 425 nm at a 1:9 ratio to 422, 422, and 422.5 nm, respectively, suggesting that a higher volume of the precursor may result in an increase in the NP size or promote the agglomeration of NPs [34]. An increase in the temperature during the biofabrication of Uv@Ag-NPs resulted in reductions in the wavelength from 422 nm at 25 °C to 420 nm at 40 °C 418.5 nm at 80 °C and 409.5 nm at 100 °C. The NP intensity was higher at both 40 and 80 °C; however, their peaks were broader, which indicated the synthesis of larger or agglomerated NPs. These data suggested that the higher temperature was an important parameter for the biogenesis of Uv@Ag-NPs. This could be attributed to the existence of algal biomolecules that become activated at higher temperatures during Uv@Ag-NPs synthesis or kinetic influence. Liu et al. reported that the reduction in NP sizes at higher temperatures could be attributed to an increase in the nucleation kinetics constant instead of the decreased growth kinetics constant, considering the concentrations of the precursors [22]. UV–Vis spectroscopy showed that acidic pH (5) resulted in a broader SPR peak with a wavelength of 408.5 nm. However, the colour of the NP suspension was transparent, suggesting a slower synthesis reaction with a low yield of UV@Ag-NPs. Moreover, the wavelength of the Uv@Ag-NPs at pH of 7 (the same as the original pH of the reaction without any adjustment) and 9 was 409.5 nm; at higher pH values (8 and 12), the wavelengths shifted from 409.5 nm to 418 and 428 nm, respectively. These data explained that the pH values of 7 and 9 were suitable for producing smaller Uv@Ag-NPs, while increasing the pH to 8 and 12 caused an increase in NP intensity with wide shifting in wavelength indicating the synthesis of larger NPs. These data could be explained by the influence of pH on the dissociation, isolation, interfacial free energy, and the net charge of NPs. For instance, in an acidic medium, the driving force of NP dissolution may be balanced by the repulsive force keeping the dispersion of NPs resulting in smaller NPs. On the contrary, the negative charge hydroxyl ions (OH^-) facilitated the reduction of silver ions to NPs by increasing the ion levels in the medium silver atoms; these tend to diffuse between adjacent adsorption sites on a surface and form bonds with nearest neighbour atoms via Brownian diffusion, resulting in the formation of larger NPs [35]. Traiwatcharanon et al. synthesised Ag-NPs using a *Pistia stratiotes* extract and studied the influence of pH on NP size [35]. They reported that acidic conditions at pH values of 4, 5, and 6 caused blue shifting in the SPR of the Ag-NPs with smaller wavelengths of 330 nm while resulting in red shifting of SPR peaks with a wavelength of 414 nm. They reported that the red shift in the basic medium suggests larger Ag-NPs with higher intensities than those generated under acidic and neutral conditions.

The data showed that the optimum illumination condition for Uv@Ag-NP synthesis was under light (409.5 nm); under dark conditions, their wavelength was 420 nm. Increasing the duration of incubation from 24 h to 72 h under illumination increased the wavelength values from 409.5 to 421.5 nm, respectively, suggesting that the duration of exposure to light influences NPs stability. This could be attributed to the photocatalytic reaction where photons produce energetic electrons that excite SPR and, as a result, reduce Ag^+ to Ag-NPs [35–37]. However, high exposure to light irritation may accelerate the agglomeration rate of NPs. Husain et al. synthesised silver nanoparticles using 30 cyanobacteria species under dark and light conditions and found that almost all species were able to generate Ag-NPs only under light conditions [38].

Based on the previous data, the optimum conditions for synthesising Uv@Ag-NPs were 1 mM $AgNO_3$, 1:1 ratio of $AgNO_3$ and algal extract, temperature of 100 °C for 1h, pH of 7, light conditions, and incubation duration of 24 h. These conditions resulted in golden brown NP suspension at a wavelength of 409.5 nm. Kusumaningruma et al. reported that the maximum SPR peak of biosynthesised Ag-NPs using *Chlorella pyrenoidosa* was at 410 nm, which confirms the nanostructure of Ag-NPs [39].

2.2.2. Uv@Ag-NPs Characterisations
TEM, SEM, EDx, and Mapping Analysis

The TEM, HR-TEM, and SEM micrographs (Figure 6) of the Uv@Ag-NPs showed that the NPs had polyform shapes, including spherical and cubic. These NPs were trapped in an algal matrix that could contain polysaccharides. Smaller spherical Ag-NPs and cubic NPs may represent the seed for generating cubic NPs [40].

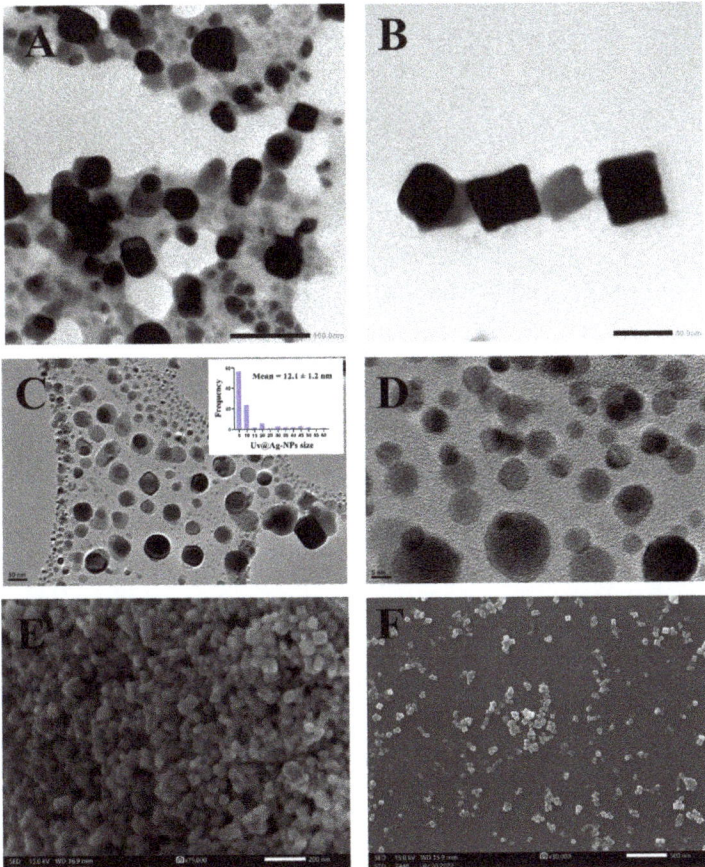

Figure 6. TEM (**A**,**B**), HR-TEM (**C**,**D**), and SEM (**E**,**F**) micrographs of Uv@Ag-NPs illustrate the uniform distribution of Uv@Ag-NPs and their spherical and cubic shapes. Scale bar = 100 nm (**A**), 50 nm (**B**,**C**), 5 nm (**D**), 200 nm (**E**), and 500 nm (**F**).

The micrographs also demonstrated that Uv@Ag-NPs were uniformly distributed without agglomeration, suggesting that Uv@Ag-NPs have good stability. The frequency distribution analysis of Uv@Ag-NPs using HR-TEM micrographs suggested that Uv@Ag-NPs are small, with a nanosize range of 5–60 nm and an average diameter of 12.1 ± 1.2 nm. Kannan et al. fabricated silver nitrate using the *Chlorophyceae Codium capitatum P.C. Silva* strain and showed that Ag-NPs have a cubic shape with a nanosize range of 3–44 nm and a mean diameter of 30 nm [41].

The elemental compositions of Uv@Ag-NPs and their distribution were determined using the EDx detector. The data showed that the main element distributed in the sample was Ag. A sharper peak was detected at 3 keV, which is a typical absorption signal of Uv@Ag-NPs with a mass percentage of 76.7%. Other elements, including carbon (6.93%), oxygen (1.81%), and chloride (12.18%), were detected while other trace elements emerged, including aluminium (0.3%), copper (1.13%), and zinc (0.91%); they may have emerged from the algal biocompounds surrounding the NPs or they existed in the polysaccharide matrix (Table 2, and Figure 7A,B) [39,42].

Table 2. EDx analysis of Uv@Ag-NPs synthesised from the unicellular *ulvophyte* sp. MBIC10591.

Element	Line	Mass%	Atom%
C	K	6.93 ± 0.02	32.29 ± 0.09
O	K	1.81 ± 0.02	6.32 ± 0.08
Al	K	0.3 ± 0.01	0.63 ± 0.02
Cl	K	12.18 ± 0.03	19.22 ± 0.05
Cu	K	1.13 ± 0.04	1.00 ± 0.04
Zn	K	0.91 ± 0.05	0.78 ± 0.04
Ag	L	76.74 ± 0.11	39.78 ± 0.06
Total		100	100

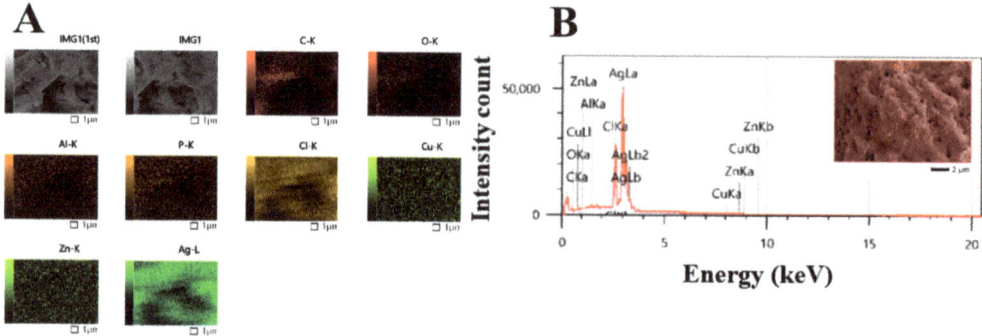

Figure 7. Map (**A**) and EDx (**B**) analysis of Uv@Ag-NPs synthesised from the unicellular *ulvophyte* sp. MBIC10591.

FTIR

The FTIR of the Uv@Ag-NPs contained 13 peaks at 3432.9 [41], 2928.7 [43], 2845.0 [44], 2130.1, 1636.5 [45], 1531.6, 1457.5 [46], 1384.4 [47], 1237.2 [48], 1085.2 [49], 889.8, 795.5 [50], and 554.0 [51] cm^{-1} (Figure 8). The IR peaks at 3432.9, 2928.7, and 2845.0 cm^{-1} corresponded to strong broad O-H stretching of alcohols or medium N-H stretching of primary amines and strong broad O-H stretching of carboxylics, broad N-H stretching of amine salts, or medium C-H stretching of alkane. However, the peaks at 2130.1, 1636.5, and 1531.6 cm^{-1} referred to the strong N=N=N stretching of azides, N=C=N stretching of carbodiimides, or N=C=S stretching of isothiocyanates or weak C≡C of alkynes; medium C=C stretching of alkenes or N-H stretching of amines; and strong N-O stretching of nitrocompounds. The peaks at 1457.5, 1384.4, 1237.2, and 1085.2 cm^{-1} were related to the medium C-H bending of alkanes; medium C-H bending of alkanes, O-H bending of alcohols, or strong S=O stretching of sulphates; strong C-O stretching of alkyls or medium C-N stretching of amines; and strong C-O stretching of primary alcohols or aliphatic ethers. The FTIR spectra at 889.8, 795.5, and 554.0 cm^{-1} were related to strong or medium C=C bending of alkenes and strong C-I stretching of halocompounds. These data may indicate that the main molecules for capping Uv@Ag-NPs were proteins and/or polysaccharides and/or alcohols, while the stabilising molecules were hydrocarbons and/or fatty acids. These data may be supported by the data of GC-MS analysis, which indicated that the main stabilising agents were fatty acids and hydrocarbons, and that phenol, alcohols, and/or amino-acid-like compounds were the reducing agents. Mahajan et al. extracellularly biofabricated Ag-NPs from silver nitrate using *Chlorella vulgaris* [52]. They analysed the functional group on the Ag-NPs using FTIR and found that the IR peaks of Ag-NPs were at 3435.88, 2092.30, 1637.82, 1559.61, 1414.42, 1037.17, and 618.16 cm^{-1}. This suggested that proteins, polysaccharides, and amides were significant passivating biomolecules for the bioreduction of AgNO$_3$ to Ag-NPs, while long-chain fatty acids were the stabilising agents.

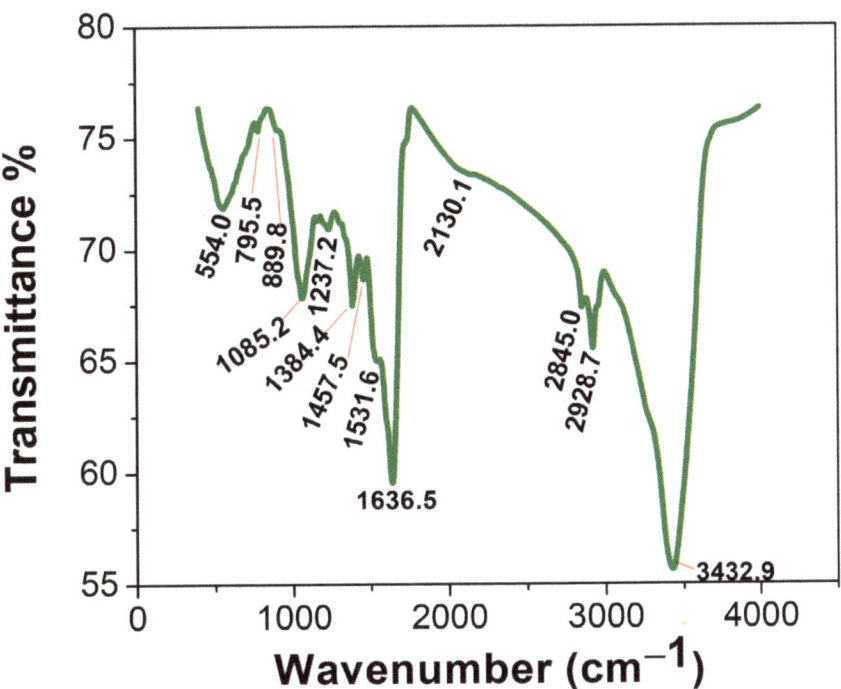

Figure 8. FTIR spectra of the Uv@Ag-NPs synthesised using the unicellular *ulvophyte* sp. MBIC10591.

DLS and Zeta Potential

The hydrodynamic diameter (HD) average of the Uv@Ag-NPs in an aqueous system was 178.1 nm with a polydispersity index of 0.38, suggesting that Uv@Ag-NPs had a polydisperse standard. The larger NP sizes than the nanosize range 5–60 nm calculated using the HR-TEM micrographs could be attributed to the algal biomaterials in the suspension and surrounding the surface of NPs; they tend to absorb water molecules on the NP surfaces, which increases the HD. The zeta potential (ZP) of the NPs is important for understanding their degree of stability in colloidal systems. NPs with higher negativity or positivity have strong repletion forces to repel each other, which prevents the agglomeration of NPs and stabilises them in a colloidal system [53]. Ardani et al. reported that the ZP value range of ±0–10 mV indicates a highly unstable colloid, while the ranges of ±10–20 mV, ±20–30 mV, and >±30 mV reveal relatively, moderately, and highly stable colloids, respectively [53]. The ZP of the Uv@Ag-NPs was −26.7 mV, indicating colloidal stability. This negative charge surrounding Uv@Ag-NPs could be normalised to those of the algal functional groups, such as hydroxyl and carboxylic groups, which surround the surfaces of NPs. Rathod et al. reported that the ZP of Ag-NPs synthesised from the *Nocardiopsis valliformis* strain OT1 was −17.1 mV, suggesting their colloidal stability (Figure 9A,B) [54].

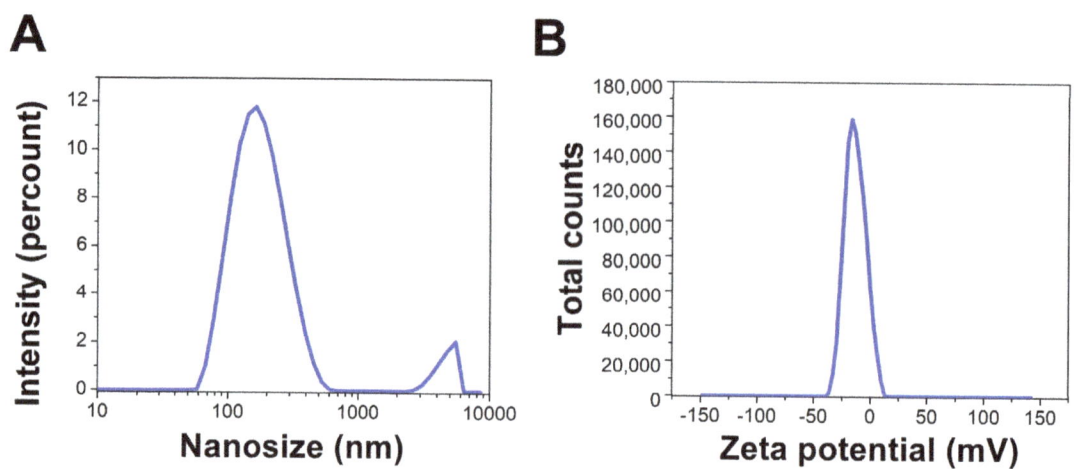

Figure 9. DLS (**A**) and zeta potential (**B**) of Uv@Ag-NPs synthesised from the unicellular *ulvophyte* sp. MBIC10591.

2.3. Antiproliferative Effect of Uv@Ag-NPs

Uv@Ag-NPs significantly reduced the proliferative activity of PC3, MDA-MB-231, T47D and HFs cell lines in a dose-dependent manner. However, MCF-7 cells responded differently to Uv@Ag-NPs. Uv@Ag-NPs drastically inhibited cellular proliferation in a dose-dependent manner from 200 to 50 µg/mL. The cell viability was non-significantly increased at 25 to 6.25 µg/mL of Uv@Ag-NPs. Interestingly, 3.13 µg/mL of Uv@Ag-NPs significantly reduced MCF-7 cell growth by 22%, whereas 1.5 µg/mL of Uv@Ag-NPs demonstrated a non-significant reduction in the malignant cell activity by 12%. This may be explained by the way that drug-responsive malignant cells behave or by the fact that smaller NPs can enter cells at lower concentrations since there are fewer aggregates present.

Similarly, the cell viability % of HFs was significantly decreased by increasing the Uv@Ag-NPs concentration from 12.5 to 200 µg/mL. Beyond 12.5 µg/mL, there was no significant activity of NPs against HFs cells. The moderate toxicity of Uv@Ag-NPs against HFs cell lines could be attributed to the algal functional groups surrounding the NPs, which have antioxidant activity, as reported in the GC-MS analysis section, increasing the NPs' biocompatibility against normal cells. These data suggested that Uv@Ag-NPs may act as potent alternative drugs for traditional therapeutic agents or pharmaceutical applications. The IC_{50} values of Uv@Ag-NPs against PC3, MDA-MB-231, T47D, MCF-7, and HFs were 27.4, 20.3, 23.8, 40.0, and 13.3 µg/mL (Figure 10). These data revealed that the most sensitive malignant cells to Uv@Ag-NPs were MDA-MB-231, followed by T47D, PC3, and MCF-7 cells. This suggested that Uv@Ag-NPs could be used as antiproliferative agents against prostate and multidrug-resistant breast cancer cell lines. The great antiproliferative activity of Uv@Ag-NPs against MDA-MB-231 cells compared to other cells may be attributed to the cellular metabolic state influencing cellular charge and their interaction with the charged NPs. On the other hand, the IC_{50} values of Ch@Ag-NPs against PC3, MDA-MB-231, T47D, MCF-7, and HFs were 111.8, 256.9, 657.0, 31.2, and 54.1 µg/mL, while the IC_{50} values of 5-FU against PC3, MDA-MB-231, T47D, MCF-7, and HFs were 10.6, 442.27, 12.75, 56.48, and 32.4 µg/mL (Figure 11). These data indicated that Uv@Ag-NPs demonstrated marked activity against tested cancer cells relative to other tested drugs, including Ch@Ag-NPs and 5-FU. The marked activity of Uv@Ag-NPs against cancer cells may be attributed to their smaller size, which facilitates penetration of cell boundaries and interactions with biomolecules, including proteins, enzymes, and antioxidants, causing cellular dysfunction and cell death [55]. Moreover, the bio-functional group derivatives from the algal components may play a significant role in enhancing the antiproliferative effects of Uv@Ag-NPs; they may facilitate the transport of NPs within cells via interactions with cellular receptors. Moreover, the negative charge on Uv@Ag-NPs may influence the therapeutic activity of NPs by increasing the attractive force between NPs and the cellular membrane and the resultant increase in the adsorption of NPs on the cellular surface. This surges the probability of these NPs moving inside cells and interacting with cell membranes [56]. Mohanta et al. synthesised Ag-NPs using *Gracilaria edulis* and found that Ag-NPs (with an average diameter of 62.7 ± 0.25 nm and a spherical shape) caused 50% death of MDA-MB-231 cells at concentrations of 344.27 ± 2.56 µg/mL, suggesting the potent antiproliferative activity of NPs [57]. Ag-NPs (with nanosize of 5–50 nm and spherical shape) synthesised from *Pleurotus djamor var. roseus* exhibited antiproliferative activity against PC3 cells with an IC$_{50}$ of 10 µg/mL [58], while Ag-NPs (with spherical shape and size range of 5–25 nm) synthesised from *Anabaena flos-aquae* reduced 50% of T47D cell growth with an IC$_{50}$ of 5 µg/mL [59]. Hexagonal Ag-NPs synthesised from the novel *Coelastrella aeroterrestrica* strain BA_Chlo4 with a diameter of 14.5 ± 0.5 nm showed marked inhibitory activity against MCF-7, MDA-MB-231, HCT-116, HepG2, HFS, and Vero with IC$_{50}$ values of 26.03, 15.92, 10.08, 5.29, 10.97, and 17.12 µg/mL, respectively [12].

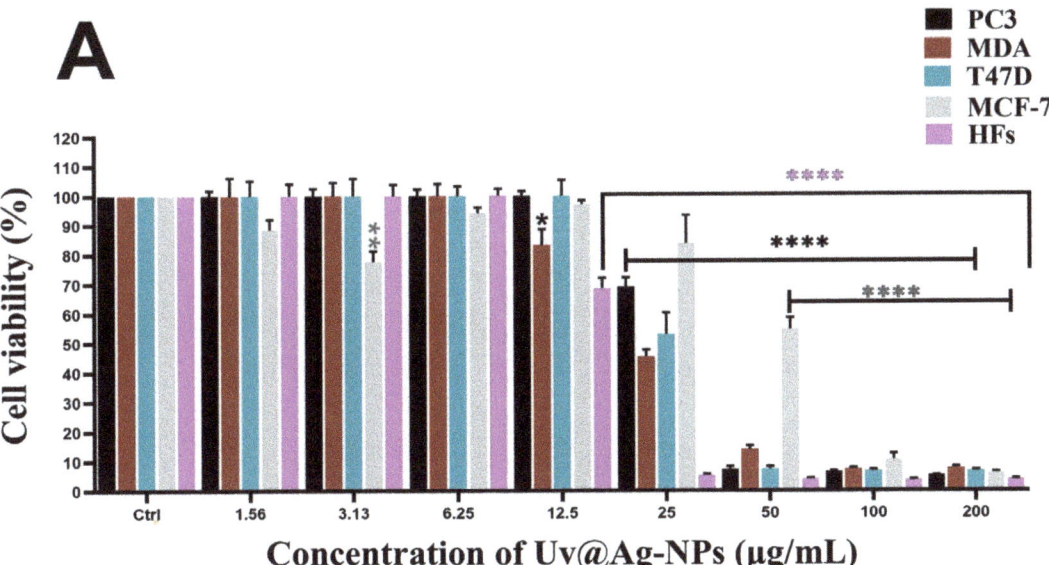

Figure 10. *Cont.*

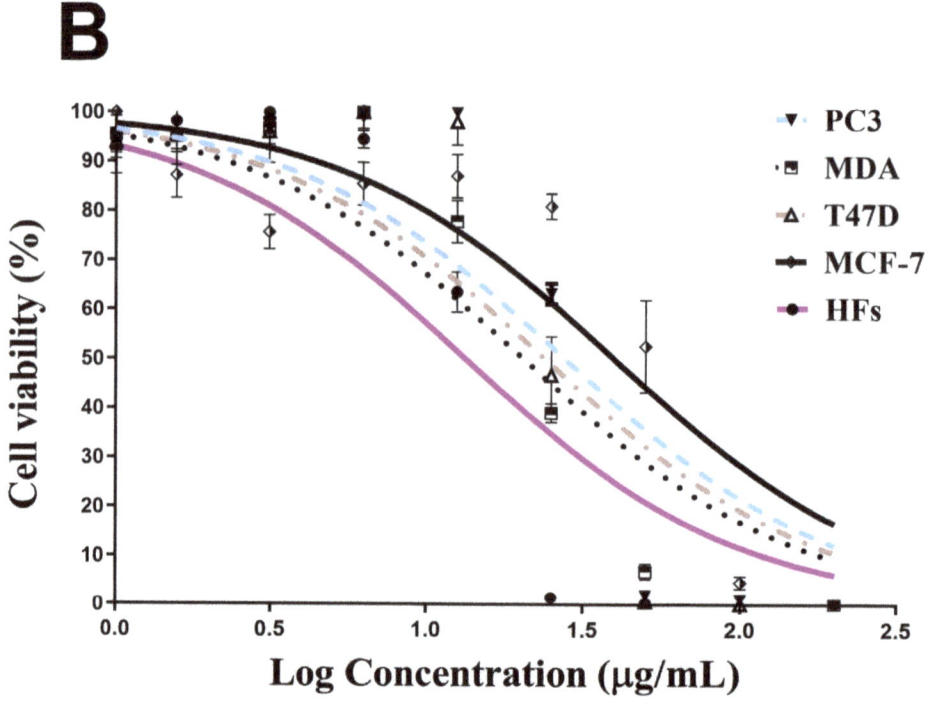

Figure 10. Antiproliferative activity (**A**,**B**) of a twofold serial dilution of 200 µg/mL of Uv@Ag-NPs synthesised from the unicellular *ulvophyte* sp. MBIC10591 against four malignant cells, PC3, MDA-MB-231, T47D, and MCF-7, and normal cells, HFs. Data are represented as mean ± SEM. *p*-values were calculated versus untreated cells: **** $p < 0.0001$, ** $p < 0.001$, and * $p < 0.01$.

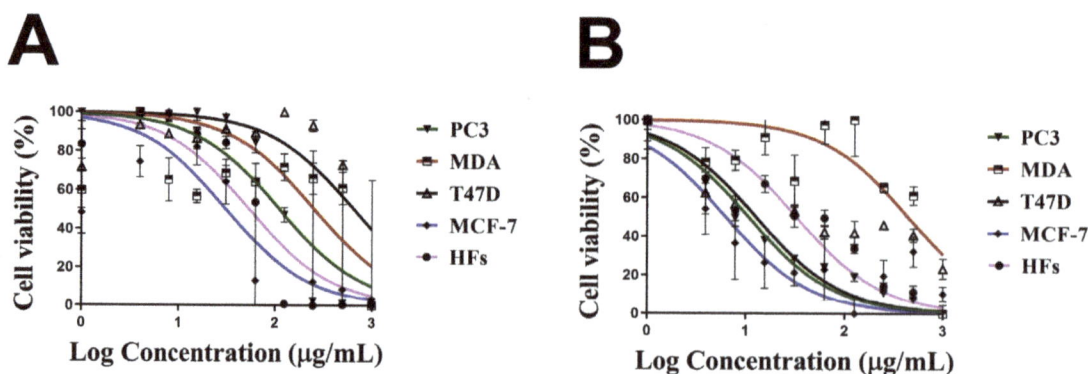

Figure 11. Cell viability of 1000 µg/mL of chemically synthesised Ag-NPs (Ch@Ag-NPs) (**A**) and 5-fluorouracil (5-FU) (**B**) against four malignant cells, PC3, MDA-MB-231, T47D, and MCF-7, and normal cells, HFs. Data are represented as mean ± SEM.

2.4. Biocidal Influence of Uv@Ag-NPs

The inhibitory effects of Uv@Ag-NPs, Ch@Ag-NPs, AgNO₃, the algal extract, and ciprofloxacin against *E. coli*, *K. pneumoniae*, *B. cereus*, and *B. subtilis* were examined using the agar well diffusion method. After excluding ciprofloxacin, the data revealed that Uv@Ag-NPs had the highest biocidal activity against the tested microbes (Figure 12 and

Table 3). The Uv@Ag-NPs showed greater activity against Gram-negative than Gram-positive bacteria. *E. coli* was the most sensitive microorganism to Uv@Ag-NPs with an IZ of 18.9 ± 0.03 mm, while *B. subtilis* showed the lowest response against Uv@Ag-NPs with an IZ of 15.1 ± 0.04 mm. The positive charge of Ag-NPs plays a significant role in enhancing their antibacterial activity via electro-attractive interactions between the negatively charged NPs and bacterial membranes [60]. Here, the Uv@Ag-NPs showed unexpected results; they had a negative charge on their surface but showed marked activity against Gram-negative bacteria, suggesting that the role of the charge in enhancing the biocidal efficiency of Uv@Ag-NPs can be overlooked. However, the biocidal activity of Uv@Ag-NPs against the tested bacteria can be attributed to their small size and large surface area and the surface chemistry of these NPs facilitating their interactions with cellular membranes and components inhibiting bacterial growth. The biocidal activity of Ag-NPs was highly dependent on the nanosize; the smaller sizes with larger surface areas allowed better contact with the cell membrane [61,62]. The Ch@Ag-NPs (negatively charged NPs), relative to Uv@Ag-NPs and AgNO$_3$, had the lowest inhibition zone; higher values (12.0 ± 0.01 mm) were recorded for *B. subtilis*, while the lower IZ was 10.1 ± 0.03 mm for *K. pneumoniae*. The lower biocidal activity of Ch@Ag-NPs compared to Uv@Ag-NPs could be attributed to the large nanosize of Ch@Ag-NPs trapping the NPs outside the bacterial wall, tackling their entrance into cells and reducing their activity against the bacterial cells. Moreover, the absence of functional groups around Ch@Ag-NPs' surface may also substantially impact the therapeutic action of Ch@Ag-NPs. It was found that the functional groups on the NPs surface mitigate their biological activity and toxicity via their interaction with cellular structures and biomolecular corona [28].

Table 3. Inhibition zones of 1 mg/mL of Uv@Ag-NPs, Ch@Ag-NPs, AgNO$_3$, algal extract, and ciprofloxacin for *E. coli, K. pneumoniae, B. cereus*, and *B. subtilis*.

Microorganisms	Drugs (µg/mL)				
	Uv@Ag-NPs	Ch@Ag-NPs	AgNO$_3$	Algal Extract	Ciprofloxacin
	IZD (mm)				
E. coli	18.9 ± 0.03	11.1 ± 0.14	15.0 ± 0.13	0.0 ± 0.0	32.5 ± 0.03
K. pneumoniae	16.0 ± 0.01	10.1 ± 0.03	14.2 ± 0.45	0.0 ± 0.0	32.0 ± 0.03
B. cereus	16.2 ± 0.18	11.4 ± 0.05	14.2 ± 0.03	0.0 ± 0.0	32.0 ± 0.06
B. subtilis	15.1 ± 0.04	12.0 ± 0.01	15.0 ± 0.19	0.0 ± 0.0	34.0 ± 0.03

The highest IZ values of AgNO$_3$ were for both *E. coli* and *B. subtilis* at 15.0 ± 0.13 and 15.0 ± 0.19 mm, respectively, while the lower values were for both *K. pneumoniae* and *B. cereus* with values of 14.2 ± 0.45 and 14.2 ± 0.03 mm, respectively. Intriguingly, 1 mg/mL of Uv@Ag-NPs and AgNO$_3$ resulted in a similar IZD value (about 15.0 mm) against *B. subtilis*. These data suggested that the biocidal activity of Uv@Ag-NPs against *B. subtilis* might be due to the nature of Ag ions rather than the algal functional groups, which might have other roles such as stabilising and charging NPs or directing the NPs to bacterial cells. Moreover, 1 mg/mL of algal extract was not enough to inhibit the bacterial growth with zero IZ against all tested microbes. These data show that Uv@Ag-NPs exhibited marked biocidal activity against the tested microbes compared with silver nitrate and Ch@Ag-NPs, suggesting that the small size with high specific surface area and functional group coating of the Uv@Ag-NPs have a significant influence on their biological activities. Ag-NPs (with a particle size of 4.06 nm) synthesised using *pu-erh tea* leaf extract inhibited the growth of *E. coli, K. pneumoniae, Salmonella Typhimurium*, and *Salmonella Enteritidis* with IZ values of 15, 10, 20, and 20 mm, respectively [63], while the Ag-NPs (with spherical shape and diameter range of 4.5 to 26 nm) synthesised from *Desertifilum IPPAS B-1220* showed antibacterial activity against *B. cereus* and *B. subtilis* with IZs of 16.33 ± 0.33 and 17.33 ± 0.33 mm, respectively [64].

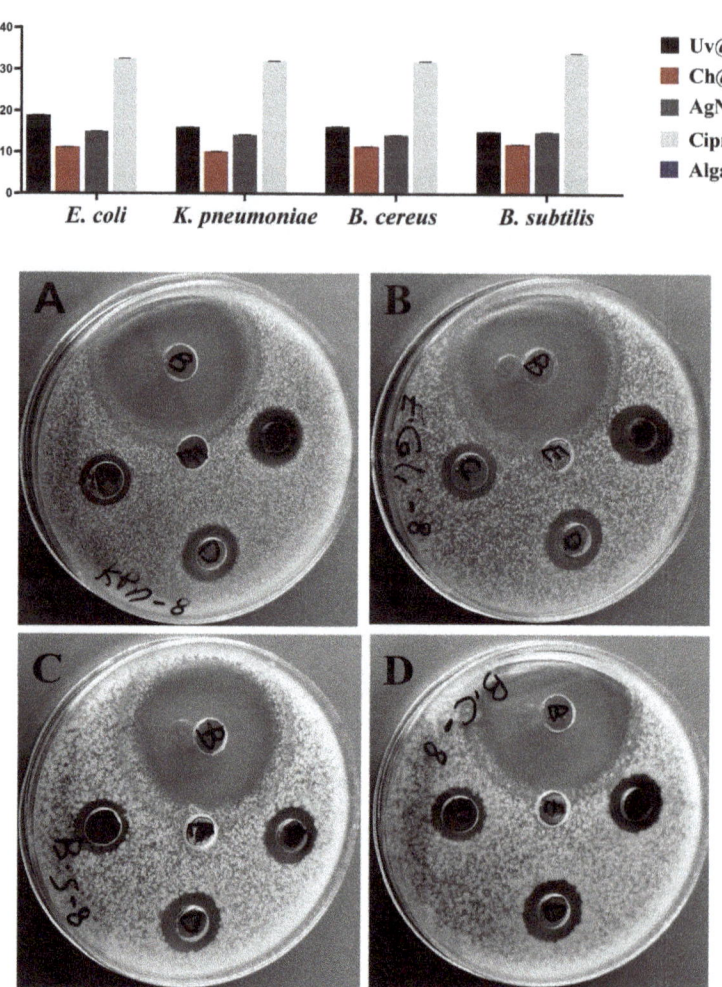

Figure 12. Inhibitory activities of 1 mg/mL of Uv@Ag-NPs, Ch@Ag-NPs, AgNO$_3$, algal extract, and ciprofloxacin against *E. coli*, *K. pneumoniae*, *B. cereus*, and *B. subtilis*. Letters written on well refer to (A) Uv@Ag-NPs, (B) ciprofloxacin, (C) Ch@Ag-NPs, (D) AgNO$_3$, and (E) algal extract.

3. Materials and Methods

3.1. Materials

Silver nitrate (AgNO$_3$); chemically synthesised NPs with nanosizes of <100 nm, spherical shapes, and 99.5% purity; and 3-(4,5-dimethylthiazol-2-yl)-2,5-diphenyltetrazolium bromide) tetrazolium reduction assay (MTT) assay and 5-fluorouracil (5-FU) were purchased from Sigma-Aldrich (St. Louis, MO, USA). The cell culture tools and media were purchased from Gibco (Thermo Fisher Scientific, Waltham, MA, USA). PC3, MDA-MB-231, T47D, MCF-7, and HFs cells were purchased from Nawah Scientific company, Egypt, who obtained the cells from the American Type Culture Collection (ATCC, Manassas, VA, USA), and microbial isolates were obtained from the Department of Microbiology, King Saud University, Riyadh, Saudi Arabia.

3.2. Unicellular ulvophyte sp. MBIC105

3.2.1. Isolation and Morphological Estimation

The microalgae were isolated from muddy soil in Riyadh, Saudi Arabia, using the serial dilution method reported by Hamida et al. [12]. The microalgae were kept in a sterile BG-11 media-containing flask in an incubator under a fluorescence lamp (2000 ± 200 Lux) with a 12:12 h dark/light cycle at room temperature for 15 days. Inverted, light, and scanning electron microscopes were used to determine the morphological appearance and purity of the microalgae. The sample was washed at least three times with water and ethanol and fixed in 70% ethanol. A small volume of the algal suspension was loaded onto a sterile glass piece fixed on a carbon paste attached to a copper stub. The sample was subsequently coated with a platinum coater (JEC-3000FC, Joel, Tokyo, Japan) for 80 s for scanning electron microscopy (SEM) (JSM-IT500HR, Joel, Japan) at 15 kV.

3.2.2. 18s rRNA Identification

The sample was identified using 18s rRNA identification. The DNA was extracted as described by Hamida et al. [12]. The PCR step with specific primers (forward primer: CCAGCAGCCGCGGTAATTCC; reverse primer: ACTTTCGTTCTTGATTAA) was performed to amplify the extracted DNA for sequencing using an ABI 3730 DNA sequencer (Thermo Fisher Scientific, USA).

3.2.3. Gas Chromatography–Mass Spectrometry (GC-MS) Analysis

The volatile components in the algal aqueous extract were screened using the Trace GC-TSQ mass spectrometer (Thermo Fisher Scientific, Austin, TX, USA) with the direct capillary column TG–5MS (30 m × 0.25 mm × 0.25 μm film thickness). Briefly, 50 mg of algal powder was soaked in 50 mL of boiled distilled water (dist. H_2O) and sonicated for 30 min. Subsequently, the sample was allowed to macerate for 24 h, followed by filtration with a syringe filter (0.22 μm). The filtrate was dried in a vacuum oven at 50 °C for 48 h. The temperature of the column oven was 50 °C initially before it was increased at a rate of 5 °C/min to 250 °C, maintained for 2 min, increased to 300 °C at a rate of 30 °C/min, and maintained for 2 more min. The injector and MS temperatures were maintained at 270 and 260 °C, respectively. Helium was utilised as a carrier gas at a constant flow rate of 1 mL/min. The solvent delay was 4 min, and diluted samples of 1 μL were injected automatically using an Autosampler AS1300 coupled with GC in the split mode. Electron ionisation mass spectra were collected at 70 eV ionisation voltages over the range of 50–650 m/z in full scan mode. The ion source temperature was set to 200 °C. Components of the algal extract were identified by comparing their mass spectra with those of the WILEY 09 and NIST 14 mass spectral databases [12].

3.2.4. Algal Aqueous Extract Preparation

The microalgae biomass was collected by centrifugation at 4700 rpm for 10 min, washed more than thrice with dist. H_2O, and lyophilised for 24 h using LYOTRAP (LTE Scientific, Greenfield, U.K.). The algal watery extract was prepared by dissolving an equal amount of algal powder with dist. H_2O and boiling at 80 °C for 30 min. Subsequently, the algal extract was spun at 4700 rpm for 10 min, and the supernatant was filtered using Whatman filter paper No. 1. The filtrate was used freshly to synthesise the Ag-NPs (Uv@Ag-NPs) [12].

3.3. Uv@Ag-NPs Synthesis

3.3.1. Optimisation Parameters for the Biofabrication of Uv@Ag-NPs

To determine the optimum conditions for Uv@Ag-NP biofabrication, various parameters were screened.

Precursor Concentrations and Ratios

Uv@Ag-NPs were produced with various concentrations (1, 2, 5, and 10 mM) of silver nitrate at a constant ratio of 1 to 9 (algal extract to silver nitrate) and a temperature of 25 °C under illumination for 24 h. Two millilitres of the synthesised Uv@Ag-NPs at each concentration was screened using UV spectroscopy (Shimadzu, Japan). After obtaining the optimum concentration, the effects of various ratios of precursor and algal extracts were determined. Four ratios were tested by mixing algal extract with 1 mM of $AgNO_3$ at ratios of 1:1, 1:2, 1:4, and 1:9, respectively, under the same constant conditions.

Temperature and pH

To estimate the influence of temperature and pH on NP biofabrication, 100 mL of 1 mM of $AgNO_3$ was mixed with 100 mL of algal extract and exposed to various temperatures of 25, 40, 60, and 100 °C for 1 h under other constant conditions. At the optimum temperature (100 °C), the pH values of $AgNO_3$ and the algal extract mixture were adjusted dropwise using 0.1 M hydrochloric acid or sodium hydroxide to 5, 7, 8, 9, and 12 under the same constant conditions for synthesis.

Illumination and Incubation Duration

Briefly, 100 mL of 1 mM of $AgNO_3$ was mixed with 100 mL of algal extract at 100 °C and a pH of 7. The mixture was incubated once in the dark and once in light (fluorescence lamp with 2000 ± 200 Lux) for 24 h. An aliquot was measured using UV spectroscopy to determine the optimum illumination conditions. The influence of the incubation duration was subsequently estimated by incubating the $AgNO_3$ and algal extract mixture under light conditions for 24, 48, and 72 h.

After obtaining the optimum conditions for biosynthesising the Uv@Ag-NPs, the NPs were synthesised on a large scale (5 L), centrifuged at 12,000 rpm for 15 min, washed at least thrice with dist. H_2O, and lyophilised for 8 h. The powder NPs were weighed and collected in sterile Eppendorf for further experiments.

3.4. Characterisation of Uv@Ag-NPs

3.4.1. UV Spectroscopy

For each optimum parameter, an aliquot (2 mL) of Uv@Ag-NPs was screened using UV spectroscopy for a wavelength range of 200–800 nm and a resolution of 1 nm.

3.4.2. Morphological and Elemental Composition Analysis of Uv@Ag-NPs

The shapes, sizes, elemental compositions, and distributions of the Uv@Ag-NPs were analysed using a high-resolution transmission electron microscope (HR-TEM), TEM, and SEM combined by an energy dispersive X-Ray analysis (EDx) detector. The Uv@Ag-NPs were collected by centrifugation at 12,000 rpm for 15 min, washed at least thrice with dist. H_2O and ethanol, and suspended in 1 mL ethanol and sonicated for 15 min. For imaging, 20 µL of the NP suspension was dropped onto the carbon-coated copper grid and air-dried to be examined by TEM (JEM-1400Flash, Joel, Tokyo, Japan) at 120 kV. Similarly, for SEM, 20 µL of the NP suspension was loaded on a sterile glass attached to a copper stub and air-dried. The sample was coated with platinum and examined at 15 kV using SEM. On the other hand, a small amount of powdered Uv@Ag-NPs was loaded onto carbon paste attached to a copper stub and coated with platinum for 80 sec to be analysed with an EDx detector (JSMIT500HR, STD-PC80, Joel, Tokyo, Japan) [12].

3.4.3. Fourier Transform Infrared Spectroscopy (FTIR) and Zeta Sizer

The surface chemistry of the Uv@Ag-NPs powders was detected in a range of 400–4000 cm^{-1} using FTIR spectroscopy (Shimadzu, Kyoto, Japan). The potential charges and hydrodynamic diameter of the Uv@Ag-NPs were determined by sonicating the NP suspension (500 µg/mL) for 15 min, diluting it 10-fold, sonicating for 1 to 2 min, and transferring it to Utype tubes at 25 °C for measurement using the zeta sizer (Malvern, U.K.).

3.5. Anticancer Activity

The antiproliferative activities of Uv@Ag-NPs, Ch@Ag-NPs, and 5-FU (as positive controls) were screened against four malignant cell lines, namely PC3, MCF-7, MDA-MB-231, and T47D, and one normal cell line, HFs, using the MTT kit. In brief, a cell density of 5×10^4 cells/mL was seeded onto a 96-well plate and incubated in a 5% CO_2 incubator for 24 h at 37 °C. At 75% confluency, the cells were subjected to serial dilution of Uv@Ag-NPs (200, 100, 50, 25, 12.5, 3.1, and 1.6 µg/mL), while the concentrations of both Ch@Ag-NPs and 5-FU were 1000, 500, 250, 125, 62.5, 31.25, 15.62, 7.81, and 3.90 µg/mL. Uv@Ag-NPs, Ch@Ag-NPs, and 5-FU were suspended in DMEM media, and the NP suspension was sonicated for 15 min. The 5-FU mixture was vortexed for 1 min. All mixtures were filtered using a 0.45 µm syringe filter for direct application to cells. The treated plates were incubated for 24 h in a 5% CO_2 incubator at 37 °C. After incubation, the media were discarded and replaced with 100 µL/well fresh media, and 10 µL/well of MTT solution (5 mg of MTT powder dissolved in 1 mL of sterile PBS, vortexed until dissolution, and filtered using a syringe filter) was added. The plates were incubated for 4 h, and the media was removed. Subsequently, 100 µL/well of DMSO was applied, and the plates were shacked at 400 rpm for 15 min to dissolve the formazan dye crystal. The plates were read on a Hercules, CA, USA) at 570 nm [3]. Cell viability (%) was estimated according to the following equation:

$$(Abs(treated)/(Abs(control)) \times 100$$

The IC_{50} (half-maximal growth inhibitory concentration) was calculated using a sigmoidal curve.

3.6. Antibacterial Activity

Escherichia coli ATCC8739, *Klebsiella pneumoniae* ATCC13883, *Bacillus cereus* ATCC9634, and *Bacillus subtilis* ATCC6633 were cultured in nutrient broth for up to 18 h at 37 °C and maintained through continuous subculturing in broth and on solid media. The inhibitory activities of 1 mg/mL of Uv@Ag-NPs, Ch@Ag-NPs, $AgNO_3$, algal extract, and 5 µg/mL ciprofloxacin were assessed against the tested bacteria using the agar well diffusion method. In brief, 4 mL of the bacterial strain was suspended in 50 mL of nutrient agar media. The mixture was poured into sterilised Petri dishes and dried at 37 °C. Four 8 mm wells were created in the agar plates using a cork borer. Subsequently, 100 µL of Uv@Ag-NPs, Ch@Ag-NPs, $AgNO_3$, algal extract, and ciprofloxacin suspensions were poured into the 8 mm wells. The plates were kept in a bacterial incubator at 37 °C for 24 h. Ch@Ag-NPs and ciprofloxacin were used as positive controls, while dist. H_2O was used as a negative control. The inhibition zone (IZ) was estimated after 24 h using a transparent ruler [65].

3.7. Statistical Analysis

All experiments were performed in triplicate independently, and the data are presented as mean ± SEM. One-way analysis of variance (ANOVA) was performed to compare differences between untreated and treated groups using graphPrism version 9.3.1 (GraphPad Software Inc., San Diego, CA, USA); $p < 0.05$ was considered statistically significant. For characterisation analysis of Uv@Ag-NPs, origin 8 (OriginLab Corporation, Northampton, MA, USA) and ImageJ (National Institutes of Health, Bethesda, MD, USA) were utilised.

4. Conclusions

These findings provide, for the first time, information about the novel microalgae unicellular *ulvophyte* sp. MBIC10591 and their potential for Ag-NP biogenesis. Herein, we report the morphological appearance of the unicellular *ulvophyte* sp. MBIC10591; the cells appeared spherical with cup-shaped spongiomorph chloroplasts with pyrenoids surrounded by several starch grains. Single cells were dominantly distributed; however, package cells with parenchyma-like structures containing more daughter cells were also found. The unicellular *ulvophyte* sp. MBIC10591 is enriched with various biomolecules, including vitamins, antioxidants, amino-acid-like compounds, organosulphur compounds, fatty acids, hydrocarbons, polysaccharides, phenol, and alcohols and may be a source of several therapeutic compounds. More investigations are needed to identify several molecules in different organic extracts of the unicellular *ulvophyte* sp. MBIC10591. These biomolecules enable the unicellular *ulvophyte* sp. MBIC10591 to biosynthesise small Ag-NPs. The Uv@Ag-NPs have UV–Vis spectra at 409.5 nm with spherical and cubic shapes. It was found that the optimum conditions for synthesising Uv@Ag-NPs include 1 mM $AgNO_3$, a ratio of 1:1 for $AgNO_3$ and algal extract, temperature of 100 °C for 1 h, and pH of 7 under light conditions for 24 h. The nanosize of these NPs was 5–60 nm with an average diameter of 12.1 ± 1.2 nm, while their HD and ZP were 178.1 nm with polydispersity index of 0.38 and −26.7 mV, respectively; these suggest their polydispersity and colloidal stability. Several functional groups were detected on Ag-NP surfaces. Proteins or/and polysaccharides or/and alcohols are responsible for reducing Ag-NPs, while fatty acids or/and hydrocarbons are the stabilising agents responsible for preventing the agglomeration of Ag-NPs. Uv@Ag-NPs exhibited marked anticancer activity against prostate cancer and multidrug resistance breast cancers with low toxicity against HFs. They also demonstrated marked inhibitory activity against Gram-negative bacteria; *E. coli* was the most susceptible to NPs, while *B. subtilis* was the most resistant. These antiproliferative activities and biocidal effects of Uv@Ag-NPs may be attributed to their unique physicochemical characteristics including their small sizes, large areas, shapes, and surface chemistry, which allow them to adsorb on cell surfaces, penetrate membranes and increase the permeability of outside walls or biomolecules such as proteins and enzymes, and interact with cellular organelles and biomolecules causing cellular dysfunction and cell death. Further study of the chemistry of the unicellular *ulvophyte* sp. MBIC10591 is recommended to discover more metabolites that can serve as drugs. Moreover, more optimisation parameters are needed to obtain more uniform shapes of Uv@Ag-NPs and assays to explore their biological activities and mechanisms inside malignant and microbial cells.

Author Contributions: Conceptualisation, R.S.H., M.A.A. (Mohamed Abdelaal Ali), and M.M.B.-M.; methodology, R.S.H., M.A.A. (Mohamed Abdelaal Ali), and M.M.B.-M.; software, R.S.H.; validation R.S.H., M.A.A. (Mohamed Abdelaal Ali), and M.M.B.-M.; formal analysis, R.S.H.; investigation, R.S.H., M.A.A. (Mohamed Abdelaal Ali), and M.M.B.-M.; resources, M.A.A. (Mohamed Abdelaal Ali), M.A.M. and M.M.B.-M.; data curation, R.S.H. and M.A.A. (Mohamed Abdelaal Ali); writing—original draft preparation, R.S.H.; writing—review and editing, R.S.H.; visualisation, R.S.H. and M.A.A. (Mohamed Abdelaal Ali); supervision, M.M.B.-M.; project administration, M.M.B.-M., H.E.A., M.A.M. and M.A.A. (Mariam Abdulaziz Alkhateeb); funding acquisition, H.E.A., M.A.A. (Mariam Abdulaziz Alkhateeb), and M.A.M. All authors have read and agreed to the published version of the manuscript.

Funding: This work was funded by the Deanship of Scientific Research at Princess Nourah bint Abdulrahman University, through the Research Groups Program (grant no. RGP-1441-0030).

Institutional Review Board Statement: Not applicable.

Informed Consent Statement: Not applicable.

Data Availability Statement: Additional data to those presented here are available from the corresponding author upon reasonable request.

Acknowledgments: This work was funded by the Deanship of Scientific Research at Princess Nourah bint Abdulrahman University, through the Research Groups Program (grant no. RGP-1441-0030).

Conflicts of Interest: The authors declare no conflict of interest.

Sample Availability: Samples of the algae or nanoparticles are available from the authors.

References

1. Singh, J.; Dutta, T.; Kim, K.-H.; Rawat, M.; Samddar, P.; Kumar, P. 'Green'synthesis of metals and their oxide nanoparticles: Applications for environmental remediation. *J. Nanobiotechnol.* **2018**, *16*, 84. [CrossRef] [PubMed]
2. Hung, Y.-P.; Chen, Y.-F.; Tsai, P.-J.; Huang, I.-H.; Ko, W.-C.; Jan, J.-S. Advances in the Application of Nanomaterials as Treatments for Bacterial Infectious Diseases. *Pharmaceutics* **2021**, *13*, 1913. [CrossRef] [PubMed]

3. Hamida, R.S.; Albasher, G.; Bin-Meferij, M.M. Oxidative Stress and Apoptotic Responses Elicited by Nostoc-Synthesized Silver Nanoparticles against Different Cancer Cell Lines. *Cancers* **2020**, *12*, 2099. [CrossRef] [PubMed]
4. He, Y.; Al-Mureish, A.; Wu, N. Nanotechnology in the treatment of diabetic complications: A comprehensive narrative review. *J. Diabetes Res.* **2021**, *2021*, 1–11. [CrossRef] [PubMed]
5. Mihai, M.M.; Dima, M.B.; Dima, B.; Holban, A.M. Nanomaterials for wound healing and infection control. *Materials* **2019**, *12*, 2176. [CrossRef] [PubMed]
6. Nath, D.; Banerjee, P. Green nanotechnology–a new hope for medical biology. *Environ. Toxicol. Pharmacol.* **2013**, *36*, 997–1014. [CrossRef] [PubMed]
7. Steinhauer, S.; Lackner, E.; Sosada-Ludwikowska, F.; Singh, V.; Krainer, J.; Wimmer-Teubenbacher, R.; Grammatikopoulos, P.; Köck, A.; Sowwan, M. Atomic-scale structure and chemical sensing application of ultrasmall size-selected Pt nanoparticles supported on SnO 2. *Mater. Adv.* **2020**, *1*, 3200–3207. [CrossRef]
8. Tao, F.F.; Nguyen, L.; Zhang, S. Introduction: Synthesis and catalysis on metal nanoparticles. In *Metal Nanoparticles for Catalysis: Advances and Applications*; Royal Society of Chemistry: Philadelphia, PA, USA, 2014.
9. Ahmed, S.F.; Mofijur, M.; Rafa, N.; Chowdhury, A.T.; Chowdhury, S.; Nahrin, M.; Islam, A.S.; Ong, H.C. Green approaches in synthesising nanomaterials for environmental nanobioremediation: Technological advancements, applications, benefits and challenges. *Environ. Res.* **2022**, *204*, 111967. [CrossRef]
10. Hamida, R.S.; Ali, M.A.; Redhwan, A.; Bin-Meferij, M.M. Cyanobacteria—A Promising Platform in Green Nanotechnology: A Review on Nanoparticles Fabrication and Their Prospective Applications. *Int. J. Nanomed.* **2020**, *15*, 6033–6066. [CrossRef]
11. Fawcett, D.; Verduin, J.J.; Shah, M.; Sharma, S.B.; Poinern, G.E.J. A review of current research into the biogenic synthesis of metal and metal oxide nanoparticles via marine algae and seagrasses. *J. Nanosci.* **2017**, *2017*, 1–15. [CrossRef]
12. Hamida, R.S.; Ali, M.A.; Almohawes, Z.N.; Alahdal, H.; Momenah, M.A.; Bin-Meferij, M.M. Green Synthesis of Hexagonal Silver Nanoparticles Using a Novel Microalgae Coelastrella aeroterrestrica Strain BA_Chlo4 and Resulting Anticancer, Antibacterial, and Antioxidant Activities. *Pharmaceutics* **2022**, *14*, 2002. [CrossRef] [PubMed]
13. Shah, S.A.A.; Hassan, S.S.U.; Bungau, S.; Si, Y.; Xu, H.; Rahman, M.H.; Behl, T.; Gitea, D.; Pavel, F.-M.; Corb Aron, R.A. Chemically diverse and biologically active secondary metabolites from marine Phylum chlorophyta. *Mar. Drugs* **2020**, *18*, 493. [CrossRef] [PubMed]
14. Singh, A.K.; Tiwari, R.; Singh, V.K.; Singh, P.; Khadim, S.R.; Singh, U.; Srivastava, V.; Hasan, S.; Asthana, R. Green synthesis of gold nanoparticles from Dunaliella salina, its characterization and in vitro anticancer activity on breast cancer cell line. *J. Drug Deliv. Sci. Technol.* **2019**, *51*, 164–176. [CrossRef]
15. Chokshi, K.; Pancha, I.; Ghosh, T.; Paliwal, C.; Maurya, R.; Ghosh, A.; Mishra, S. Green synthesis, characterization and antioxidant potential of silver nanoparticles biosynthesized from de-oiled biomass of thermotolerant oleaginous microalgae Acutodesmus dimorphus. *RSC Adv.* **2016**, *6*, 72269–72274. [CrossRef]
16. Mawed, S.A.; Centoducati, G.; Farag, M.R.; Alagawany, M.; Abou-Zeid, S.M.; Elhady, W.M.; El-Saadony, M.T.; Di Cerbo, A.; Al-Zahaby, S.A. Dunaliella salina Microalga Restores the Metabolic Equilibrium and Ameliorates the Hepatic Inflammatory Response Induced by Zinc Oxide Nanoparticles (ZnO-NPs) in Male Zebrafish. *Biology* **2022**, *11*, 1447. [CrossRef]
17. Caliskan, G.; Mutaf, T.; Agba, H.C.; Elibol, M. Green Synthesis and Characterization of Titanium Nanoparticles Using Microalga, Phaeodactylum tricornutum. *Geomicrobiol. J.* **2022**, *39*, 83–96. [CrossRef]
18. Princy, K.; Gopinath, A. Optimization of physicochemical parameters in the biofabrication of gold nanoparticles using marine macroalgae Padina tetrastromatica and its catalytic efficacy in the degradation of organic dyes. *J. Nanostructure Chem.* **2018**, *8*, 333–342. [CrossRef]
19. Alves, M.F.; Murray, P.G. Biological Synthesis of Monodisperse Uniform-Size Silver Nanoparticles (AgNPs) by Fungal Cell-Free Extracts at Elevated Temperature and pH. *J. Fungi* **2022**, *8*, 439. [CrossRef]
20. Hamida, R.S.; Ali, M.A.; Abdelmeguid, N.E.; Al-Zaban, M.I.; Baz, L.; Bin-Meferij, M.M. Lichens—A Potential Source for Nanoparticles Fabrication: A Review on Nanoparticles Biosynthesis and Their Prospective Applications. *J. Fungi* **2021**, *7*, 291. [CrossRef]
21. Khan, S.; Singh, S.; Gaikwad, S.; Nawani, N.; Junnarkar, M.; Pawar, S.V. Optimization of process parameters for the synthesis of silver nanoparticles from Piper betle leaf aqueous extract, and evaluation of their antiphytofungal activity. *Environ. Sci. Pollut. Res.* **2020**, *27*, 27221–27233. [CrossRef]
22. Liu, H.; Zhang, H.; Wang, J.; Wei, J. Effect of temperature on the size of biosynthesized silver nanoparticle: Deep insight into microscopic kinetics analysis. *Arab. J. Chem.* **2020**, *13*, 1011–1019. [CrossRef]
23. Aziz, N.; Faraz, M.; Pandey, R.; Shakir, M.; Fatma, T.; Varma, A.; Barman, I.; Prasad, R. Facile algae-derived route to biogenic silver nanoparticles: Synthesis, antibacterial, and photocatalytic properties. *Langmuir* **2015**, *31*, 11605–11612. [CrossRef] [PubMed]
24. Rai, M.; Yadav, A.; Gade, A. Silver nanoparticles as a new generation of antimicrobials. *Biotechnol. Adv.* **2009**, *27*, 76–83. [CrossRef] [PubMed]
25. Dos Santos, C.A.; Seckler, M.M.; Ingle, A.P.; Gupta, I.; Galdiero, S.; Galdiero, M.; Gade, A.; Rai, M. Silver nanoparticles: Therapeutical uses, toxicity, and safety issues. *J. Pharm. Sci.* **2014**, *103*, 1931–1944. [CrossRef]
26. Quintero-Quiroz, C.; Acevedo, N.; Zapata-Giraldo, J.; Botero, L.E.; Quintero, J.; Zárate-Triviño, D.; Saldarriaga, J.; Pérez, V.Z. Optimization of silver nanoparticle synthesis by chemical reduction and evaluation of its antimicrobial and toxic activity. *Biomater. Res.* **2019**, *23*, 27. [CrossRef]

27. Yin, I.X.; Zhang, J.; Zhao, I.S.; Mei, M.L.; Li, Q.; Chu, C.H. The Antibacterial Mechanism of Silver Nanoparticles and Its Application in Dentistry. *Int. J. Nanomed.* **2020**, *15*, 2555. [CrossRef]
28. Hamida, R.S.; Ali, M.A.; Goda, D.A.; Khalil, M.I.; Al-Zaban, M.I. Novel Biogenic Silver Nanoparticle-Induced Reactive Oxygen Species Inhibit the Biofilm Formation and Virulence Activities of Methicillin-Resistant Staphylococcus aureus (MRSA) Strain. *Front. Bioeng. Biotechnol.* **2020**, *8*, 433. [CrossRef]
29. Sommer, V.; Mikhailyuk, T.; Glaser, K.; Karsten, U. Uncovering unique green algae and cyanobacteria isolated from biocrusts in highly saline potash tailing pile habitats, using an integrative approach. *Microorganisms* **2020**, *8*, 1667. [CrossRef]
30. Huda, J.A.-T.; Mohammed, Y.H.; Imad, H.H. Phytochemical analysis of Urtica dioica leaves by fourier-transform infrared spectroscopy and gas chromatography-mass spectrometry. *J. Pharmacogn. Phytother.* **2015**, *7*, 238–252. [CrossRef]
31. Wang, Y.-B.; Qin, J.; Zheng, X.-Y.; Bai, Y.; Yang, K.; Xie, L.-P. Diallyl trisulfide induces Bcl-2 and caspase-3-dependent apoptosis via downregulation of Akt phosphorylation in human T24 bladder cancer cells. *Phytomedicine* **2010**, *17*, 363–368. [CrossRef]
32. Junwei, L.; Juntao, C.; Changyu, N.; Peng, W. Molecules and functions of rosewood: Pterocarpus cambodianus. *Arab. J. Chem.* **2018**, *11*, 763–770. [CrossRef]
33. Olasehinde, T.A.; Odjadjare, E.C.; Mabinya, L.V.; Olaniran, A.O.; Okoh, A.I. Chlorella sorokiniana and Chlorella minutissima exhibit antioxidant potentials, inhibit cholinesterases and modulate disaggregation of β-amyloid fibrils. *Electron. J. Biotechnol.* **2019**, *40*, 1–9. [CrossRef]
34. Ahmad, N.; Ang, B.C.; Amalina, M.A.; Bong, C.W. Influence of precursor concentration and temperature on the formation of nanosilver in chemical reduction method. *Sains Malays.* **2018**, *47*, 157–168.
35. Traiwatcharanon, P.; Timsorn, K.; Wongchoosuk, C. Flexible room-temperature resistive humidity sensor based on silver nanoparticles. *Mater. Res. Express* **2017**, *4*, 085038. [CrossRef]
36. Hamal, D.B.; Klabunde, K.J. Synthesis, characterization, and visible light activity of new nanoparticle photocatalysts based on silver, carbon, and sulfur-doped TiO2. *J. Colloid Interface Sci.* **2007**, *311*, 514–522. [CrossRef]
37. Ball, R.; Weitz, D.; Witten, T.; Leyvraz, F. Universal kinetics in reaction-limited aggregation. *Phys. Rev. Lett.* **1987**, *58*, 274. [CrossRef]
38. Husain, S.; Sardar, M.; Fatma, T. Screening of cyanobacterial extracts for synthesis of silver nanoparticles. *World J. Microbiol. Biotechnol.* **2015**, *31*, 1279–1283. [CrossRef]
39. Kusumaningrum, H.P.; Zainuri, M.; Marhaendrajaya, I.; Subagio, A. Nanosilver microalgae biosynthesis: Cell appearance based on SEM and EDX methods. *J. Phys. Conf. Ser.* **2018**, *1025*, 012084. [CrossRef]
40. Kemper, F.; Beckert, E.; Eberhardt, R.; Tünnermann, A. Light filter tailoring–the impact of light emitting diode irradiation on the morphology and optical properties of silver nanoparticles within polyethylenimine thin films. *RSC Adv.* **2017**, *7*, 41603–41609. [CrossRef]
41. Kannan, R.; Stirk, W.; Van Staden, J. Synthesis of silver nanoparticles using the seaweed *Codium capitatum* P.C. Silva (Chlorophyceae). *South Afr. J. Bot.* **2013**, *86*, 1–4. [CrossRef]
42. Borah, D.; Das, N.; Das, N.; Bhattacharjee, A.; Sarmah, P.; Ghosh, K.; Chandel, M.; Rout, J.; Pandey, P.; Ghosh, N.N. Alga-mediated facile green synthesis of silver nanoparticles: Photophysical, catalytic and antibacterial activity. *Appl. Organomet. Chem.* **2020**, *34*, e5597. [CrossRef]
43. Al-Zahrani, S.; Astudillo-Calderón, S.; Pintos, B.; Pérez-Urria, E.; Manzanera, J.A.; Martín, L.; Gomez-Garay, A. Role of synthetic plant extracts on the production of silver-derived nanoparticles. *Plants* **2021**, *10*, 1671. [CrossRef] [PubMed]
44. Ajeboriogbon, A.; Adewuyi, B.; Talabi, H. Synthesis of silver nanoparticles from selected plants extract. *Acta Tech. Corviniensis-Bull. Eng.* **2019**, *12*, 55–58.
45. Rani, K.; Lekha, C. Green synthesis and characterization of silver nanoparticles using leaf and stem aqueous extract of Pauzolzia Bennettiana and their antioxidant activity. *J. Pharmacogn. Phytochem.* **2018**, *7*, 129–132.
46. Yadav, A.; Theivasanthi, T.; Paul, P.; Upadhyay, K. Extracellular biosynthesis of silver nanoparticles from plant growth promoting rhizobacteria *Pseudomonas* sp. *arXiv* **2015**, arXiv:1511.03130.
47. Oliveira, G.Z.S.; Lopes, C.A.P.; Sousa, M.H.; Silva, L.P. Synthesis of silver nanoparticles using aqueous extracts of Pterodon emarginatus leaves collected in the summer and winter seasons. *Int. Nano Lett.* **2019**, *9*, 109–117. [CrossRef]
48. Huq, M. Green Synthesis of Silver Nanoparticles Using Pseudoduganella eburnea MAHUQ-39 and Their Antimicrobial Mechanisms Investigation against Drug Resistant Human Pathogens. *Int. J. Mol. Sci.* **2020**, *21*, 1510. [CrossRef]
49. El-Naggar, N.E.-A.; Hussein, M.H.; El-Sawah, A.A. Phycobiliprotein-mediated synthesis of biogenic silver nanoparticles, characterization, in vitro and in vivo assessment of anticancer activities. *Sci. Rep.* **2018**, *8*, 8925. [CrossRef]
50. Mukaratirwa-Muchanyereyi, N.; Gusha, C.; Mujuru, M.; Guyo, U.; Nyoni, S. Synthesis of Silver nanoparticles using plant extracts from Erythrina abyssinica aerial parts and assessment of their anti-bacterial and anti-oxidant activities. *Results Chem.* **2022**, *4*, 100402. [CrossRef]
51. Shameli, K.; Ahmad, M.B.; Jazayeri, S.D.; Sedaghat, S.; Shabanzadeh, P.; Jahangirian, H.; Mahdavi, M.; Abdollahi, Y. Synthesis and characterization of polyethylene glycol mediated silver nanoparticles by the green method. *Int. J. Mol. Sci.* **2012**, *13*, 6639–6650. [CrossRef]
52. Mahajan, A.; Arya, A.; Chundawat, T.S. Green synthesis of silver nanoparticles using green alga (*Chlorella vulgaris*) and its application for synthesis of quinolines derivatives. *Synth. Commun.* **2019**, *49*, 1926–1937. [CrossRef]

53. Ardani, H.; Imawan, C.; Handayani, W.; Djuhana, D.; Harmoko, A.; Fauzia, V. Enhancement of the stability of silver nanoparticles synthesized using aqueous extract of Diospyros discolor Willd. leaves using polyvinyl alcohol. *IOP Conf. Ser. Mater. Sci. Eng.* **2017**, *188*, 012056. [CrossRef]
54. Rathod, D.; Golinska, P.; Wypij, M.; Dahm, H.; Rai, M. A new report of Nocardiopsis valliformis strain OT1 from alkaline Lonar crater of India and its use in synthesis of silver nanoparticles with special reference to evaluation of antibacterial activity and cytotoxicity. *Med. Microbiol. Immunol.* **2016**, *205*, 435–447. [CrossRef] [PubMed]
55. Shang, L.; Nienhaus, K.; Nienhaus, G.U. Engineered nanoparticles interacting with cells: Size matters. *J. Nanobiotechnol.* **2014**, *12*, 5. [CrossRef]
56. Fröhlich, E. The role of surface charge in cellular uptake and cytotoxicity of medical nanoparticles. *Int. J. Nanomed.* **2012**, *7*, 5577–5591. [CrossRef]
57. Mohanta, Y.K.; Mishra, A.K.; Nayak, D.; Patra, B.; Bratovcic, A.; Avula, S.K.; Mohanta, T.K.; Murugan, K.; Saravanan, M. Exploring Dose-Dependent Cytotoxicity Profile of Gracilaria edulis-Mediated Green Synthesized Silver Nanoparticles against MDA-MB-231 Breast Carcinoma. *Oxidative Med. Cell. Longev.* **2022**, *2022*, 1–15. [CrossRef]
58. Raman, J.; Reddy, G.R.; Lakshmanan, H.; Selvaraj, V.; Gajendran, B.; Nanjian, R.; Chinnasamy, A.; Sabaratnam, V. Mycosynthesis and characterization of silver nanoparticles from Pleurotus djamor var. roseus and their in vitro cytotoxicity effect on PC3 cells. *Process Biochem.* **2015**, *50*, 140–147. [CrossRef]
59. Ebrahimzadeh, Z.; Salehzadeh, A.; Naeemi, A.; Jalali, A. Silver nanoparticles biosynthesized by Anabaena flos-aquae enhance the apoptosis in breast cancer cell line. *Bull. Mater. Sci.* **2020**, *43*, 92. [CrossRef]
60. Abbaszadegan, A.; Ghahramani, Y.; Gholami, A.; Hemmateenejad, B.; Dorostkar, S.; Nabavizadeh, M.; Sharghi, H. The effect of charge at the surface of silver nanoparticles on antimicrobial activity against gram-positive and gram-negative bacteria: A preliminary study. *J. Nanomater.* **2015**, *2015*, 1–8. [CrossRef]
61. Kaur, A.; Preet, S.; Kumar, V.; Kumar, R.; Kumar, R. Synergetic effect of vancomycin loaded silver nanoparticles for enhanced antibacterial activity. *Colloids Surf. B Biointerfaces* **2019**, *176*, 62–69. [CrossRef] [PubMed]
62. Bruna, T.; Maldonado-Bravo, F.; Jara, P.; Caro, N. Silver nanoparticles and their antibacterial applications. *Int. J. Mol. Sci.* **2021**, *22*, 7202. [CrossRef]
63. Loo, Y.Y.; Rukayadi, Y.; Nor-Khaizura, M.-A.-R.; Kuan, C.H.; Chieng, B.W.; Nishibuchi, M.; Radu, S. In vitro antimicrobial activity of green synthesized silver nanoparticles against selected gram-negative foodborne pathogens. *Front. Microbiol.* **2018**, *9*, 1555. [CrossRef] [PubMed]
64. Hamida, R.S.; Abdelmeguid, N.E.; Ali, M.A.; Bin-Meferij, M.M.; Khalil, M.I. Synthesis of silver nanoparticles using a novel cyanobacteria Desertifilum sp. extract: Their antibacterial and cytotoxicity effects. *Int. J. Nanomed.* **2020**, *15*, 49. [CrossRef] [PubMed]
65. Hamida, R.S.; Ali, M.A.; Goda, D.A.; Al-Zaban, M.I. Lethal Mechanisms of Nostoc-Synthesized Silver Nanoparticles Against Different Pathogenic Bacteria. *Int. J. Nanomed.* **2020**, *15*, 10499. [CrossRef]

Disclaimer/Publisher's Note: The statements, opinions and data contained in all publications are solely those of the individual author(s) and contributor(s) and not of MDPI and/or the editor(s). MDPI and/or the editor(s) disclaim responsibility for any injury to people or property resulting from any ideas, methods, instructions or products referred to in the content.

Article

Synthesis, Characterization, and Antibacterial Activity of Mg-Doped CuO Nanoparticles

Russul M. Adnan [1], Malak Mezher [2], Alaa M. Abdallah [3], Ramadan Awad [3,4] and Mahmoud I. Khalil [2,5,*]

1. Department of Chemistry, Faculty of Science, Beirut Arab University, Beirut P.O.Box 11-5020, Lebanon
2. Department of Biological Sciences, Faculty of Science, Beirut Arab University, Beirut P.O.Box 11-5020, Lebanon
3. Department of Physics, Faculty of Science, Beirut Arab University, Beirut P.O.Box 11-5020, Lebanon
4. Department of Physics, Faculty of Science, Alexandria University, Alexandria 21568, Egypt
5. Molecular Biology Unit, Department of Zoology, Faculty of Science, Alexandria University, Alexandria 21568, Egypt
* Correspondence: mahmoud_ibrahim@alexu.edu.eg; Tel.: +20-1223256303

Abstract: This study aims to investigate the effect of magnesium (Mg) doping on the characteristics and antibacterial properties of copper oxide (CuO) nanoparticles (NPs). The Mg-doped CuO NPs were fabricated by the co-precipitation method. NPs were characterized by X-ray Powder Diffraction (XRD), Transmission Electron Microscope (TEM), Energy Dispersive X-ray (EDX) analysis, Fourier Transform Infrared Spectroscopy (FTIR), and Photoluminescence (PL). Broth microdilution, agar-well diffusion, and time-kill assays were employed to assess the antibacterial activity of the NPs. XRD revealed the monoclinic structure of CuO NPs and the successful incorporation of Mg dopant to the $Cu_{1-x}Mg_xO$ NPs. TEM revealed the spherical shape of the CuO NPs. Mg doping affected the morphology of NPs and decreased their agglomeration. EDX patterns confirmed the high purity of the undoped and Mg-doped CuO NPs. FTIR analysis revealed the shifts in the Cu–O bond induced by the Mg dopant. The position, width, and intensity of the PL bands were affected as a result of Mg doping, which is an indication of vacancies. Both undoped and doped CuO NPs exhibited significant antibacterial capacities. NPs inhibited the growth of Gram-positive and Gram-negative bacteria. These results highlight the potential use of Mg-doped CuO NPs as an antibacterial agent.

Keywords: CuO nanoparticles; bi-metallic NPs; Mg-dopant; co-precipitation; antibacterial; EDX; photoluminescence

1. Introduction

Nanoparticles (NPs) revolutionized the industrial world. This revolution is due to their outstanding performance and remarkable optical, electrical, catalytic, and corrosion resistance, in addition to their antibacterial properties [1]. Copper oxide (CuO) has a vital role in multi-functional applications [2]. CuO is an important inorganic p-type semiconductor with a band gap of around 1.2–1.8 eV [3,4]. The most stable phases of copper oxide are cubic cuprous oxide (Cu_2O) and monoclinic cupric oxide (CuO) [5]. Its applicability ranges between catalysis, photovoltaics, an electrode for lithium-ion batteries, solar energy conversion, supercapacitors, corrosion inhibition, antimicrobial, and anticancer applications [2,5,6].

CuO NPs serve as a good template for multi-functional applications. Its performance can be enhanced by implementing dopants into the CuO lattice. Since Mg dopants have enhanced the structural and antibacterial properties of CuO NPs, and they have a comparable ionic radius (72 pm) to the ionic radius of Cu^{2+} ions (73 pm) [7], the Mg-doped CuO NPs may be promising candidates for numerous applications, especially antibacterial applications [8–10].

Previous studies showed the antibacterial activity of doped CuO NPs. Doped NPs synthesized by co-precipitation revealed that the doping elements promote the release of

Cu^{2+} from the doped CuO NPs [8–10]. Furthermore, the doped CuO NPs possess better antibacterial activity against Gram-positive bacteria than Gram-negative bacteria, especially against *S. aureus*. The 5% Mg-doped CuO NPs exhibited bactericidal activity at very low concentrations and their bacteriostatic rate reached 99.9% [7]. The high antibacterial activity of doped CuO NPs may be attributed to the inactivation of proteins in the cell wall of bacteria. This activity may be due to the binding of Cu^{2+} ions to the surface of the bacterial cell. Numerous previous studies showed the bactericidal action of doped CuO NPs, especially against *E. coli* and *S. aureus* [8–10]. This inhibitory action may be dependent on the structure of CuO NPs, as reported previously [8].

Doped NPs were shown to exhibit a better inhibitory effect on bacterial growth than that pure CuO NPs. However, other studies reported that pure NPs exhibited better inhibitory effects. The 5%, 7%, and 10% Mg-doped CuO NPs, at low concentrations, exhibited the same antibacterial activity as that of the pure CuO NPs at higher concentrations [10]. It was reported that the antibacterial activity of CuO NPs is enhanced by Mg^{2+} doping. The increased release of Cu^{2+} and Mg^{2+} ions from doped CuO with the increase of Mg^{2+} doping content may explain the inhibition of the growth of bacteria [7]. Besides, undoped CuO and Mg-doped CuO NPs showed considerable antimicrobial activity versus several bacterial pathogens, especially *S. aureus*, *P. aeruginosa*, and *E. coli* [10]. Furthermore, the doped NPs possess significant antibacterial activity against many bacterial isolates. Thus, such fabrications may provide a potential alternative to the standard methods of bacterial inhibition.

In addition, the capping of CuO NPs with Ethylenediamine tetra-acetic acid (EDTA) was shown to enhance the antibacterial activity. EDTA, which is a water-soluble polymer, acts by stabilizing the surfaces and modifying the growth (size) during NP synthesis. This enhances the antibacterial action of NPs, due to the ability of EDTA to reduce the size of NPs, which in turn increases the action of NPs against bacteria [11].

Here the impact of undoped and Mg-doped CuO NPs was explored on the inhibition of various bacterial isolates. Previous studies have shown that undoped CuO and Mg-doped CuO NPs showed considerable antimicrobial activity against several bacterial pathogens [10]. In this regard, the objective of this study is to investigate the impact of undoped and Mg-doped CuO NPs capped with EDTA on their structural, morphological, and inhibition capacities against various bacteria isolated from the Lebanese sewage sludge, including Gram-positive bacteria (*S. aureus* and *E. faecium*) and Gram-negative bacteria (*E. coli* and *S. maltophilia*).

2. Results

2.1. Characterization of NPs

2.1.1. X-ray Diffraction

The XRD patterns of undoped and Mg-doped CuO NPs ($Cu_{1-x}Mg_xO$ NPs) are shown in Figure 1. The patterns of the undoped CuO NPs show that all the peaks reflect the planes of monoclinic CuO, which are (110), (002), (111), ($\bar{1}$12), ($\bar{2}$02), (112), (020), (202), ($\bar{1}$13), (022), ($\bar{3}$11), (113), (311), and (004). Given that no other secondary phases, relating to impurities or the Cu_2O phase are found. Moreover, when CuO is doped with magnesium, the diffraction peaks of CuO NPs are not altered. To assure this observation, the XRD patterns of all the samples are refined via the MAUD program (Figure 1). The refinements checked the possible formation of MgO as a secondary phase. However, all the samples are well fitted to the pure CuO phase, without any presence of MgO. The obtained patterns for undoped and Mg-doped CuO NPs are similar to the previously reported literature [7,12,13]. Lv et al. [7] synthesized Mg^{2+}, Zn^{2+}, and Ce^{4+}-doped CuO nanoparticles by the hydrothermal method. They obtained a pure phase of CuO in the XRD patterns with a doping concentration of less than 7%, indicating the total incorporation of the dopants into the lattice, without the formation of secondary phases. However, beyond 7%, at 10% doping percentage, MgO, ZnO, and CeO_2 phases were formed in the CuO lattice. This further aligns with the present study, as all the prepared Mg-doped CuO NPs formed pure CuO phase, as the doping percentage ranged between 0.5% ($x = 0.005$) to 2% ($x = 0.020$).

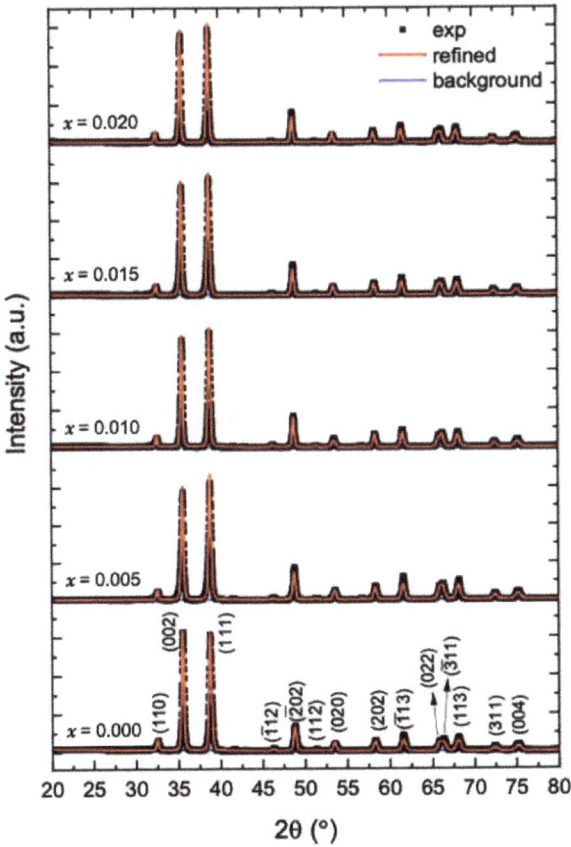

Figure 1. Refinements of the XRD patterns of undoped and Mg-doped CuO NPs ($Cu_{1-x}Mg_xO$ NPs).

2.1.2. Transmission Electron Microscope

The morphology and size of the undoped and Mg-doped CuO NPs ($Cu_{1-x}Mg_xO$ NPs) were determined using the Transmission Electron Microscope (TEM) technique. The TEM images demonstrate spherical NPs for the three selected samples, as shown in Figure 2. The Mg doping led to some noticeable changes in the NP morphology. The TEM images showed agglomeration for the undoped CuO NPs ($x = 0.000$), whereas, with the doping of Mg at concentrations of $x = 0.005$ and $x = 0.020$, the agglomeration of the doped CuO NPs is reduced, showing more uniform shapes, as depicted in Figure 2. The average particle sizes of the synthesized samples are determined from the particle size distribution which is extracted from the TEM images, using ImageJ software. This distribution is fitted by a Gaussian function, from which the average particle sizes are determined along with the standard deviation (SD), as shown in Figure 2. The obtained average particle size for undoped CuO NPs is 41.31 ± 1.76 nm. Upon Mg doping with $x = 0.005$, the particle size decreased to 27.14 ± 6.70 nm and increased slightly to 33.78 ± 8.54 nm with $x = 0.020$. These alterations in the average grain sizes with the increase in the concentration of Mg-doping may be attributed to the dissimilarity in Pauling electronegativity that affected the growth rate of Mg-doped CuO nanoparticles. The host Cu ions have a Pauling electronegativity of 1.9, which is higher than that of the doped Mg ions (1.31). This dissimilarity proves the decrease in the growth rate at low concentrations of Mg-doped CuO NPs [13]. However, at higher concentrations, the Mg-doped ions may incorporate into the lattice, not only filling substitutional sites but also occupying interstitial sites, that yield larger grains, as seen in the sample with $x = 0.020$.

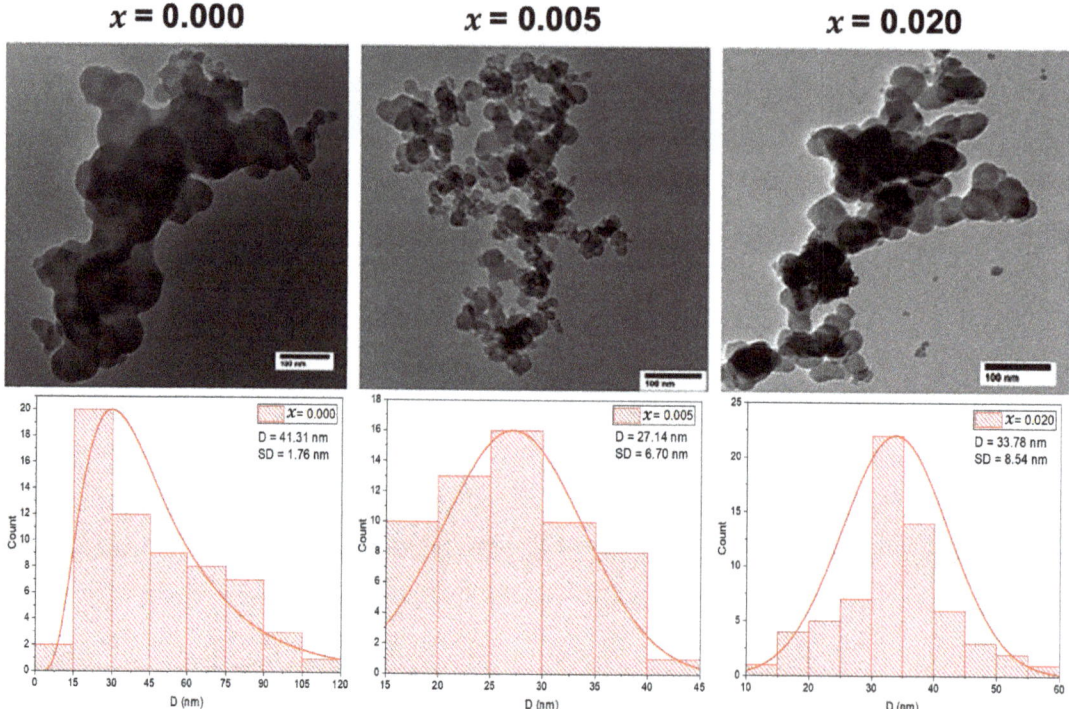

Figure 2. TEM images with the grain size distribution of the undoped and Mg-doped CuO NPs ($Cu_{1-x}Mg_xO$ NPs).

2.1.3. Scanning Electron Microscope and Energy Dispersive X-ray

The morphology of the nanoparticles is further studied by the scanning electron microscope, as shown in Figure 3. The SEM micrographs assure the nanocrystalline nature of the undoped and Mg-doped CuO NPs, without agglomeration. It is also noticed that the average grain size decreased with Mg-doping, re-assuring the TEM analysis. The average grain size, extracted from the SEM images, is found to be around 25.7 nm for undoped CuO NPs and 22.5 nm for Mg-doped CuO NPs. The chemical composition was studied with an energy-dispersive X-ray (EDX) technique. The presence of copper (Cu) and oxygen (O) elements in the undoped CuO NPs is confirmed by the EDX pattern, shown in Figure 3 ($x = 0.000$). No traces of precursors were detected. The EDX pattern of Mg-doped CuO NPs with $x = 0.020$ is exhibited in Figure 3. In addition to Cu and O, Mg is detected in the pattern, confirming the presence of the Mg-doped in the CuO NPs. The average atomic percent (at%) of copper and oxygen were microstructures of three different regions of the samples and demonstrated in the insets of (Figure 3) as pie charts. In undoped CuO NPs, the ratio between Cu and O is 0.82 while it is equal to 0.97 in Mg-doped CuO NPs ($x = 0.020$). This indicates that the stoichiometric nature of the samples is affected by Mg dopants. These variations may be due to surface crystalline defects [14]. Moreover, the ratio of the atomic percentage of Mg/Cu for $x = 0.020$ samples is calculated to be 0.0255, further confirming the successful synthesis of Mg-doped CuO with matching experimental and theoretical values.

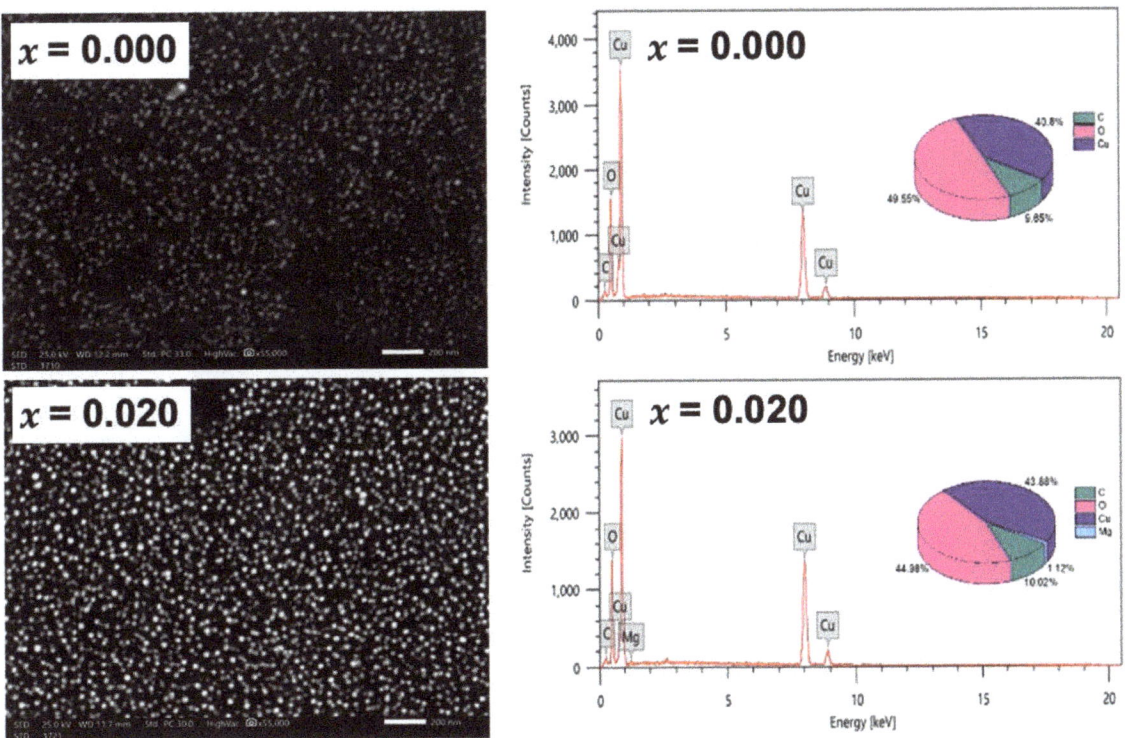

Figure 3. SEM and EDX patterns for undoped, and Mg-doped CuO NPs ($Cu_{1-x}Mg_xO$ NPs).

2.1.4. Fourier Transform Infrared (FTIR)

The FTIR spectra, represented in Figure 4, of the undoped and Mg-doped CuO NPs, demonstrated different vibrational bands. A broad absorption band ranged between 3200 and 3600 cm^{-1}. The adsorbed water is present in all spectra. A small peak is detected at 2344–2354 c, and a peak centered around 1618–1656 cm^{-1} is observed. The main peaks, ranging between 700–400 cm^{-1}, are displayed in Figure 4. Three peaks are detected, which are centered around 480–486, 521–530, and 580–584 cm^{-1}.

It is noticed that the peak position is slightly affected by Mg concentration in the CuO lattice. The peak attributed to the symmetric vibration of Cu–O fluctuated around 483 ± 3 cm^{-1} with Mg-doping concentrations. However, Cu–O asymmetric stretching and wagging peaks of the pure CuO Nps are shifted monotonously to lower wavenumber with the Mg-doping from 530 and 584 cm^{-1} to 521 and 580 cm^{-1} with $x = 0.020$. Pramothkumar et al. [15] reported the same pattern of variation upon Mn, Co, and Ni doping to CuO NPs, and explained these shifts according to the dopant effect, which can in turn affect the surface area and defects in the samples. Singh et al. [16] reported the successful doping of Zn to CuO, which led to the increase in Cu–O bond length in the samples with the increase of Zn dopant concentration.

2.1.5. Photoluminescence (PL)

Figure 5a shows the room temperature photoluminescence (PL) emission spectra for $Cu_{1-x}Mg_xO$ NPs with an excitation wavelength of 200 nm. A prominent UV peak appeared at 310 ± 1 nm in all samples, with the highest intensity, as compared to other peaks. Furthermore, the visible emissions in the PL spectra are deconvoluted by four Voigt functions to elucidate the origin of these emissions. The position of the fitted peaks is listed in Table 1, along with the position of the UV peak. It is noticed that the increase in the concentration of the Mg doping in CuO NPs did not affect the position of the peaks, however, it affects their intensity. This is similar to the reported literature, where the doping concentration does not affect the position of the peaks in the visible part of the PL spectra [17–20]. The deconvolution of the PL spectra of $Cu_{1-x}Mg_xO$ NPs yielded violet (391 ± 1 nm), blue (452 ± 5.5 nm), green (536 nm), and orange-red (628 ± 5 nm) emission peaks.

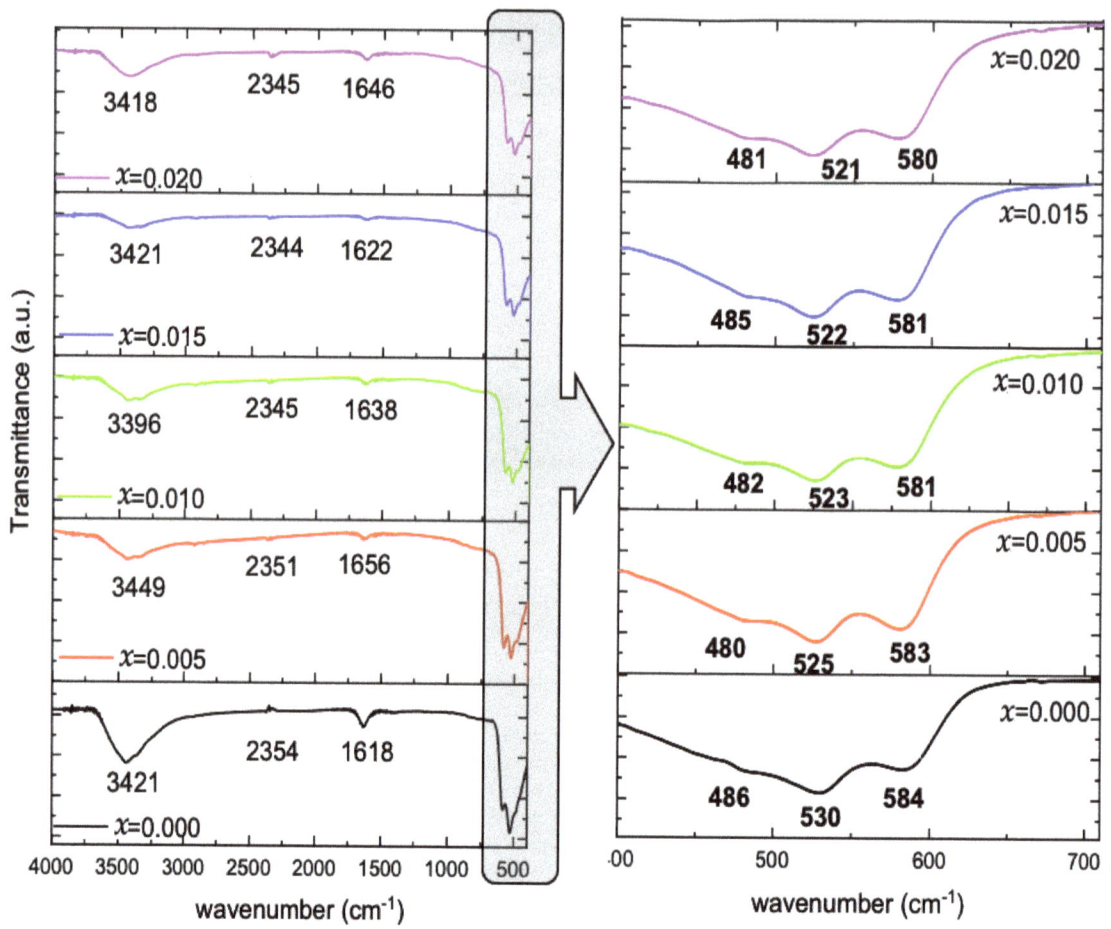

Figure 4. FTIR spectra of $Cu_{1-x}Mg_xO$ NPs, and enlarged view for Cu-O vibrations.

2.2. Antibacterial Activity of the Undoped and Mg-Doped CuO NPs

2.2.1. MIC and MBC

The four bacterial isolates were tested for their susceptibility against the undoped and Mg-doped CuO NPs. Undoped CuO NPs had a bactericidal effect on Gram-positive bacteria while having a bacteriostatic effect on Gram-negative bacteria. The most sensitive bacterium was *E. faecium* (MIC = 0.375 mg/mL and MBC = 0.75 mg/mL), followed by *S. aureus* (MIC = 1.5 mg/mL and MBC = 3 mg/mL). *E. coli* and *S. maltophilia* were sensitive at the highest NP concentration used (3 mg/mL). Similarly, Mg-doped NPs exhibited bactericidal activity on Gram-positive bacteria and bacteriostatic activity on Gram-negative bacteria.

E. faecium was the most sensitive bacterium (MIC = 1.5 mg/mL and MBC = 3 mg/mL) and *E. coli* was the most resistant bacterium. Collectively, undoped and Mg-doped NPs had a better effect on Gram-positive bacteria than that on Gram-negative bacteria. The MIC and MBC results are shown in Table 2 and Figures A1–A3 (Appendix A).

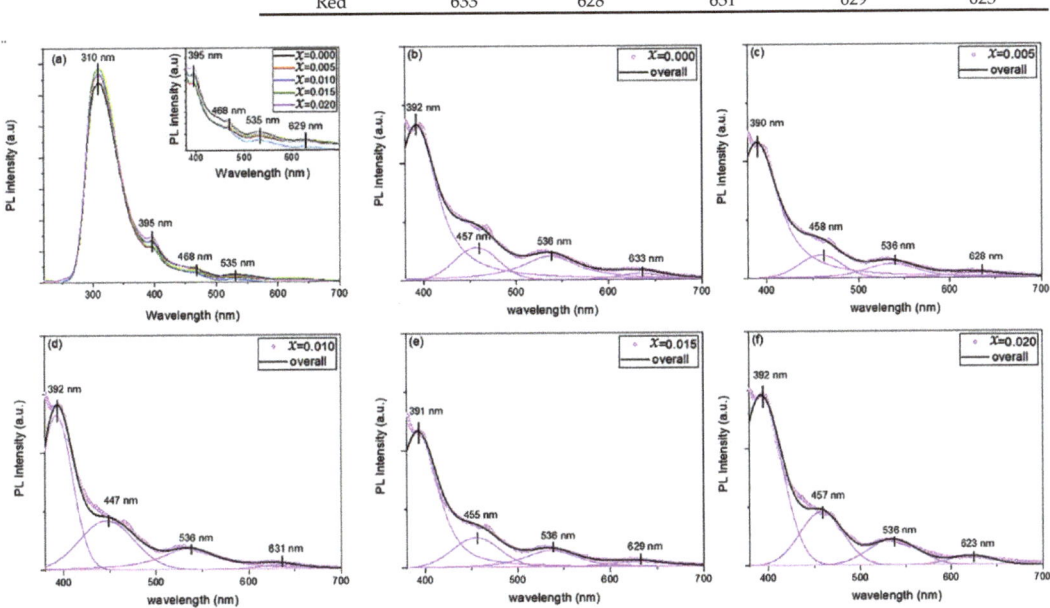

Figure 5. (**a**) Photoluminescence spectra of $Cu_{1-x}Mg_xO$ NPs, and (**b**–**f**) their deconvolution for undoped and Mg-doped CuO NPs.

Table 1. Position of peaks by deconvolution of PL spectra for $Cu_{1-x}Mg_xO$ NPs.

Peaks Color	Wavelength (nm)				
	$x = 0.000$	$x = 0.005$	$x = 0.010$	$x = 0.015$	$x = 0.020$
UV	310	310	311	309	309
Violet	392	390	392	391	392
Blue	457	458	447	455	457
Green	536	536	536	536	536
Red	633	628	631	629	623

Table 2. MIC and MBC of the undoped and Mg-doped ($Cu_{1-x}Mg_xO$) NPs against four bacterial isolates.

Bacterial Isolates	MICs and MBCs (mg/mL)														
	$x = 0.000$			$x = 0.005$			$x = 0.010$			$x = 0.015$			$x = 0.020$		
	MIC	MBC	MIC/MBC	MIC	MBC	MIC/MBC	MIC	MBC	MIC/MBC	MIC	MBC	MIC/MBC	MIC	MBC	MIC/MBC
	Gram-positive bacteria														
S. aureus	1.5	3	2	1.5	3	2	1.5	3	2	3	>3	nd	1.5	3	2
E. faecium	0.375	0.75	2	1.5	3	2	1.5	3	2	1.5	3	2	1.5	3	2
	Gram-negative bacteria														
E. coli	3	>3	nd	3	>3	nd	3	>3	nd	3	>3	nd	3	>3	nd
S. maltophilia	3	>3	nd	1.5	3	2	1.5	3	2	1.5	3	2	3	>3	nd

nd: not determined, MIC: minimum inhibitory concentration, MBC: minimum bactericidal concentration.

It was reported that when the MBC/MIC ratio < 4, this will reflect a bactericidal effect [21]. The antibacterial results against the four bacterial isolates shown in Table 2 revealed that the undoped and Mg-doped NPs exhibited bactericidal effects against *S. aureus* and *E. faecium* with MBC/MIC ratio = 2 and a bacteriostatic effect against *E. coli* and *S. maltophilia*.

2.2.2. Agar Well Diffusion

All isolates showed sensitivity to the undoped CuO NP ($x = 0.000$). *E. faecium* was the most sensitive. It showed a sensitivity against the lowest NP concentration (0.1875 mg/mL). The other three bacteria showed a sensitivity against an NP concentration of 1.5 mg/mL. Sensitivity was considered positive for ZOI diameters > 7 mm [17]. All the bacterial isolates were sensitive to the Mg-doped CuO NP with $x = 0.005$. *E. coli* and *S. maltophilia* showed a sensitivity to the NP at a concentration of 0.75 mg/mL. *E. faecium* and *S. aureus* showed sensitivity against a concentration of 1.5 mg/mL. All investigated bacteria showed sensitivity to Mg-doped NPs with $x = 0.010$. *S. aureus* showed susceptibility against a concentration of 0.75 mg/mL. *E. faecium* showed a sensitivity against a concentration of 1.5 mg/mL. *S. aureus*, *E. coli*, and *S. maltophilia* exhibited a sensitivity at the highest concentration (3 mg/mL). All isolates were sensitive to the Mg-doped CuO NP with $x = 0.015$. *E. faecium* was the most sensitive. It showed a sensitivity starting from the lowest concentration (0.1875 mg/mL). *S. aureus*, *E. coli*, and *S. maltophilia* were sensitive at concentrations starting from 0.375 mg/mL. All isolates were sensitive at the highest NP concentration (3 mg/mL). For Mg-doped CuO NP with $x = 0.020$, all bacteria exhibited sensitivity only at the highest NP concentration (3 mg/mL). The agar well diffusion results of the undoped and Mg-doped CuO NPs are shown in Table 3 and Figure A4 (Appendix A). All results were significant with a p-value < 0.05 shown in Table A1 (Appendix A).

Table 3. Agar well diffusion of undoped and Mg-doped CuO NPs against four bacterial isolates.

		Bacterial Isolates			
	Nanoparticles	Gram-Positive Bacteria		Gram-Negative Bacteria	
		S. aureus	*E. faecium*	*E. coli*	*S. maltophilia*
Sample	Concentration (mg/mL)	ZOI ± SEM (mm)			
$x = 0.000$	0.1875	7 ± 0.0	7.6 ± 0.2	0 ± 0.0	0 ± 0.0
	0.375	7 ± 0.0	8.6 ± 0.2	0 ± 0.0	0 ± 0.0
	0.75	7 ± 0.0	9 ± 0.7	0 ± 0.0	0 ± 0.0
	1.5	8 ± 0.4	12.3 ± 1.4	19 ± 0.9	14 ± 0.4
	3	12.3 ± 0.2	13.3 ± 0.7	22.3 ± 0.3	17.3 ± 0.4
$x = 0.005$	0.1875	7 ± 0.0	7 ± 0.0	0 ± 0.0	0 ± 0.0
	0.375	7 ± 0.0	7 ± 0.0	0 ± 0.0	0 ± 0.0
	0.75	7 ± 0.0	7 ± 0.0	10 ± 0.0	8.3 ± 0.2
	1.5	9.6 ± 0.2	8 ± 0.4	15.6 ± 0.7	10.6 ± 0.2
	3	13.6 ± 0.2	12.3 ± 0.2	20 ± 0.4	11.6 ± 0.5
$x = 0.010$	0.1875	7 ± 0.0	7 ± 0.0	0 ± 0.0	7 ± 0.0
	0.375	7 ± 0.0	7 ± 0.0	0 ± 0.0	7 ± 0.0
	0.75	7.6 ± 0.2	7 ± 0.0	0 ± 0.0	7 ± 0.0
	1.5	7.6 ± 0.2	8.6 ± 0.5	0 ± 0.0	7 ± 0.0
	3	10 ± 0.9	10.6 ± 0.7	16.6 ± 0.7	7 ± 0.0
$x = 0.015$	0.1875	7 ± 0.0	7.6 ± 0.2	0 ± 0.0	0 ± 0.0
	0.375	7 ± 0.0	7.6 ± 0.2	0 ± 0.0	0 ± 0.0
	0.75	10.3 ± 0.2	7.6 ± 0.2	0 ± 0.0	0 ± 0.0
	1.5	11.3 ± 0.2	7.6 ± 0.2	11.6 ± 0.5	0 ± 0.0
	3	17.6 ± 1.18	7.6 ± 0.2	16.3 ± 0.9	10.6 ± 0.5

Table 3. Cont.

		Bacterial Isolates			
	Nanoparticles	Gram-Positive Bacteria		Gram-Negative Bacteria	
		S. aureus	E. faecium	E. coli	S. maltophilia
Sample	Concentration (mg/mL)	ZOI ± SEM (mm)			
x = 0.020	0.1875	0 ± 0.0	0 ± 0.0	0 ± 0.0	0 ± 0.0
	0.375	0 ± 0.0	0 ± 0.0	0 ± 0.0	0 ± 0.0
	0.75	0 ± 0.0	0 ± 0.0	0 ± 0.0	0 ± 0.0
	1.5	0 ± 0.0	0 ± 0.0	0 ± 0.0	0 ± 0.0
	3	9.6 ± 0.7	9.3 ± 0.5	13.6 ± 0.7	7.6 ± 0.2
Dox	0.25	33.6 ± 0.7	19.3 ± 0.7	27.6 ± 0.2	34.6 ± 0.2
Amo	0.25	0 ± 0.0	0 ± 0.0	0 ± 0.0	0 ± 0.0

Dox: Doxycycline, Amo: Amoxicillin, ZOI: zone of inhibition, SEM: standard error of the mean.

2.2.3. Time-Kill Results

The time-kill test was performed using the MICs of the undoped and Mg-doped CuO NPs against four bacterial isolates to detect the time needed for each NP to exert its antibacterial effect. All bacterial isolates were sensitive to all tested NPs after 2 h of incubation. The activities were sustained till 24 h of incubation. The time-kill results for the different bacterial isolates are shown in Figure 6.

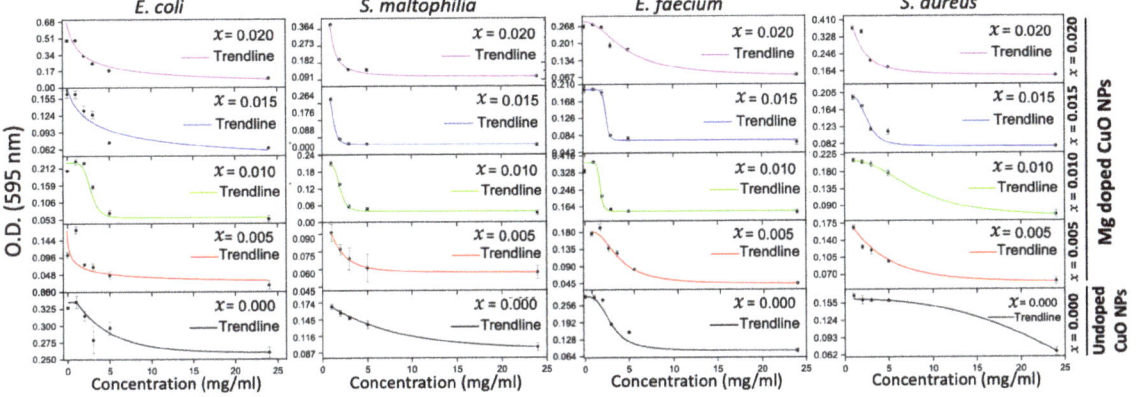

Figure 6. Time-kill results of the undoped and Mg-doped NPs against four bacterial isolates.

3. Discussion

The peaks of the undoped CuO NPs shown by the XRD correspond to the primary defining peaks of the monoclinic CuO phase with a space group of C2/c [22]. The absence of a secondary phase suggests that the samples exhibit a highly single phase [23]. The refinements indicate the total incorporation of Mg dopant in the $Cu_{1-x}Mg_xO$ NPs.

The TEM results demonstrated the change in the morphology and size of the NPs after doping, indicating that the increase in the doping concentration increases the uniformity of the NPs and in turn decreases their size.

The composition of the NPs was detected by EDX. The absence of precursors indicates the purity of the formed CuO NPs. Noting that the emergence of carbon may be due to the use of carbon tape in the measurements, or some residues from EDTA [24]. In addition, the ratio between Cu and O in the undoped and Mg-doped NPs indicates that the stoichiometric nature of the samples is affected by Mg dopants. These variations may be due to surface crystalline defects [14].

Infrared spectroscopy can be used as a fingerprint to identify different molecules by comparing vibration bands. The broad absorption band observed by the FTIR is associated with the hydroxyl (O–H) stretching vibration mode of the water molecule [25]. The adsorbed water observed is due to the physical adsorption of water from the atmosphere [26]. The small peak observed at 2344–2354 cm^{-1} is related to the vibration of CO$_2$ in the air [26] and the peak centered around 1618–1656 cm^{-1} is due to H–O–H bending vibrations of water molecules [25]. The three main peaks are attributed to Cu–O symmetric stretching, Cu–O asymmetric stretching, and Cu–O wagging, respectively. These peaks validate the successful formation of CuO NPs [5]. Pramothkumar et al. [15] reported the same pattern of variation observed upon Mn, Co, and Ni doping to CuO NPs, and explained these shifts according to the dopant effect, which can in turn affect the surface area and defects in the samples. Singh et al. [16] reported the successful doping of Zn to CuO, which led to the increase in Cu–O bond length in the samples with the increase of Zn dopant concentration.

The PL spectra detect the imperfections and defects within the samples, where the prevalence of the imperfections and surface states varies depending on the synthesizing circumstances, particle size and shape, types of dopants, and concentrations [27,28]. The origin of the UV peak is directly related to the recombination of electron–hole pair, near the band gap transition [17,20]. It is noticed that the position of the UV peak is slightly invariant with the doping concentration, however, its intensity increased with Mg-doping. This enhancement of the intensity may be related to the passivation of surface defects that generate radiative recombination [29]. Additionally, the intensity of the UV peak is affected by the electron density and the variation of the morphology and size of the nanoparticles, with the increase of the doping concentration [30]. The visible emissions are highly sensitive to the change in the synthesis conditions, accounting for the type of dopant and its concentration, the size of the nanoparticle, and its morphology [18]. The size of Mg-doped CuO NPs decreased with the increase of the doping concentration, as noted from TEM and SEM analysis. Hence, the large surface-to-volume ratio stimulates more surface-defect states, as vacancies and interstitials, creating trap levels that radiate visible emissions [18]. Mainly, the intensity of the visible emissions is quenched with the increment of the doping concentration in $Cu_{1-x}Mg_xO$ NPs, as can be noticed from the inset of Figure 5a. This decrement in the intensity may be due to the trapping of the photoexcited electron from the conduction band of CuO NPs by the formed deep-level centers from Mg doping [17,30]. The violet and blue emissions are mostly attributed to deep-level defects, indicating the existence of Cu vacancies in the lattice [17]. The green emission was reported to originate from the recombination of single ionized electrons with a photogenerated hole in the valence band, noting the presence of singly ionized oxygen vacancies or dangling bonds of copper [19]. The orange–red emission ascends from the recombination of an electron bound to donor and free holes [18].

The reported antibacterial properties of NPs, especially CuO NPs, made their usage efficient against bacteria [1,2]. In this regard, CuO NPs were used against four bacterial pathogens isolated from sewage sludge. All bacterial isolates, except *S. maltophilia*, are frequently present in Lebanon, especially in the feces of animals. *S. maltophilia* is a rare Gram-negative bacterium in Lebanon [21,31,32]. In this investigation, the antibiotic Dox was used as a reference antibiotic. It belongs to the tetracycline family of antibiotics. It acts by inhibiting protein synthesis by binding to the 30S ribosomal subunit, leading to the destruction of the bacterial cells [33–35]. This study showed that the undoped and Mg-doped CuO NPs exerted antibacterial activities against all bacterial isolates. Using the agar well diffusion assay, the NPs had better effects on Gram-positive bacteria than on Gram-negative bacteria. The results are consistent with previous studies that showed that Gram-negative bacteria are more resistant to NPs, due to the rigidity of their cell wall [36–38]. This activity depends on the metal oxides present in the NPs. The latter could penetrate the cell wall of bacteria, leading to cell autolysis [10]. Among the investigated NPs, the undoped and Mg-doped CuO NPs with $x = 0.005$ and $x = 0.010$ were efficient as antibacterial agents. They were effective at low NPs concentrations against all bacterial isolates. In contrast, the Mg-doped CuO NPs with $x = 0.015$ and $x = 0.020$ had lower antibacterial activity against the investigated bacterial isolates. They were effective at higher concentrations. Regarding bacterial susceptibility, S. aureus and E. faecium were the most sensitive. Their growth was inhibited by all tested NPs at significantly low concentrations. These results are consistent with previous studies that have shown the sensitivity of Gram-positive bacteria, especially S. aureus, to NPs [8,9,39]. On the other hand, E. coli and S. maltophilia were more resistant. The inhibition of their growth required higher concentrations of the NPs. This could be attributed to the shape of NPs. Their spherical shape, demonstrated by the TEM results reflects their significant inhibitory activity against Gram-positive bacteria [8,40]. Spherical NPs are shown to have good antibacterial activity due to the sphere prisms that can penetrate easily into the bacterial cell membrane [41].

The MIC results confirmed the results of the agar well diffusion assay. All the investigated NPs, except the Mg-doped CuO NPs with $x = 0.020$, had bactericidal activities. The Mg-doped CuO NP with $x = 0.020$ had bacteriostatic activity only. The bacteriostatic and bactericidal effects of NPs depend on their metal oxide content and their morphology, and the architecture of the bacteria [42,43]. This means that the metal oxides of the tested NPs can react with the bacterial cell wall through special mechanisms, which are still not very specific. However, previous studies reported the following mechanisms: disruption of the bacterial cell wall, generating reactive oxygen species, and binding with specific cytoplasmic targets and production of metabolites, leading to these bacteriostatic and bactericidal effects [42,44]. Kumer et al. showed that ZnO NPs exhibited good antibacterial activity against Gram-positive bacteria. This activity was better than that of Ag-doped ZnO NPs [45]. In addition, Prakash et al. reported that TiO_2 NPs with doped antibacterial activity were better than the undoped TiO_2 NPs [46]. The bactericidal properties of the NPs depend on the shape, the surface area of the particle, the type of metal ions, and the chemically reactive functional groups [8,10,35]. The high bactericidal effect is attributed to the different shapes (spherical in our case) of the particles, which help them penetrate the bacterial cell membrane. Moreover, the large surface area permits the production of reactive oxygen. This induces oxidative stress on bacteria, which interrupts the electron flow in the inner membrane, thus causing cell damage [35,36]. On the other hand, the observed bacteriostatic effect could be due to the low number of metal ions coming from the metal oxides, which prevents the Cu ions from interacting with the bacterial cell wall. So, the main bactericidal mechanism may rely on the damage of the cell membrane by the metal oxides [35,37].

Time-kill results have shown that the most frequent inhibition time of the tested NPs started at 2 h of incubation. NPs can prevent the adaptation and duplication of bacteria [47]. This effect could be attributed to the limiting effect of NPs on the nutrient uptake by the bacteria, which eventually will lead to cell lysis. This is consistent with previous studies that showed that NPs affect the metabolic activities and division of bacterial cells. Metal oxides may lead to nutrient deprivation [8,10,36]. In addition, the size of the particles and the surface area may specify the time needed for the interaction between the NP and the bacterial cell wall. When the size of the particle is smaller, the interaction becomes faster, thus decreasing the time needed for the inhibition of bacterial growth [8,10,36]. This slows metabolic processes and leads to cell death.

Collectively, previous studies reported that the size of the NPs reflects their antibacterial effect [48]. In addition, the variation in the antibacterial activity depends on the morphology of the NPs. Furthermore, the variation in the intensity of PL accompanied by the variation in oxygen interferes with the antibacterial activity. So, the shape and the morphology of NPs play a vital role in the inhibition of bacterial growth.

4. Materials and Methods

4.1. Synthesis of NPs

The undoped and Mg-doped CuO NPs were prepared by the co-precipitation method, with the chemical formula of $Cu_{1-x}Mg_xO$ ($x = 0.000, 0.005, 0.010, 0.015,$ and 0.020). The

CuO NPs were synthesized using copper (II) chloride dehydrate (Merk), magnesium chloride hexahydrate (Sigma-Aldrich, ≥99.0%), and ethylenediamine tetra-acetic acid (EDTA) (0.1 M). The weighed reagents were prepared with a molarity of 1 M in de-ionized water and stirred for 15 minutes. The solution was then titrated with sodium hydroxide NaOH (2 M). NaOH was added slowly under vigorous stirring until pH reached 12. After that, the precipitate was heated at 60 °C for 2 h, then sonicated for 10 minutes. The black precipitate obtained was washed with de-ionized water several times until pH reached 7. Finally, the washed precipitate was dried at 100 °C for 16 h and ground into fine powders. The powders were sintered at 600 °C for 4 h.

4.2. Characterization of Mg-Doped CuO NPs

The structural properties of both undoped and doped CuO NPs were studied by XRD using the X-ray Bruker D8 Focus power diffractometer with Cu K_α radiation, operated at 40 kV and 40 mA, in the range $20 \leq 2\theta° \leq 80$. Material Analysis Using Diffraction (MAUD) software was then used to check for the presence of CuO and MgO phases in the resultant NPs using the CIF files downloaded from the Crystallography Open Database (COD). The morphology of the prepared CuO NPs was investigated using the JEM 100 CX Transmission Electron microscope (TEM), operated at 80 kV. The main functional groups of the synthesized samples were detected using the Nicolet iS5 Fourier Transform Infra-Red (FTIR) spectra after preparing potassium bromide (KBr) pellets mixed with the undoped and doped CuO NPs (1:100). The purity of the $Cu_{1-x}Mg_xO$ NPs was studied using energy dispersive X-ray (EDX), operated at a voltage of 20 kV with laser power of 5 mW and magnification objective of 50x. The Photoluminescence (PL) spectra were studied by a Jasco FP-8600 spectrofluorometer with Xenon (Xe) laser at 200 nm excitation wavelength for $Cu_{1-x}Mg_xO$ nanoparticles, dispersed in ethanol.

4.3. Isolation of Bacteria

Briefly, *E. faecium, S. aureus, E. coli,* and *S. maltophilia* were isolated from wastewater by streaking 100 μL of the samples on different selective media (blood agar, chocolate agar, MacConkey agar, mannitol salt agar (MSA), eosin methylene blue (EMB) agar, and cetrimide agar). The plates were incubated at 37 °C for 24 h. After isolation, bacteria were Gram stained to differentiate between Gram-positive and Gram-negative bacteria. Bacteria were further identified by VITEK assay.

4.4. Broth Microdilution Assay: Minimum Inhibitory Concentration (MIC) and Minimum Bactericidal Concentration (MBC)

The MICs of the undoped and Mg-doped CuO NPs were determined against four bacteria employing the microwell dilution method. The test was performed in sterile 96-well microplates by dispensing into each well 90 μL of nutrient broth and 10 μL of bacterial suspensions adjusted to 0.5 McFarland. Then, 100 μL of each NP (0.1875–3 mg/mL) was added to the wells. The plates were incubated at 37 °C for 24 h and the optical density (O.D.) was measured at 595 nm, using an ELISA microtiter plate reader. The MIC is defined as the lowest concentration of the NPs that inhibits the visible growth of the tested bacteria in the wells. Doxycycline (Dox) was used as a reference antibiotic. After incubation, 10 μL of the clear wells was transferred to Muller Hinton agar (MHA) plates and incubated at 37 °C for 24 h to detect the MBC [16]. The MBC is defined as the lowest concentration that inhibits the visible growth of bacteria on the plates. All experiments were repeated at least three times.

4.5. Agar Well Diffusion Assay

Agar well diffusion assays were performed in triplicate for the undoped and Mg-doped CuO NPs on four bacterial isolates using MHA. A standard inoculum was prepared for each tested bacterial isolate as described in the MIC and MBC broth microdilution assay. The plates were inoculated with 100 μL of each bacterial suspension, which was spread evenly over the entire surface of the agar. Plates were then punched with a 6 mm cork-borer. A total of 100 µL of each NP (0.1875–3 mg/mL) was pipetted into the wells and the plates were incubated at 37 °C for 24 h. Dox and Amoxicillin (Amo) were used as reference antibiotics. For each well, the diameter of the zone of inhibition (ZOI) was measured. ZOI of diameter > 7 mm was considered a significant inhibitory effect [49].

4.6. Time-Kill Test

Time-kill studies were performed to detect the time needed by the undoped and Mg-doped CuO NPs to inhibit bacterial growth. The test was performed in sterile 96-well microplates by dispensing into each well 90 µL of nutrient broth and 10 µL of the bacterial suspensions adjusted to 0.5 McFarland. Then, 100 µL of each NP's MIC was added to the wells. The plates were incubated at 37 °C and the O.D. was measured at 595 nm, using an ELISA microtiter plate reader at different time points (0–24 h) [47]. All experiments were repeated at least three times.

4.7. Statistical Analysis

All statistical tests were done in Excel software, and graphs were drawn on Origin software. Statistical significance was determined by t-test. Differences with p-value < 0.05 were considered statistically significant.

5. Conclusions

Pure and Mg-doped CuO NPs were fabricated via the co-precipitation method. The XRD patterns with their refinements confirmed the total incorporation of Mg dopant in the $Cu_{1-x}Mg_xO$ NPs and the production of CuO NPs without any impurities. Besides, the morphology was changed upon Mg doping, in which the NPs showed a uniform shape with less agglomeration. FTIR spectra demonstrated the main vibrational modes of undoped and doped CuO NPs. The Cu–O bond was shifted as the Mg concentration for doping increased, confirming the incorporation of the dopant and its effect in modifying the surface area and defects. The EDX patterns further confirmed the purity of CuO NPs and the inclusion of Mg inside the NPs successfully. PL studies proved the enhancement of visible emissions of CuO nanoparticles associated with Mg doping. This study showed significant antibacterial activity of undoped and Mg-doped NPs. The results showed that the NPs had significant antibacterial activity against different Gram-positive and Gram-negative bacteria. Thus, the undoped and Mg-doped CuO NPs exhibited a significant impact on the structural, morphological, and inhibition capacities against *S. aureus, E. faecium, E. coli,* and *S. maltophilia,* isolated from the Lebanese wastewater. These results may provide an approach to using CuO NPs as antibacterial agents to prevent bacterial contaminations.

Author Contributions: Conceptualization, R.A. and M.I.K.; Data curation, R.M.A. and M.M.; Formal analysis, R.M.A., M.M. and A.M.A.; Investigation, R.M.A., M.M., R.A. and M.I.K.; Methodology, R.M.A., M.M., A.M.A., R.A. and M.I.K.; Resources, R.A. and M.I.K.; Software, M.M. and A.M.A.; Supervision, R.A. and M.I.K.; Validation, R.A. and M.I.K.; Writing—original draft, R.M.A. and M.M.; Writing—review & editing, R.M.A., M.M., R.A. and M.I.K. All authors have read and agreed to the published version of the manuscript.

Funding: This research received no external funding.

Institutional Review Board Statement: Not applicable.

Informed Consent Statement: Not applicable.

Data Availability Statement: The data supporting the reported results are available with the corresponding author and will be provided upon request.

Conflicts of Interest: The authors declare no conflict of interest.

Appendix A

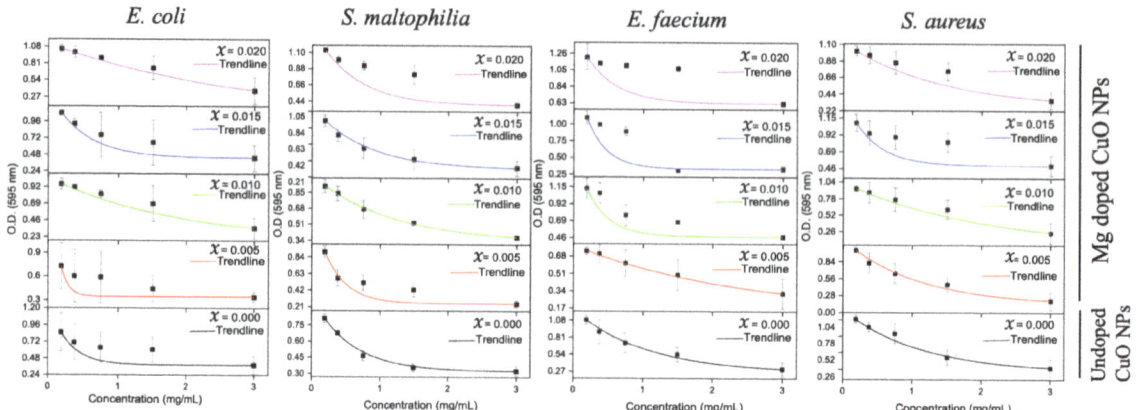

Figure A1. MICs of the undoped and Mg-doped NPs against four bacterial isolates.

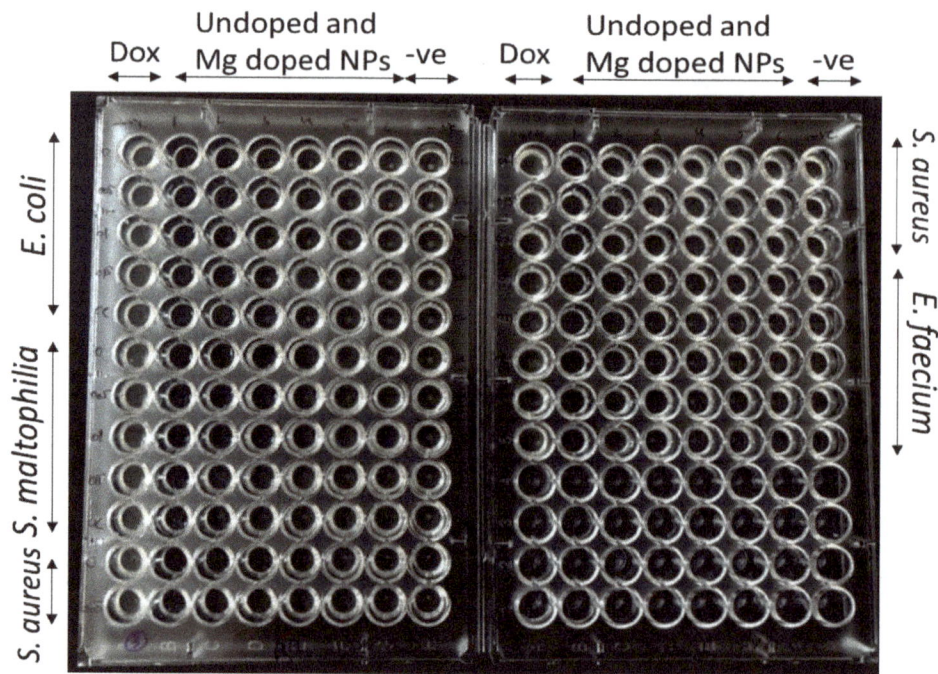

Figure A2. MIC results of the undoped and Mg-doped CuO NPs against four bacterial isolates.

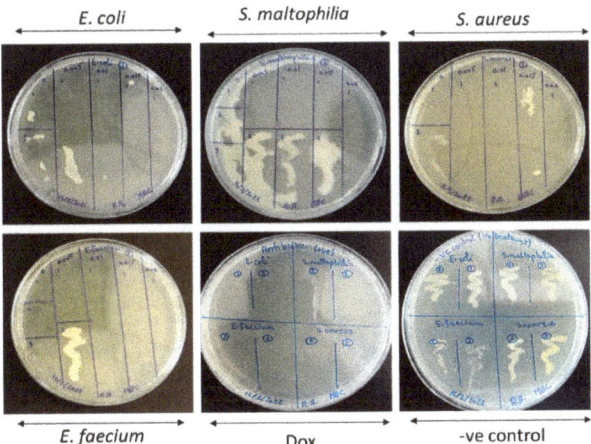

Figure A3. MBC results of the undoped and Mg-doped CuO NPs against four bacterial isolates.

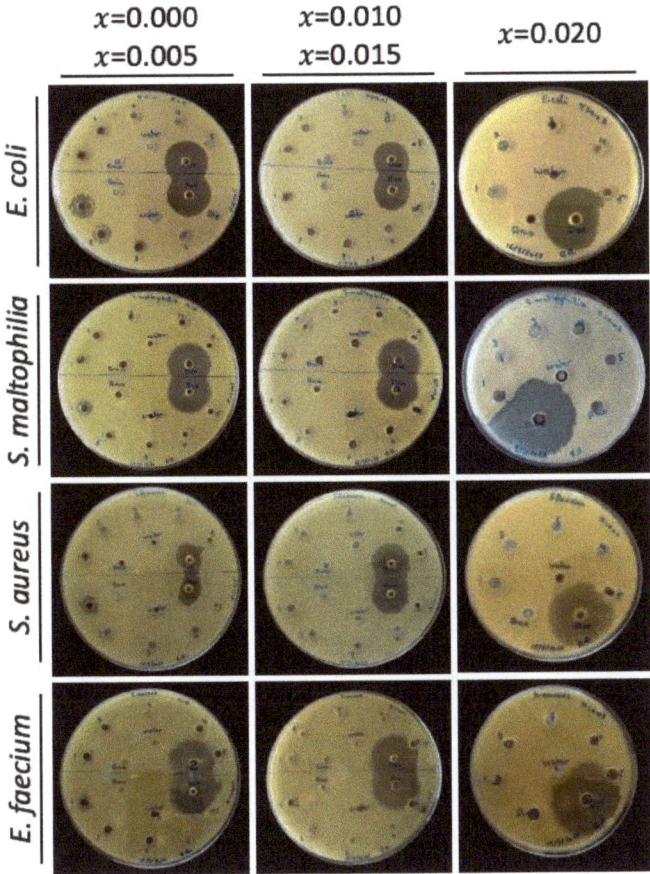

Figure A4. Agar well diffusion results of the undoped and Mg-doped CuO NPs against four bacterial isolates.

Table A1. Calculated p-value and significance levels of the agar well diffusion results of the undoped and Mg-doped CuO NPs.

Nanoparticles		Bacterial Isolates							
		Gram-Positive Bacteria				Gram-Negative Bacteria			
		S. aureus	Significance	E. faecium	Significance	E. coli	Significance	S. maltophilia	Significance
S	(mg/mL)				p-Values (vs. Dox)				
$x = 0.000$	0.1875	0.005	**	0.003	**	<0.001	***	<0.001	***
	0.375	0.005	**	0.006	**	<0.001	***	<0.001	***
	0.75	0.005	**	0.004	**	<0.001	***	<0.001	***
	1.5	0.01	**	0.010	**	0.01	**	0.001	***
	3	0.007	**	0.007	**	0.02	*	<0.001	***
$x = 0.005$	0.1875	0.001	***	0.005	**	<0.001	***	<0.001	***
	0.375	0.001	***	0.005	**	<0.001	***	<0.001	***
	0.75	0.001	***	0.005	**	<0.001	***	<0.001	***
	1.5	<0.001	***	0.010	**	0.002	**	<0.001	***
	3	<0.001	***	0.006	**	0.001	***	0.001	***
$x = 0.010$	0.1875	0.001	***	0.005	**	<0.001	***	<0.001	***
	0.375	0.001	***	0.005	**	<0.001	***	<0.001	***
	0.75	0.001	***	0.005	**	0.030	*	<0.001	***
	1.5	0.001	***	0.010	**	0.040	*	<0.001	***
	3	0.002	**	0.020	*	0.002	**	<0.001	***
$x = 0.015$	0.1875	0.001	***	0.003	**	<0.001	***	<0.001	***
	0.375	0.001	***	0.003	**	<0.001	***	<0.001	***
	0.75	0.002	**	0.003	**	<0.001	***	<0.001	***
	1.5	<0.001	***	0.003	**	0.003	**	<0.001	***
	3	0.001	***	0.003	**	0.006	**	0.001	***
$x = 0.020$	0.1875	<0.001	***	0.002	**	<0.001	***	<0.001	***
	0.375	<0.001	***	0.002	**	<0.001	***	<0.001	***
	0.75	<0.001	***	0.002	**	<0.001	***	<0.001	***
	1.5	<0.001	***	0.002	**	<0.001	***	<0.001	***
	3	0.002	**	0.02	*	0.001	***	<0.001	***

S: samples. p-values were calculated such that: * $p < 0.05$, ** $p < 0.01$, *** $p < 0.001$.

References

1. Chandrappa, K.G.; Venkatesha, T.V. Generation of Nanostructured CuO by Electrochemical Method and Its Zn–Ni–CuO Composite Thin Films for Corrosion Protection. *Mater. Corros.* **2013**, *64*, 831–839. [CrossRef]
2. Surendhiran, S.; Gowthambabu, V.; Balamurugan, A.; Sudha, M.; Senthil Kumar, V.B.; Suresh, K.C. Rapid Green Synthesis of CuO Nanoparticles and Evaluation of Its Photocatalytic and Electrochemical Corrosion Inhibition Performance. *Mater. Today Proc.* **2021**, *47*, 1011–1016. [CrossRef]
3. Butte, S.M.; Waghuley, S.A. Optical Properties of Cu_2O and CuO. *AIP Conf. Proc.* **2020**, *2220*, 020093. [CrossRef]
4. Koshy, J.; Samuel, M.S.; Chandran, A.; George, K.C. Optical Properties of CuO Nanoparticles. *AIP Conf. Proc.* **2011**, *1391*, 576–578. [CrossRef]
5. Arun, K.; Batra, A.; Krishna, A.; Bhat, K.; Aggarwal, M.; Francis, J. Surfactant Free Hydrothermal Synthesis of Copper Oxide Nanoparticles. *Am. J. Mater. Sci.* **2015**, *2015*, 36–38. [CrossRef]
6. Singh, S.; Goswami, N.; Mohapatra, S.R.; Singh, A.K.; Kaushik, S.D. Significant Magnetic, Dielectric and Magnetodielectric Properties of CuO Nanoparticles Prepared by Exploding Wire Technique. *Mater. Sci. Eng. B* **2021**, *271*, 115301. [CrossRef]
7. Lv, Y.; Li, L.; Yin, P.; Lei, T. Synthesis and Evaluation of the Structural and Antibacterial Properties of Doped Copper Oxide. *Dalton Trans.* **2020**, *49*, 4699–4709. [CrossRef]
8. Khalid, A.; Ahmad, P.; Alharthi, A.I.; Muhammad, S.; Khandaker, M.U.; Rehman, M.; Faruque, M.R.I.; Din, I.U.; Alotaibi, M.A.; Alzimami, K.; et al. Structural, Optical, and Antibacterial Efficacy of Pure and Zinc-Doped Copper Oxide Against Pathogenic Bacteria. *Nanomaterials* **2021**, *11*, 451. [CrossRef]

9. Raba-Páez, A.M.; Malafatti, J.O.D.; Parra-Vargas, C.A.; Paris, E.C.; Rincón-Joya, M. Effect of Tungsten Doping on the Structural, Morphological and Bactericidal Properties of Nanostructured CuO. *PLoS ONE* **2020**, *15*, e0239868. [CrossRef]
10. Ahamed, M.; Alhadlaq, H.A.; Khan, M.A.M.; Karuppiah, P.; Al-Dhabi, N.A. Synthesis, Characterization, and Antimicrobial Activity of Copper Oxide Nanoparticles. *J. Nanomater.* **2014**, *2014*, 17. [CrossRef]
11. Jayaprakash, J.; Srinivasan, N.; Chandrasekaran, P. Surface Modifications of CuO Nanoparticles Using Ethylene Diamine Tetra Acetic Acid as a Capping Agent by Sol–Gel Routine. *Spectrochim. Acta. A. Mol. Biomol. Spectrosc.* **2014**, *123*, 363–368. [CrossRef] [PubMed]
12. Raveesha, H.R.; Sudhakar, M.S.; Pratibha, S.; Ravikumara, C.R.; Nagaswarupa, H.P.; Dhananjaya, N. Costus Pictus Leaf Extract Mediated Biosynthesis of Fe and Mg Doped CuO Nanoparticles: Structural, Electrochemical and Antibacterial Analysis. *Mater. Res. Express* **2019**, *6*, 1150.e5. [CrossRef]
13. Din, S.U.; Sajid, M.; Imran, M.; Iqbal, J.; Shah, B.A.; Shah, S. One Step Facile Synthesis, Characterization and Antimicrobial Properties of Mg-Doped CuO Nanostructures. *Mater. Res. Express* **2019**, *6*, 085022. [CrossRef]
14. Scimeca, M.; Bischetti, S.; Lamsira, H.K.; Bonfiglio, R.; Bonanno, E. Energy Dispersive X-Ray (EDX) Microanalysis: A Powerful Tool in Biomedical Research and Diagnosis. *Eur. J. Histochem. EJH* **2018**, *62*, 2841. [CrossRef] [PubMed]
15. Pramothkumar, A.; Senthilkumar, N.; Mercy Gnana Malar, K.C.; Meena, M.; Vetha Potheher, I. A Comparative Analysis on the Dye Degradation Efficiency of Pure, Co, Ni and Mn-Doped CuO Nanoparticles. *J. Mater. Sci. Mater. Electron.* **2019**, *30*, 19043–19059. [CrossRef]
16. Singh, H.; Yadav, K.L. Structural, Dielectric, Vibrational and Magnetic Properties of Sm Doped BiFeO$_3$ Multiferroic Ceramics Prepared by a Rapid Liquid Phase Sintering Method. *Ceram. Int.* **2015**, *41*, 9285–9295. [CrossRef]
17. Singh, S.J.; Chinnamuthu, P. Highly Efficient Natural-Sunlight-Driven Photodegradation of Organic Dyes with Combustion Derived Ce-Doped CuO Nanoparticles. *Colloids Surf. Physicochem. Eng. Asp.* **2021**, *625*, 126864. [CrossRef]
18. Siddiqui, H.; Parra, M.R.; Malik, M.M.; Haque, F.Z. Structural and Optical Properties of Li Substituted CuO Nanoparticles. *Opt. Quantum Electron.* **2018**, *50*, 260. [CrossRef]
19. Siddiqui, H.; Parra, M.R.; Qureshi, M.S.; Malik, M.M.; Haque, F.Z. Studies of Structural, Optical, and Electrical Properties Associated with Defects in Sodium-Doped Copper Oxide (CuO/Na) Nanostructures. *J. Mater. Sci.* **2018**, *53*, 8826–8843. [CrossRef]
20. Jamal, M.; Billah, M.M.; Ayon, S.A. Opto-Structural and Magnetic Properties of Fluorine Doped CuO Nanoparticles: An Experimental Study. *Ceram. Int.* **2022**. [CrossRef]
21. Dagher, L.A.; Hassan, J.; Kharroubi, S.; Jaafar, H.; Kassem, I.I. Nationwide Assessment of Water Quality in Rivers across Lebanon by Quantifying Fecal Indicators Densities and Profiling Antibiotic Resistance of *Escherichia coli*. *Antibiotics* **2021**, *10*, 883. [CrossRef] [PubMed]
22. Singh, S.J.; Lim, Y.Y.; Hmar, J.J.L.; Chinnamuthu, P. Temperature Dependency on Ce-Doped CuO Nanoparticles: A Comparative Study via XRD Line Broadening Analysis. *Appl. Phys. Mater. Sci. Process.* **2022**, *128*, 188. [CrossRef]
23. Gupta, D.; Meher, S.R.; Illyaskutty, N.; Alex, Z.C. Facile Synthesis of Cu$_2$O and CuO Nanoparticles and Study of Their Structural, Optical and Electronic Properties. *J. Alloys Compd.* **2018**, *743*, 737–745. [CrossRef]
24. Kumar, P.; Nene, A.G.; Punia, S.; Kumar, M.; Abbas, Z.; Thakral, F.; Tuli, H.S. SYNTHESIS, CHARACTERIZATION AND ANTIBACTERIAL ACTIVITY OF CUO NANOPARTICLES. *Int. J. Appl. Pharm.* **2020**, *12*, 17–20. [CrossRef]
25. Kayani, Z.N.; Umer, M.; Riaz, S.; Naseem, S. Characterization of Copper Oxide Nanoparticles Fabricated by the Sol–Gel Method. *J. Electron. Mater.* **2015**, *44*, 3704–3709. [CrossRef]
26. Muthukumaran, S.; Gopalakrishnan, R. Structural, FTIR and Photoluminescence Studies of Cu Doped ZnO Nanopowders by Co-Precipitation Method. *Opt. Mater.* **2012**, *34*, 1946–1953. [CrossRef]
27. Wang, H.; Xu, J.-Z.; Zhu, J.-J.; Chen, H.-Y. Preparation of CuO Nanoparticles by Microwave Irradiation. *J. Cryst. Growth* **2002**, *244*, 88–94. [CrossRef]
28. Chandrasekar, M.; Subash, M.; Logambal, S.; Udhayakumar, G.; Uthrakumar, R.; Inmozhi, C.; Al-Onazi, W.A.; Al-Mohaimeed, A.M.; Chen, T.-W.; Kanimozhi, K. Synthesis and Characterization Studies of Pure and Ni Doped CuO Nanoparticles by Hydrothermal Method. *J. King Saud Univ. Sci.* **2022**, *34*, 101831. [CrossRef]
29. Siddiqui, H.; Parra, M.R.; Haque, F.Z. Optimization of Process Parameters and Its Effect on Structure and Morphology of CuO Nanoparticle Synthesized via the Sol–gel Technique. *J. Sol-Gel Sci. Technol.* **2018**, *87*, 125–135. [CrossRef]
30. Abdallah, A.M.; Awad, R. Mixed Magnetic Behavior in Gadolinium and Ruthenium Co-Doped Nickel Oxide Nanoparticles. *Phys. Scr.* **2022**, *97*, 015802. [CrossRef]
31. Jisr, N.; Younes, G.; El Omari, K.; Hamze, M.; Sukhn, C.; El-Dakdouki, M.H. Levels of Heavy Metals, Total Petroleum Hydrocarbons, and Microbial Load in Commercially Valuable Fish from the Marine Area of Tripoli, Lebanon. *Environ. Monit. Assess.* **2020**, *192*, 705. [CrossRef]
32. Romanos, D.; Nemer, N.; Khairallah, Y.; Thérèse Abi Saab, M. Assessing the Quality of Sewage Sludge as an Agricultural Soil Amendment in Mediterranean Habitats. *Int. J. Recycl. Org. Waste Agric.* **2019**, *8*, 377–383. [CrossRef]
33. Arenz, S.; Wilson, D.N. Bacterial Protein Synthesis as a Target for Antibiotic Inhibition. *Cold Spring Harb. Perspect. Med.* **2016**, *6*, a025361. [CrossRef] [PubMed]
34. Speer, B.S.; Shoemaker, N.B.; Salyers, A.A. Bacterial Resistance to Tetracycline: Mechanisms, Transfer, and Clinical Significance. *Clin. Microbiol. Rev.* **1992**, *5*, 387–399. [CrossRef] [PubMed]

35. Yoon, B.K.; Jackman, J.A.; Valle-González, E.R.; Cho, N.-J. Antibacterial Free Fatty Acids and Monoglycerides: Biological Activities, Experimental Testing, and Therapeutic Applications. *Int. J. Mol. Sci.* **2018**, *19*, 1114. [CrossRef] [PubMed]
36. EL-Mekkawi, D.M.; Selim, M.M.; Hamdi, N.; Hassan, S.A.; Ezzat, A. Studies on the Influence of the Physicochemical Characteristics of Nanostructured Copper, Zinc and Magnesium Oxides on Their Antibacterial Activities. *J. Environ. Chem. Eng.* **2018**, *6*, 5608–5615. [CrossRef]
37. Mersian, H.; Alizadeh, M.; Hadi, N. Synthesis of Zirconium Doped Copper Oxide (CuO) Nanoparticles by the Pechini Route and Investigation of Their Structural and Antibacterial Properties. *Ceram. Int.* **2018**, *44*, 20399–20408. [CrossRef]
38. Moniri Javadhesari, S.; Alipour, S.; Mohammadnejad, S.; Akbarpour, M.R. Antibacterial Activity of Ultra-Small Copper Oxide (II) Nanoparticles Synthesized by Mechanochemical Processing against *S. aureus* and *E. coli*. *Mater. Sci. Eng. C* **2019**, *105*, 110011. [CrossRef]
39. Rutgersson, C.; Ebmeyer, S.; Lassen, S.B.; Karkman, A.; Fick, J.; Kristiansson, E.; Brandt, K.K.; Flach, C.-F.; Larsson, D.G.J. Long-Term Application of Swedish Sewage Sludge on Farmland Does Not Cause Clear Changes in the Soil Bacterial Resistome. *Environ. Int.* **2020**, *137*, 105339. [CrossRef]
40. Thakur, N.; Anu; Kumar, K.; Kumar, A. Effect of (Ag, Zn) Co-Doping on Structural, Optical and Bactericidal Properties of CuO Nanoparticles Synthesized by a Microwave-Assisted Method. *Dalton Trans.* **2021**, *50*, 6188–6203. [CrossRef]
41. Dang, T.M.D.; Le, T.T.T.; Fribourg-Blanc, E.; Dang, M.C. Synthesis and Optical Properties of Copper Nanoparticles Prepared by a Chemical Reduction Method. *Adv. Nat. Sci. Nanosci. Nanotechnol.* **2011**, *2*, 015009. [CrossRef]
42. Maruthapandi, M.; Saravanan, A.; Luong, J.H.T.; Gedanken, A. Antimicrobial Properties of Polyaniline and Polypyrrole Decorated with Zinc-Doped Copper Oxide Microparticles. *Polymers* **2020**, *12*, 1286. [CrossRef] [PubMed]
43. Harikumar, P.S. Antibacterial Activity of Copper Nanoparticles and Copper Nanocomposites against *Escherichia coli* Bacteria. *Int. J. Sci.* **2016**, *2*, 83–90. [CrossRef]
44. Ren, E.; Zhang, C.; Li, D.; Pang, X.; Liu, G. Leveraging Metal Oxide Nanoparticles for Bacteria Tracing and Eradicating. *VIEW* **2020**, *1*, 20200052. [CrossRef]
45. Kumar, V.; Prakash, J.; Singh, J.P.; Chae, K.H.; Swart, C.; Ntwaeaborwa, O.M.; Swart, H.C.; Dutta, V. Role of Silver Doping on the Defects Related Photoluminescence and Antibacterial Behaviour of Zinc Oxide Nanoparticles. *Colloids Surf. B Biointerfaces* **2017**, *159*, 191–199. [CrossRef] [PubMed]
46. Prakash, J.; Samriti; Kumar, A.; Dai, H.; Janegitz, B.C.; Krishnan, V.; Swart, H.C.; Sun, S. Novel Rare Earth Metal–Doped One-Dimensional TiO_2 Nanostructures: Fundamentals and Multifunctional Applications. *Mater. Today Sustain.* **2021**, *13*, 100066. [CrossRef]
47. Iseppi, R.; Tardugno, R.; Brighenti, V.; Benvenuti, S.; Sabia, C.; Pellati, F.; Messi, P. Phytochemical Composition and In Vitro Antimicrobial Activity of Essential Oils from the *Lamiaceae* Family against *Streptococcus agalactiae* and *Candida albicans* Biofilms. *Antibiotics* **2020**, *9*, 592. [CrossRef]
48. Bomila, R.; Srinivasan, S.; Venkatesan, A.; Bharath, B.; Perinbam, K. Structural, Optical and Antibacterial Activity Studies of Ce-Doped ZnO Nanoparticles Prepared by Wet-Chemical Method. *Mater. Res. Innov.* **2018**, *22*, 379–386. [CrossRef]
49. Vasireddy, L.; Bingle, L.E.H.; Davies, M.S. Antimicrobial Activity of Essential Oils against Multidrug-Resistant Clinical Isolates of the *Burkholderia cepacia* Complex. *PLoS ONE* **2018**, *13*, e0201835. [CrossRef]

Disclaimer/Publisher's Note: The statements, opinions and data contained in all publications are solely those of the individual author(s) and contributor(s) and not of MDPI and/or the editor(s). MDPI and/or the editor(s) disclaim responsibility for any injury to people or property resulting from any ideas, methods, instructions or products referred to in the content.

Article

Biosynthesis of Gold Nanoparticles and Its Effect against *Pseudomonas aeruginosa*

Syed Ghazanfar Ali [1], Mohammad Jalal [1], Hilal Ahmad [2], Khalid Umar [3,*], Akil Ahmad [4,*], Mohammed B. Alshammari [4] and Haris Manzoor Khan [1]

[1] Department of Microbiology, Jawaharlal Nehru Medical College, Aligarh Muslim University, Aligarh 202002, India
[2] SRM Institute of Science and Technology, Kattankulathur, Chennai 603203, India
[3] School of Chemical Sciences, Universiti Sains Malaysia, Pulau Pinang 11800, Malaysia
[4] Department of Chemistry, College of Science and Humanities in Al-Kharj, Prince Sattam bin Abdulaziz University, Al-Kharj 11942, Saudi Arabia
* Correspondence: khalidumar4@gmail.com (K.U.); aj.ahmad@psau.edu.sa (A.A.)

Abstract: Antimicrobial resistance has posed a serious health concern worldwide, which is mainly due to the excessive use of antibiotics. In this study, gold nanoparticles synthesized from the plant *Tinospora cordifolia* were used against multidrug-resistant *Pseudomonas aeruginosa*. The active components involved in the reduction and stabilization of gold nanoparticles were revealed by gas chromatography–mass spectrophotometry(GC-MS) of the stem extract of *Tinospora cordifolia*. Gold nanoparticles (TG-AuNPs) were effective against *P. aeruginosa* at different concentrations (50,100, and 150 µg/mL). TG-AuNPs effectively reduced the pyocyanin level by 63.1% in PAO1 and by 68.7% in clinical isolates at 150 µg/mL; similarly, swarming and swimming motilities decreased by 53.1% and 53.8% for PAO1 and 66.6% and 52.8% in clinical isolates, respectively. Biofilm production was also reduced, and at a maximum concentration of 150 µg/mL of TG-AuNPs a 59.09% reduction inPAO1 and 64.7% reduction in clinical isolates were observed. Lower concentrations of TG-AuNPs (100 and 50 µg/mL) also reduced the pyocyanin, biofilm, swarming, and swimming. Phenotypically, the downregulation of exopolysaccharide secretion from *P. aeruginosa* due to TG-AuNPs was observed on Congo red agar plates

Keywords: biofilm; GC-MS; gold nanoparticles; *Pseudomonas aeruginosa*; pyocyanin

1. Introduction

The emergence of antimicrobial resistance has become a serious health concern worldwide since the multidrug resistance in microorganisms has increased the morbidity and mortality rates worldwide [1–4]. The major problem with antibiotic therapy is that microorganisms develop resistance against antibiotics within a short span both in hospital- as well as in community-acquired infections. [5]. The resistance developed in microorganisms against the antibiotics poses a serious health challenge to treat infectious diseases, resulting in increased mortality [6]. Moreover, it is challenging to develop new antimicrobials or alternative therapeutics within a short span of time to treat pathogens [7–9]. *Pseudomonas aeruginosa* is one such kind of pathogen which develops resistance to antimicrobials and has been included in the list of ESKAPE pathogens, i.e., those pathogens which even surpass antibiotic treatment, therefore listed as critical priority pathogens [10,11]. A statistical survey report of 2019 from the United States Center for Disease Control and Prevention (CDC) states 32,600 cases and 2700 deaths from multidrug-resistant *P. aeruginosa* [12].

P. aeruginosa spreads its pathogenesis through different virulence factors such as pyocyanin, biofilm formation, and motility (pili and flagella).These virulence factors are responsible for attachment, colonization, and invasion into the host tissue, resulting in life threatening infection [13]. Pyocyanin cytotoxicity has already been reported, which involves pro-inflammation and free radical production, which cause cellular damage and necrosis [14–16]. The motility helps the microorganism to strive better in harsh environmental conditions, and it is an important virulence factor since it is necessary for proliferation, colonization, and infection [17]. Swarming and swimming in *P. aeruginosa* are different types of motilities [18]. Another virulence factor associated with *P. aeruginosa* is biofilm. A report from the United States National Institute of Health states that 80% of microbial infections are caused due to biofilm in the human body [19]. Biofilm can be formed on the respiratory system, reproductive organs, medical devices, etc. [20,21]. Exopolysaccharide (EPS) plays a crucial role in the development of biofilm; the EPS production allows irreversible attachment of *P. aeruginosa* on the surface, and it also allows social interaction, enhances gene transfer, and provides protection against antimicrobials [22]. Biofilm provides protection to microorganisms from harsh external environment, making them resistant. The main function of biofilm is to protect the microorganisms present within it from the harsh external environment and make them resistant [23].

Nanotechnology is an emerging field that is currently not only confined to physics or chemistry but has shown its promising applications in the field of medicine, specifically against microbial resistance. Nanomaterials are small-sized particles that have a large surface area to volume ratio. Due to the large surface area to volume ratio, metal nanoparticles possess unique properties, some of which are of human interest, viz., treatment against bacterial infection [24]; some other biomedical applications include diagnostics, photothermal therapy, and electrical and optical sensing [25]. Gold nanoparticles are less toxic in nature and possess good compatibility with human cells in addition to being antimicrobial in nature [26]. Anticancer properties of gold nanoparticles have also been reported [27]. Enzymes such as acetylcholinesterase and butyrlcholinesterase when released in excess block the function of acetylcholine, which results in dementia. Some studies have claimed that gold nanoparticles downregulate the enzymes acetylcholinesterase and butyrlcholinesterase [28,29]. There are different methods of synthesis of nanoparticles, but the green method is preferred over chemical methods since chemical processes use harmful chemicals for reduction as well as for stabilization; moreover, the chemicals used pose a serious threat to the environment [30]. On the other hand, green synthesis, which uses green plants or parts of the plants, is an eco-friendly synthesis that does not use any chemicals [31]. The phytoconstituents from the plants act as reducing and stabilizing agents. Moreover, the use of plants does not pose any serious challenge, since their availability is abundant without any harmful effects.

Tinospora cordifolia (Willd.) Miers is a medicinal plant. The plant has been used traditionally for the treatment of fever, jaundice, chronic diarrhea, cancer, etc. [32]. The stem of *T. cordifolia* has antidiabetic effects, since it regulates the blood glucose level in the body [33]. The extract from the roots of *T. cordifolia* possesses the ability to scavenge free radicals which are generated during aflatoxicosis [34].

Green synthesized nanoparticles (silver, zinc, etc.) from different plants possessing antibacterial, antivirulence, and antibiofilm potential have been well documented [35–39]. The synergistic effect of metal/metal oxide nanoparticles showing antibiosis has also been reported [40].

In view of the beneficial role of plants and medicinal properties of the stem of *Tinospora cordifolia*, we synthesized gold nanoparticles from the stem of *Tinospora cordifolia* plant [41] and further checked for antimicrobial activity and antivirulence against *P. aeruginosa*.

2. Result

The formation of gold nanoparticles from the stem extract of *Tinopora cordifolia* is represented by the equation below. The formation of gold nanoparticles in detailed view is shown as a flowchart diagram and attached as Supplementary Figure S1.

$$\textit{TinosporaCordifolia} \text{ stem Extract} + 1 \text{ mM AuCl}_3 \xrightarrow{\text{After 24 h}} \text{Gold Nanoparticles} \quad (1)$$

2.1. SEM, TEM, and XRD Analyses

The TG-AuNPs as analyzed by SEM were poly dispersed and were of varying shape, but the majority of particles seemed to be spherical, whereas TEM analysis indicated the average particle size to be 16.25 nm (Figures 1 and 2).

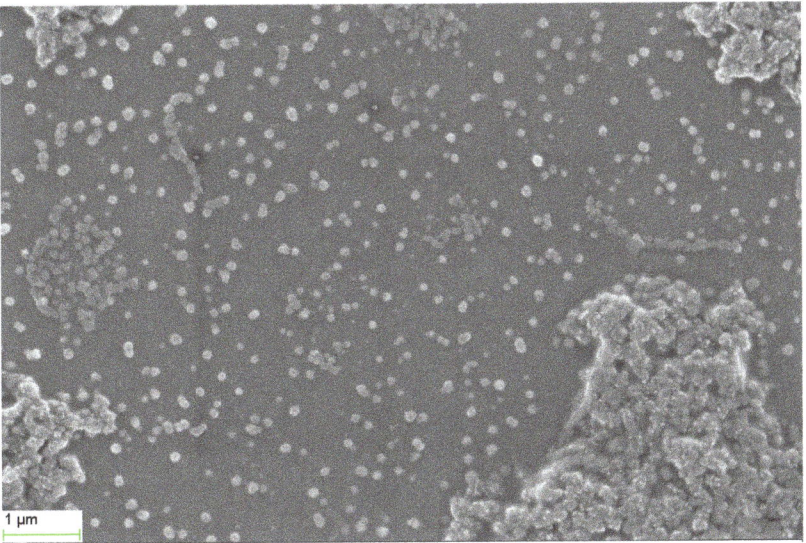

Figure 1. SEM image of TG-AuNPs.

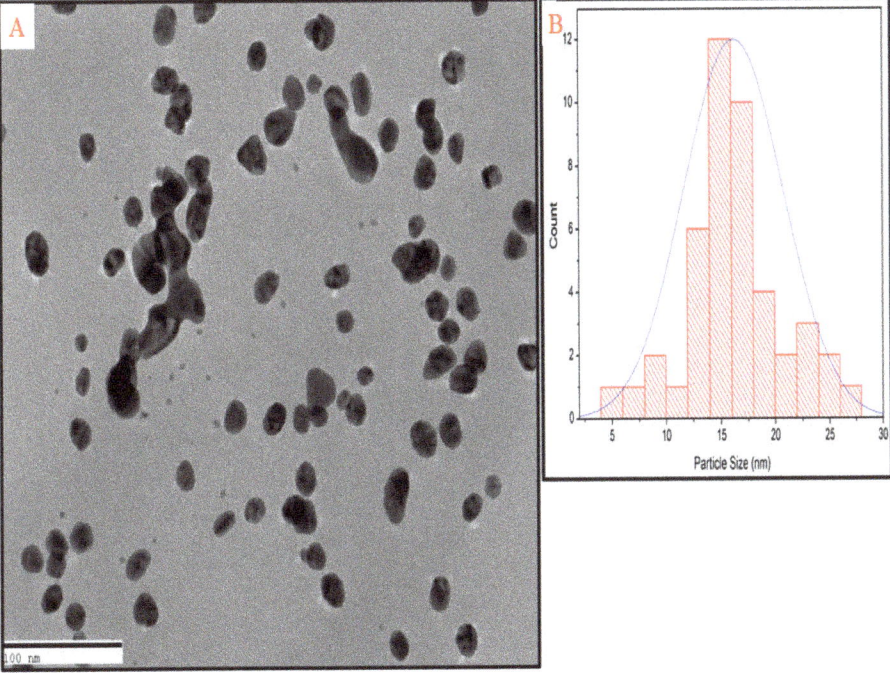

Figure 2. (**A**) TEM image of TG-AuNPs; (**B**) Particle size distribution of TG-AuNPs.

XRD analysis confirmed the crystalline nature of the gold nanoparticles. The respective diffraction peaks at 38.2°, 44.5°, 64.74°, and 77.6°, relating to (111), (200), (220), and (311) facets of the face-centered cubic (FCC) crystal lattice, correspond to pure gold (Figure 3) (JCPDS card no. 04-0784).

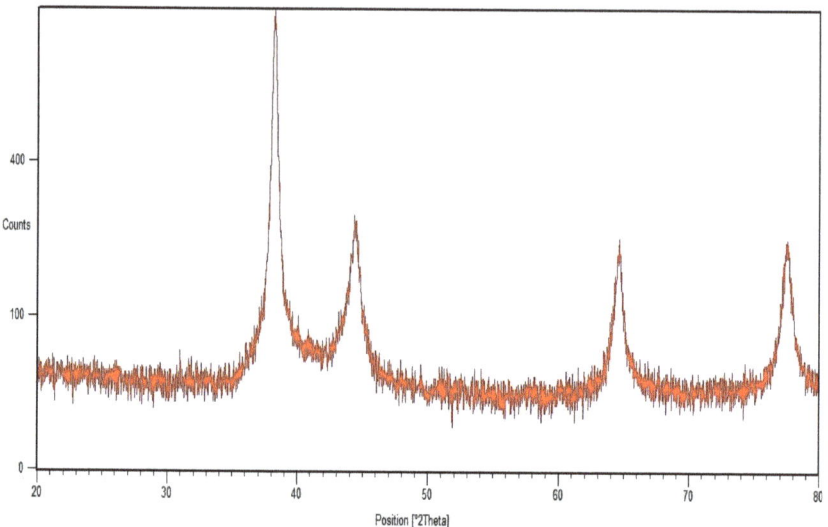

Figure 3. XRD of TG-AuNPs.

2.2. GC-MS of Tinospora Cordifolia Stem Extract

The GC-MS of the methanolic stem extract of *Tinospora cordifolia* revealed 7-Tetradecanal (12.95%), n-Hexadecanoic acid (11.32%), 9–12Octadecadienoic acid (10.39%), Benzene (5.97%), Pregna-5,16-dien-20-one,3-(acetyloxy)-16-methyle (3.85%), and Octadecanoic acid (3.40%) as the major components. The detailed analysis of GC-MS along with other compounds is shown in Table 1. The chromatogram reflecting different peaks obtained in the GC-MS analysis is shown in Figure 4.

Table 1. Major components of GC-MS analysis of *Tinospora cordifolia* stem extract.

Peak	R. Time	Area	Area%	Name
1	23.767	27437724	12.95	7-Tetradecenal, (Z)-
2	21.652	23992314	11.32	n-Hexadecanoic acid, methyl ester
3	23.692	22020115	10.39	9,12-octadecadienoic acid (Z,Z)-
4	32.034	12640436	5.97	BENZENE, (2-ETHYL-4-METHYLE-1,3-PENTADIENYL)-
5	31.137	8164505	3.85	Pregna-5,16-dien-20-one, 3-(acetyloxy)-16-methyle-, (3.beta.)
6	24.007	7197068	3.40	Octadecanoic acid
7	33.871	6441213	3.04	Octacosanol
8	38.278	6350228	3.00	.gamma.-Sitosterol
9	37.853	5967832	2.82	1,4-METHANOAZULENE, DECAHYDRO-4,8,8-TRIMET
10	17.570	5954103	2.81	Inositol, 1-deoxy-

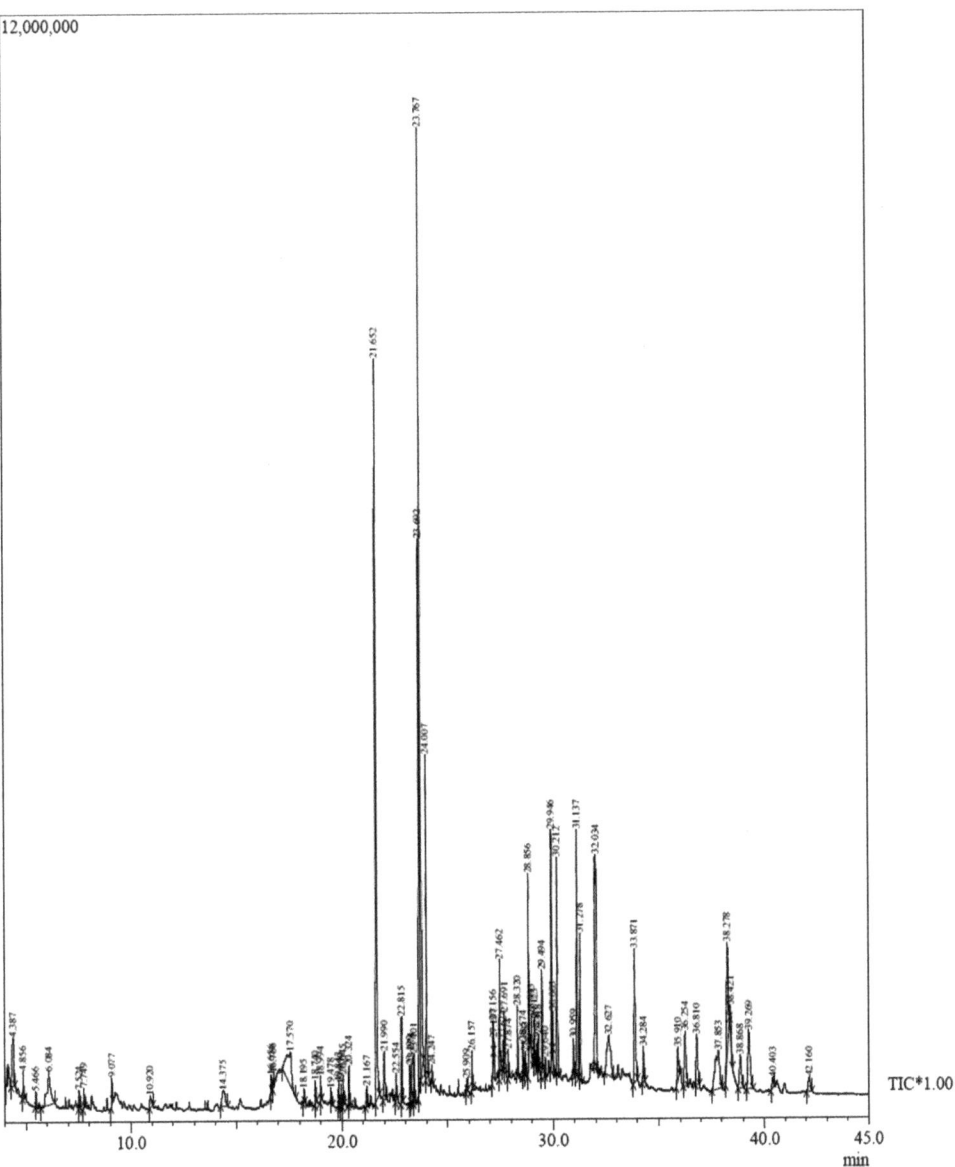

Figure 4. Representative GC-MS chromatogram of stem extract of *Tinospora cordifolia*.

2.3. Antibiotic Profile

P. aeruginosa (n = 10) were resistant to different antibiotics, and the details of antibiotics are the following: tobramycin (Tob, 10 µg,), piperacillin (Pi, 100 µg), nitrofurantoin (Nit, 300 µg), piperacillin-tazobactam (Pit, 100/10 µg), cefepime (Cpm, 30 µg),imipenem (Ipm, 10 µg), amikacin (Ak, 30 µg),ceftazidime (Caz, 30 µg),levofloxacin (Le, 5 µg), and sparfloxacin (Spx, 5 µg)

2.4. MIC of TG-AuNPs

The MIC of PAO1 was found to be 1000 µg/mL, whereas for all 10 clinical isolates the MICs varied:20% of the isolates showed an MIC of 1000 µg/mL, 50% of isolates showed an MIC of 1500 µg/mL; and 30% of isolates showed an MIC of 1800 µg/mL (Table 2). Three different concentrations, viz., 150,100, and 50 µg/mL, were considered for further antivirulence approaches.

Table 2. MIC of PAO1 and clinical isolates of *P. aeruginosa*.

Standard (N = 1)		Clinical Isolate (N = 10)	
Isolate	MIC (µgmL^{-1})	Isolates	MIC (µgmL^{-1})
PAO1	1000	20%	1000
		50%	1500
		30%	1800

2.4.1. Effect of TG-AuNPs on Pyocyanin of *P. aeruginosa*

Gold nanoparticles (TG-AuNPs) effectively downregulated the virulence of *P. aeruginosa*. In PAO1, a 63.1% reduction in the level of pyocyanin was observed at 150 µg/mL, whereas a similar concentration (150 µg/mL) of TG-AuNPs decreased the level of pyocyanin from 57.1% to 68.7% in clinical isolates. The lower concentration of 100 µg/mL caused a 43.9% reduction and 41.6% to 55.3% reduction in the level of pyocyanin for PAO1 and clinical isolates, respectively. The lowest concentration, i.e., 50 µg/mL of TG-AuNPs, caused 23.5% and 41.7% to 28.3% reductions in pyocyanin level for PAO1 and clinical isolates, respectively(Figures 5A and 6).

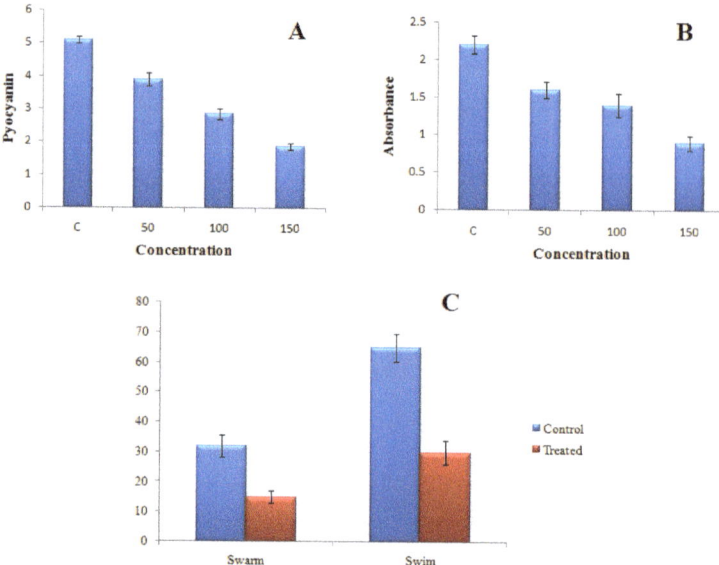

Figure 5. Representative of treated and untreated culture of PAO1 with TG-AuNPs: (**A**) pyocyanin; (**B**) biofilm; (**C**) motility (swarm and swim). For pyocyanin and biofilm 50, 100, and 150 µg/mL concentrations of TG-AuNPs were considered, whereas for motility only a 150 µg/mL of concentration of TG-AuNP was considered. Pyocyanin expressed as µg/mL. Absorbance measured at 595 nm. Swarm and Swim expressed as zone size in mm.

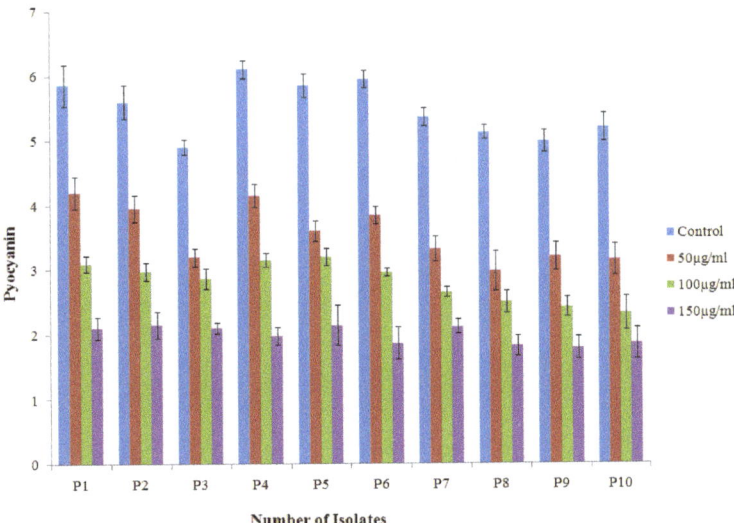

Figure 6. Bar graphs representative of level of pyocyanin after treatment of clinical isolates of *P. aeruginosa* with TG-AuNPs at 50, 100 and 150 µg/mL, along with control (untreated). Pyocyanin expressed as µg/mL.

2.4.2. Effect of TG-AuNPs on Swarming and Swimming Motilities

The swarming and swimming motilities were also affected by the TG-AuNPs. Swarming and swimming motilities of PAO1 were reduced by 53.1% and 53.8% in the case of TG-AuNPs at 150 µg/mL (Figure 5C). Similar observations were also recorded for the clinical isolates. The reductions from 50% to 66.6% in swarming and 41.5 to 52.8% in swimming were observed at 150 µg/mL (Figures 7–9).

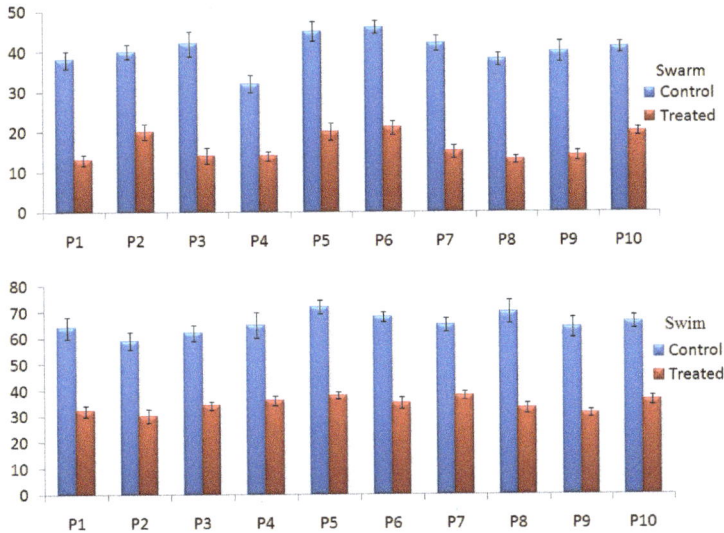

Figure 7. Bar graphs representative of swarm and swim after treatment of clinical isolates of *P. aeruginosa* with TG-AuNPsat150 µg/mL, along with control (untreated). Swarm and Swim expressed as zone size in mm.

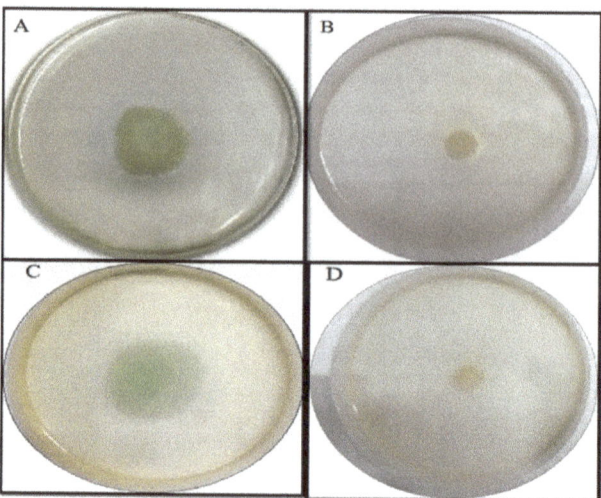

Figure 8. Representative of swarming of *P. aeruginosa*. (**A**) Swarm of PAO1. (**B**) Swarm of PAO1 after treatment with 150 µg/mL of TG-AuNPs. (**C**) Swarm of clinical isolate of *P. aeruginosa*. (**D**) Swarm of clinical isolate of *P. aeruginosa* after treatment with 150 µg/mL of TG-AuNPs.

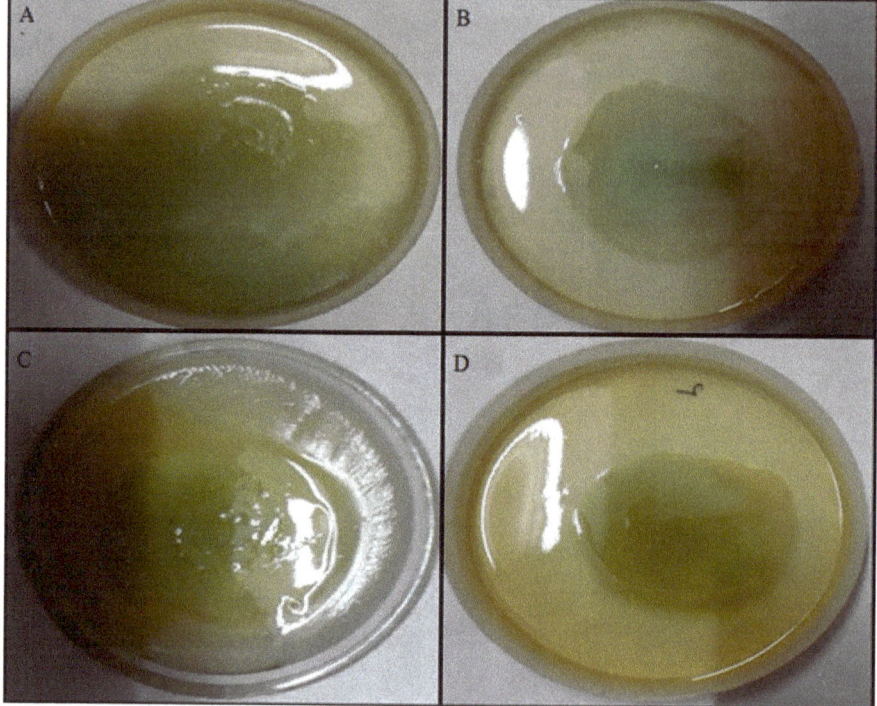

Figure 9. Representative of swimming of *P. aeruginosa*. (**A**) Swim of PAO1. (**B**) Swim of PAO1 after treatment with 150 µg/mL of TG-AuNPs. (**C**) Swim of clinical isolate of *P. aeruginosa*. (**D**) Swim of clinical isolate of *P. aeruginosa* after treatment with 150 µg/mL of TG-AuNPs.

2.4.3. Effect of TG-AuNPs on the Biofilm by Crystal Violet Assay

Biofilm formation was also reduced at all three concentrations for PAO1, as well as for clinical isolates of *P. aeruginosa*. In PAO1, a 59.09% reduction in biofilm was observed at 150 μg/mL of TG-AuNPs, whereas a 49.1% to 64.7% reduction in biofilm formation was observed for clinical isolates of *P. aeruginosa* (Fig 5B). A lower concentration, i.e., 100 μg/mL, caused 36.3% and 29.9% to 47.1% reductions in biofilm formation forPAO1 and clinical isolates, respectively. Further, the lowest concentration, i.e., 50 μg/mL, effectively reduced the biofilm by 27.2% and 14.6% to 35.1% in PAO1 and clinical isolates, respectively (Figure 10).

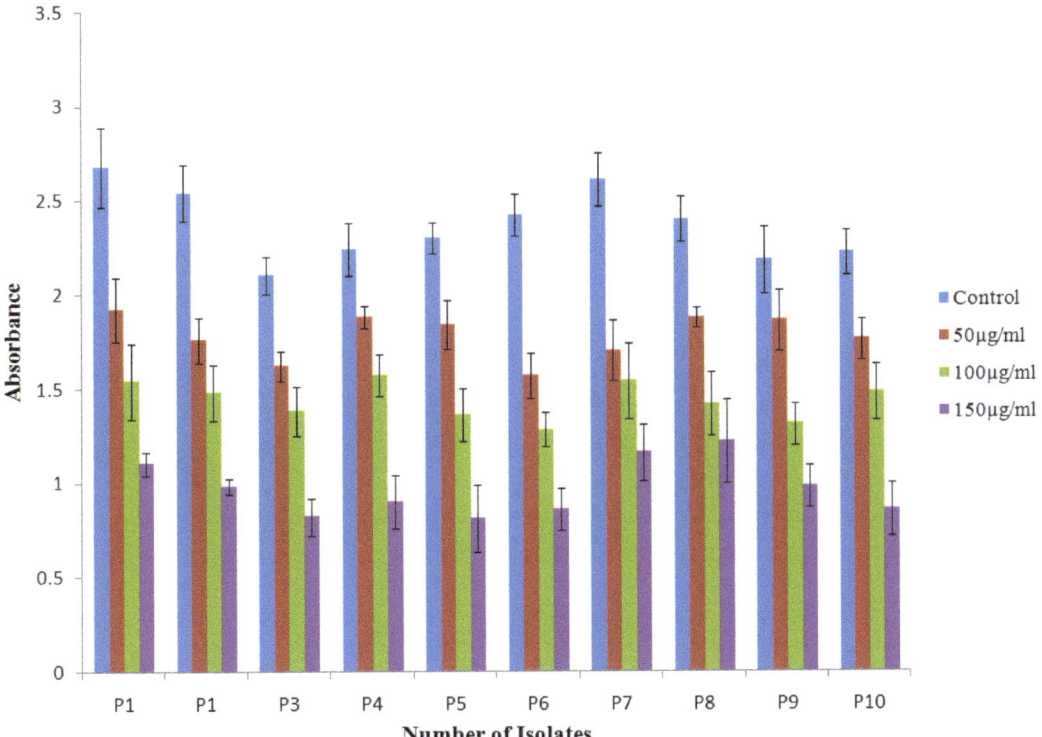

Figure 10. Bar graphs representative of biofilm after treatment of clinical isolates of *P. aeruginosa* with TG-AuNPs at 50, 100, and 150 μg/mL, along with control (untreated). Absorbance measured at 595 nm.

2.4.4. Effect of TG-AuNPs Using Congo Red Agar (CRA) Method

TG-AuNPs at 150 μg/mL effectively reduced the exopolysaccharide production, which can be observed by the loss of black consistencies in colonies on Congo red agar plates amended with TG-AuNPs. The loss of black consistencies in PAO1 and clinical isolates of *P. aeruginosa* can be clearly seen when compared with the control (plates without TG-AuNPs) (Figure 11).

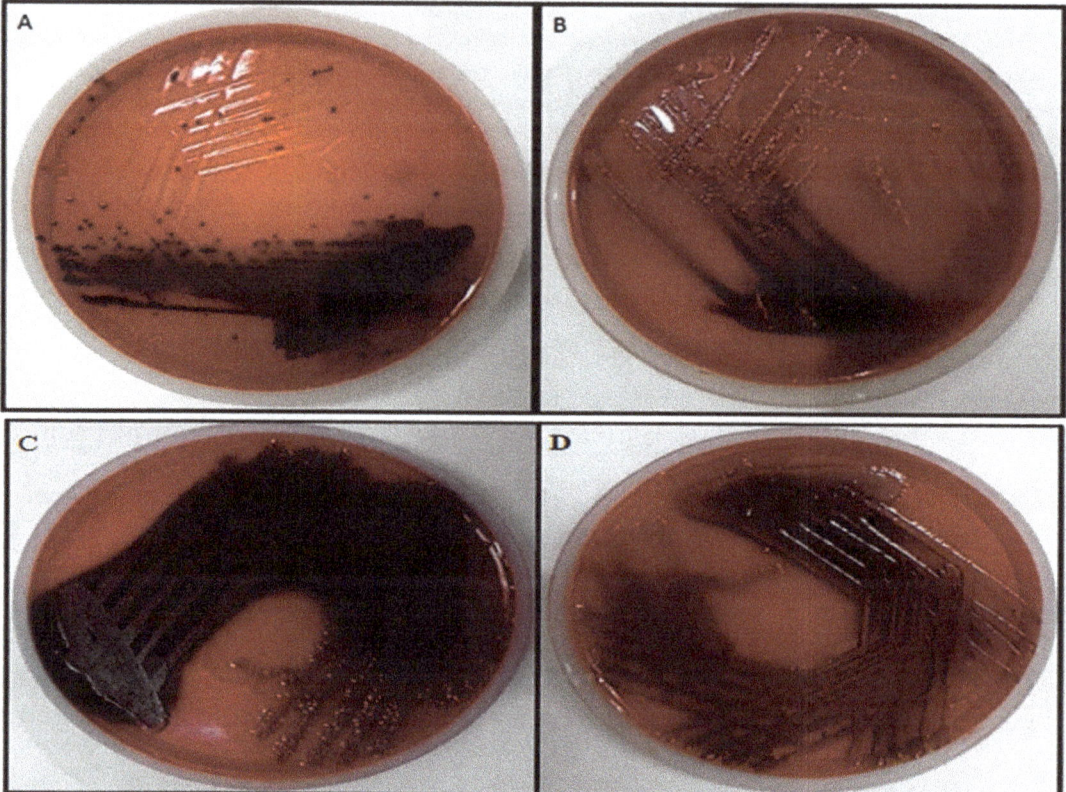

Figure 11. Representative of biofilm of *P. aeruginosa* on Congo red agar. Black coloration represents production of exopolysaccharide. (**A**) Biofilm of PAO1. (**B**) Biofilm of PAO1 after treatment with 150 µg/mL of TG-AuNPs. (**C**) Biofilm of clinical isolate of *P. aeruginosa*. (**D**) Biofilm of clinical isolate of *P. aeruginosa* after treatment with 150 µg/mL of TG-AuNPs.

3. Discussion

The SEM analysis revealed that particles were polydispersed and not agglomerated. Since we can observe the surface morphology of nanoparticles through SEM, in order to better understand the size of nanoparticles TEM was performed, and it revealed the average particle size to be 16.25 nm. The histogram in Figure 2B represents the particle size distribution, which shows the varying size of nanoparticles. The methanolic stem extract of *Tinospora cordifolia* further revealed the presence of 7-Tetradecanal (12.95%), followed by n –Hexadecanoic acid (11.32%), 9,12-octadecadienoic acid (Z,Z) (10.39%), Benzene (5.97%), and Pregna-5,16-dien-20-one (3.85%). Some of the major components are shown in Table 1. We are of the opinion that 7 Tetradecanal and n–Hexadecanoic acid could be the major components responsible for the reduction inprecursor salt and stabilization of nanoparticles, although other components could also be responsible for the reduction and stabilization. Phytochemicals present in the plants reduce the metal ions, and the reduced metal ions are linked using atmospheric oxygen or from degrading phytochemicals. The phytochemicals also prevent the agglomeration of metal nanoparticles [42,43].

In our study, three different concentrations of TG-AuNPs (50, 100 and 150 µg/mL) were considered, which were lower than the MIC for PAO1 as well as for multidrug-resistant clinical isolates.

Pyocyanin, a major component involved in the pathogenesis of *P. aeruginosa*, allows the *P. aeruginosa* to coordinate and respond according to the change in environmental conditions [44]. In our study, the pyocyanin level was decreased for both PAO1 and multidrug-resistant clinical isolates. The maximum reductions of 63.1% for PAO1 and 57.1–68.7% for clinical isolates of *P. aeruginosa* for pyocyanin were observed at 150 μg/mL of TG-AuNPs. Lower concentrations, i.e., 100 and 50 μg/mL of TG-AuNPs, also caused reductions in the level of pyocyanin. Our results are in agreement with the previous studies, where 40–88% and 20–82% reductions were observed for the pyocyanin level at $1/2$ and $1/4$ MIC of gold nanoparticles [45].

Swarming is a movement of bacteria (motility) that helps in colonization on the surface and helps in biofilm formation [46]. In addition to representing motility, the differentiation of swarm cells results in the alteration of metabolic bias and gene expression, indicating complex lifestyle adaptation [47,48]. When motility is regarding an aqueous solution, it is called swimming. The decrease in swarming and swimming motilities were also observed at 150 μg/mL. The decrease in the swarming and swimming motilities of *P. aeruginosa* both in PAO1 and clinical isolates are clearly observed in Figures 8 and 9. In the plates without TG-AuNPs, more movement was observed in both swarm and swim, but in plates with TG-AuNPs restricted movement was seen. Swarming and swimming motility decreased by 53.1% and 53.8% for PAO1, whereas 50–66.6% and 41.5–52.8% reductions in swarming and swimming motility were observed for clinical isolates, respectively. At lower concentrations of 100 and 50 μg/mL of TG-AuNPs, zones of swarm and swim were not easy to measure, since they were equivalent to the control (untreated); therefore, we included only the 150 μg/mL concentration. Our results are supported by previous studies, where a complete reduction in swimming and approximately 30% and 50% reductions in swarming at 32 and 256 μg/mL of TG-AuNPs were observed [49]

One of the most important aspects of pathogenesis in *P. aeruginosa* is the formation of biofilm, through which the bacteria avoid the host immune response [50,51]. Biofilm is the aggregation of microbial communities and the site for the spread of infection. Further, the exopolysaccharide secretion forms the mask and does not allow the antimicrobial to penetrate [52].

Biofilm formation of PAO1 reduced by 59.09%, whereas a 49.1% to 64.7% reduction was observed for clinical isolates of *P aeruginosa* at 150 μg/mL. Lower concentrations of 100 and 50 μg/mL also caused a reduction in biofilm, both in PAO1 and clinical isolates. Our results are also in agreement with the previous studies of Elshaer and shaaban [45], where they have shown the downregulation of biofilm formation by 26–68% and 21–37% at $1/2$ and $1/4$ MIC levels of gold nanoparticles. The loss of black consistency on the Congo red agar plate is the benchmark showing the decrease in EPS production. Our results showed the decrease in black consistency on Congo red agar plates amended with 150 μg/mL of TG-AuNPs both for PAO1 and for clinical isolates of *P. aeruginosa*, which is an indication of the loss of exopolysaccharide secretion (Figure 11). Our results are also in agreement with the previous studies, where baicalein fabricated nanoparticles reduced the exopolysaccharide secretion on Congo red agar plates [53]. Similar results showing the loss of exopolysaccharide production have been shown by Qais et al. [54].

4. Materials and Methods

All chemicals used are of 'AR' grade

4.1. Materials Used with Specification

- Stem of *Tinospora cordifolia*—for obtaining extract.
- Gold chloride ($AuCl_3$), Sigma Aldrich (Germany)—salt for preparing gold nanoparticles.
- Methanol, Merck (Germany)—solvent used for extraction during GC-MS.
- Nutrient broth, Hi media (India)—liquid media for growth of bacteria.
- Nutrient agar, Hi media (India)—solid media for growth of bacteria.
- Chloroform, Merck (Germany)—used in pyocyanin extraction.

- Hydrochloric acid (HCl), Rankem (India)—used in pyocyanin extraction.
- Glucose, Rankem (India)—inoculated with nutrient media for swarming and swimming assay.
- Bacteriological agar, Hi media (India)—for solidifying liquid media.
- Crystal violet, Merck (Germany)—used in biofilm assay.
- Brain heart infusion, Hi media (India)—media used in Congo red biofilm assay.
- Sucrose, Rankem (India)—for analyzing biofilm using Congo red assay, since sucrose provides extra nutrients for growth of microorganisms.
- Congo red, Merck (Germany)—dye used in biofilm assay.

4.2. Synthesis of AuNPs

The gold nanoparticles were synthesized as previously described [41]. The part of the plant, i.e., stem, was collected from the nearby area of Aligarh, Uttar Pradesh, India. The stem consists of an outer husk, which was removed and sun-dried for few days until it became hard. The stem was then ground into powder form; the powder (10 gm) was then mixed with water (100 mL) and purified using filter paper. Furthermore, the centrifugation at 1200 rpm for 5 min allowed the removal of heavy biomaterials. The aqueous extract (10 mL) was mixed with 90 mL $AuCl_3$ and left for 24 h.

4.3. Characterization of Nanoparticles

4.3.1. Scanning Electron Microscopy (SEM)

The green synthesized gold nanoparticles (TG-AuNPs) were characterized using SEM (JSM 6510 LV) for analyzing morphology, as described by Ali et al. [41]. In brief, a drop of green synthesized gold nanoparticles (TG-AuNPs) was initially placed on the glass coverslip. The drop was allowed to dry on the glass coverslip at room temperature. After drying, the samples were placed under SEM and analyzed at an accelerating voltage of 15 kv and viewed on the screen attached to the SEM.

4.3.2. Transmission Electron Microscopy (TEM)

TEM was used to analyze the size of TG-AuNPs, as previously described [41]. Briefly, a drop of gold nanoparticles (TG-AuNPs) was placed on a copper grid and left at room temperature for drying. After drying, the sample was placed in the TEM. Before viewing the vacuum was created, and the sample was illuminated with electronic radiations inside the TEM. The beam of the electron transmitted in the TEM allowed the detection of the sample on screen.

4.3.3. X-ray Diffraction (XRD)

Gold nanoparticles were examined for crystalline or amorphous nature using XRD (Rigaku, Pittsburg, PA, USA) with a scanning 2 theta angle from $20°$ to $80°$ at 40 KeV.

4.4. GC-MS for Bioactive Compounds in Plant Extract

The GC-MS for bioactive compounds in plant extract was performed using a Shimadzu GC-MS-QP 2010 Plus fitted with an RTX-5 capillary column (60 m $\times$ 0.25 mm $\times$ 0.25 µm). Helium gas was used at 40.9 cm/s linear velocity. The oven temperature which was programmed at 90 °C was increased to 280 °C with a ramp rate of 10 °C/min. The total running time of GC was 50 min. The electron impact ionization method was applied with the ion source set at 230 °C. Methanol was the solvent used.

4.5. Bacterial Isolates

P. aeruginosa ($n = 10$) were isolated from the routine patient samples received in the Department of Microbiology J N Medical College & Hospital and were further identified using biochemical tests. The isolates were further tested for antibiotic sensitivity following the Clinical and Laboratory Standards institute guideline [55]. PAO1 was used as a control sample.

4.6. Determination of Minimum Inhibitory Concentration (MIC)

MIC was determined using the broth dilution method as previously described [56]. Briefly, overnight grown cultures of *P. aeruginosa* (PAO1 and clinical isolates) (2×10^6 CFU/mL) were allowed to inoculate the nutrient broth with or without different concentrations of nanoparticles and were incubated at 37 °C for 24 h.

4.6.1. Effect of TG-AuNPs on Pyocyanin

P. aeruginosa were inoculated with 5 mL nutrient broth in presence or absence of varying concentrations of TG-AuNPs at 150 rpm at 37 °C for 16 h in shaking incubator. Pyocyanin from *P. aeruginosa* treated or untreated with TG-AuNPs was extracted using 3 mL chloroform and then further re-extracted into 1 mL 0.2 NHCl until the color of the solution turned pink to deep red. Optical density at 520 nm multiplied by 17.070 determined the pyocyanin/mL of culture supernatant [57].

4.6.2. Effect of TG-AuNPs on the Swarming Motility

Swarming of *P. aeruginosa* was analyzed by the procedure described by Chelvam et al. [58]. Semi-solid agar plates were prepared using nutrient broth and glucose (0.5%) mixed with bacteriological agar (0.5%). Before the pouring of media into plates, TG-AuNPs were added to the cooled media. Plates without TG-AuNPs were considered as control. After drying the plates, *P. aeruginosa* was spot inoculated on both the plates (with or without nanoparticles) and further incubated at 37 °C for 24 h.

4.6.3. Effect of TG-AuNPs on Swimming Motility

Swimming was also checked by the procedure described by Chelvam et al. [58]. Semi-solid agar media constituting nutrient broth along with 0.25% bacteriological agar and 0.5% glucose were mixed, then autoclaved and cooled. TG-AuNPs were added before the pouring of media into the plates, and control plates were without TG-AuNPs. After drying, the spot inoculation of overnight grown *P. aeruginosa* was completed on the semi-solid agar plates including the plate without TG-AuNPs and incubated at 37 °C for 24 h.

4.7. Antibiofilm Potential of TG-AuNPs

4.7.1. Effect of TG-AuNPs Using Crystal Violet Assay

Biofilm formation of *P. aeruginosa* by crystal violet assay was evaluated as previously described [59]. Briefly, 100µL (1×10^7 CFU/mL) of mid-exponential *P. aeruginosa* culture was used to inoculate the tubes (2 mL) with or without TG-AuNPs. After inoculation, tubes were incubated at 70 rev/min for 24 h in shaking incubator. Tubes were then washed and stained with 0.1% *w/v* crystal violet for 30 min and then again washed three times, and finally filled with absolute ethanol and absorbance was recorded at 595 nm.

4.7.2. Effect of TG-AuNPs Using Congo Red Assay

Antibiofilm efficacy of TG-AuNPs was observed by the method as described [38]. Briefly, brain heart infusion broth (37 g/L), sucrose (50 g/L), and bacteriological agar (10 g/L) were mixed and autoclaved, whereas Congo red agar solution (0.8 g/L) was autoclaved separately. After autoclaving and cooling, the Congo red agar solution was mixed with the brain heart infusion solution along with the desired concentration of TG-AuNPs and poured into the plates. Control plates were not amended with TG-AuNPs. *P. aeruginosa* was streaked on the control plates as well as on the plates amended with TG-AuNPs and incubated at 37 °C for 24 h.

5. Conclusions

In this paper, the green synthesized gold nanoparticles were used to target the virulence of multidrug-resistant *P aeruginosa*. The TG-AuNPs at very low concentrations (50, 100, and 150µg/mL) were effective against the virulence factors of P. aeruginosa, viz., pyocyanin, swarming, swimming, and biofilm. The TG-AuNPs downregulated the pyocyanin production, along with the decrease in swarming and swimming motilities. The TG-AuNPs also lowered the biofilm formation, since it decreased the EPS production, which is a necessary requirement for biofilm. Finally, the GC-MS analysis of the plant extract showed the active component involved in the reduction and stabilization of TG-AuNPs. Finally, we are of the opinion that gold nanoparticles can be used as an alternative therapy at a very low concentration against multidrug-resistant microorganisms. Although the gold nanoparticles have shown their antivirulence effect at very low concentrations, extensive research on the toxicological aspect still needs to be conducted to better understand the effect of nanoparticles on different organs before they can be used inhuman applications.

Supplementary Materials: The following supporting information can be downloaded at: https://www.mdpi.com/article/10.3390/molecules27248685/s1, Figure S1: Flowchart for stepwise formation of gold nanoparticles.

Author Contributions: Conceptualization, S.G.A., M.J. and H.A.; Methodology, writing-original draft preparation, visualization, investigation, S.G.A., A.A. and H.M.K.; Writing Reviewing and Editing, M.B.A., A.A., K.U. and H.A.; Software, formal analysis, S.G.A., H.A. and K.U.; Supervision K.U. and H.M.K. All authors have read and agreed to the published version of the manuscript.

Funding: This research received no external funding.

Institutional Review Board Statement: Not applicable.

Informed Consent Statement: Not applicable.

Data Availability Statement: Not applicable.

Acknowledgments: We appreciate the support of Prince Sattam bin Abdulaziz University, Al-Kharj, Saudi Arabia.

Conflicts of Interest: The authors declare no conflict of interest.

Sample Availability: Samples of the compounds are not available from the authors.

References

1. Founou, R.C.; Founou, L.L.; Essack, S.Y. Clinical and economic impact of antibiotic resistance in developing countries: A systematic review and meta-analysis. *PLoS ONE* **2017**, *12*, e0189621. [CrossRef] [PubMed]
2. Tillotson, G.S.; Zinner, S.H. Burden of antimicrobial resistance in an era of decreasing susceptibility. *Expert Rev. Anti-Infect. Ther.* **2017**, *15*, 663–676. [CrossRef] [PubMed]
3. Abadi, A.T.B.; Rizvanov, A.A.; Haertlé, T.; Blatt, N.L. World Health Organization report: Current crisis of antibiotic resistance. *BioNanoScience* **2019**, *9*, 778–788. [CrossRef]
4. De Oliveira, D.M.; Forde, B.M.; Kidd, T.J.; Harris, P.N.; Schembri, M.A.; Beatson, S.A.; Paterson, D.L.; Walker, M.J. Antimicrobial resistance in ESKAPE pathogens. *Clin. Microbiol. Rev.* **2020**, *33*, e00181-19. [CrossRef]
5. Ventola, C.L. The antibiotic resistance crisis: Part 1: Causes and threats. *Pharm. Ther.* **2015**, *40*, 277–283.
6. Marston, H.D.; Dixon, D.M.; Knisely, J.M.; Palmore, T.N.; Fauci, A.S.J.J. Antimicrobial resistance. *JAMA* **2016**, *316*, 1193–1204. [CrossRef]
7. Mizar, P.; Arya, R.; Kim, T.; Cha, S.; Ryu, K.-S.; Yeo, W.-S.; Bae, T.; Kim, D.W.; Park, K.H.; Kim, K.K.; et al. Total Synthesis of Xanthoangelol B and Its Various Fragments: Toward Inhibition of Virulence Factor Production of Staphylococcus aureus. *J. Med. Chem.* **2018**, *61*, 10473–10487. [CrossRef]
8. Yeo, W.-S.; Arya, R.; Kim, K.K.; Jeong, H.; Cho, K.H.; Bae, T. The FDA-approved anti-cancer drugs, streptozotocin and floxuridine, reduce the virulence of *Staphylococcus aureus*. *Sci. Rep.* **2018**, *8*, 2521. [CrossRef]
9. Imdad, S.; Chaurasia, A.K.; Kim, K.K. Identification and validation of an antivirulence agent targeting HlyU-regulated virulence in Vibrio vulnificus. *Front. Cell Infect. Microbiol.* **2018**, *8*, 152. [CrossRef]
10. Rice, L.B. *Federal Funding for the Study of Antimicrobial Resistance in Nosocomial Pathogens: No ESKAPE*; The University of Chicago Press: Chicago, IL, USA, 2008; Volume 197, pp. 1079–1081.
11. Tacconelli, E.; Magrini, N.; Kahlmeter, G.; Singh, N. *Global Priority List of Antibiotic-Resistant Bacteria to Guide Research, Discovery, and Development of New Antibiotics*; World Health Organization: Geneva, Switzerland, 2017; Volume 27, pp. 318–327.
12. CDC. *Antibiotic Resistance Threats in the United States*; Department of Health and Human Services; CDC: Atlanta, GA, USA, 2019; pp. 1–150.
13. Rajkowska, K.; Otlewska, A.; Guiamet, P.S.; Wrzosek, H.; Machnowski, W. Pre-Columbian Archeological Textiles: A Source of Pseudomonas aeruginosa with Virulence Attributes. *Appl. Sci.* **2020**, *10*, 116. [CrossRef]

14. Britigan, B.E.; Roeder, T.L.; Rasmussen, G.T.; Shasby, D.M.; McCormick, M.L.; Cox, C.D. Interaction of the *Pseudomonas aeruginosa* secretory products pyocyanin and pyochelin generates hydroxyl radical and causes synergistic damage to endothelial cells. Implications for Pseudomonas-associated tissue injury. *J. Clin. Investig.* **1992**, *90*, 2187–2196. [CrossRef] [PubMed]
15. Denning, G.M.; Wollenweber, L.A.; Railsback, M.A.; Cox, C.D.; Stoll, L.L.; Britigan, B.E. Pseudomonas pyocyanin increases interleukin-8 expression by human airway epithelial cells. *Infect. Immun.* **1998**, *66*, 5777–5784. [CrossRef] [PubMed]
16. Lau, G.W.; Ran, H.; Kong, F.; Hassett, D.J.; Mavrodi, D. *Pseudomonas aeruginosa* pyocyanin is critical for lung infection in mice. *Infect. Immun.* **2004**, *72*, 4275–4278. [CrossRef] [PubMed]
17. Behzadi, P.; Baráth, Z.; Gajdács, M. It's Not Easy Being Green: A Narrative Review on the Microbiology, Virulence and Therapeutic Prospects of Multidrug-Resistant *Pseudomonas aeruginosa*. *Antibiotics* **2021**, *10*, 42. [CrossRef] [PubMed]
18. Khan, F.; Pham, D.T.N.; Oloketuyi, S.F.; Kim, Y.-M. Regulation and controlling the motility properties of Pseudomonas aeruginosa. *Appl. Microbiol. Biotechnol.* **2020**, *104*, 33–49. [CrossRef] [PubMed]
19. Khatoon, Z.; McTiernan, C.D.; Suuronen, E.J.; Mah, T.-F.; Alarcon, E.I. Bacterial Biofilm Formation on Implantable Devices and Approaches to its Treatment and Prevention. *Heliyon* **2018**, *4*, e01067. [CrossRef] [PubMed]
20. Ramage, G.; Williams, C. The Clinical Importance of Fungal Biofilms. *Adv. Appl. Microbiol.* **2013**, *84*, 27–83. [CrossRef]
21. Shakibaie, M.R. Bacterial Biofilm and its Clinical Implications. *Ann. Microbiol. Res.* **2018**, *2*, 45–50. [CrossRef]
22. Flemming, H.-C.; Wingender, J.; Szewzyk, U.; Steinberg, P.; Rice, S.A.; Kjelleberg, S. Biofilms: An emergent form of bacterial life. *Nat. Rev. Microbiol.* **2016**, *14*, 563–575. [CrossRef]
23. Yin, W.; Wang, Y.; Liu, L.; He, J. Biofilms: The Microbial "Protective Clothing" in Extreme Environments. *Int. J. Mol. Sci.* **2019**, *20*, 3423. [CrossRef]
24. Huh, A.J.; Kwon, Y.J. "Nanoantibiotics": A new paradigm for treating infectious diseases using nanomaterials in the antibiotics resistant era. *J. Control. Release* **2011**, *156*, 128–145. [CrossRef] [PubMed]
25. Bhardwaj, V.; Kaushik, A. Biomedical Applications of Nanotechnology and Nanomaterials. *Micromachines* **2017**, *8*, 298. [CrossRef] [PubMed]
26. Suriyakala, G.; Sathiyaraj, S.; Babujanarthanam, R.; Alarjani, K.M.; Hussein, D.S.; Rasheed, R.A.; Kanimozhi, K. Green Synthesis of Gold Nanoparticles Using Jatropha s Jacq. Flower Extract and Their Antibacterial Activity. *J. King Saud. Univ. Sci.* **2022**, *34*, 101830. [CrossRef]
27. Botteon, C.E.A.; Silva, L.B.; Ccana-Ccapatinta, G.V.; Silva, T.S.; Ambrosio, S.R.; Veneziani, R.C.S.; Bastos, J.K.; Marcato, P.D. Biosynthesis and characterization of gold nanoparticles using Brazilian red propolis and evaluation of its antimicrobial and anticancer activities. *Sci. Rep.* **2021**, *11*, 1974. [CrossRef]
28. Zainab; Saeed, K.; Ammara; Ahmad, S.; Ahmad, H.; Ullah, F.; Sadiq, A.; Uddin, A.; Khan, I.; Ahmad, M. Green Synthesis, Characterization and Cholinesterase Inhibitory Potential of Gold Nanoparticles. *J. Mex. Chem. Soc.* **2021**, *65*, 416–423.
29. Ahmad, S.; Zainab; Ahmad, H.; Khan, I.; Alghamdi, S.; Almehmadi, M.; Ali, M.; Ullah, A.; Hussain, H.; Khan, N.M.; et al. Green synthesis of gold nanaoparticles using Delphinium Chitralense tuber extracts, their characterization and enzyme inhibitory potential. *Braz. J. Biol.* **2022**, *82*, e257622. [CrossRef] [PubMed]
30. Tagad, C.K.; Dugasani, S.R.; Aiyer, R.; Park, S.; Kulkarni, A.; Sabharwal, S. Green Synthesis of Silver Nanoparticles and Their Application for the Development of Optical Fiber Based Hydrogen Peroxide Sensors. *Sens. Actuators B Chem.* **2013**, *183*, 144–149. [CrossRef]
31. Abdel-Halim, E.S.; El-Rafie, M.H.; Al-Deyab, S.S. Polyacrylamide/Guar Gum Graft Copolymer for Preparation of Silver Nanoparticles. *Carbohydr. Polym.* **2011**, *85*, 692–697. [CrossRef]
32. Parthipan, M.; Aravindhan, V.; Rajendran, A. Medico-botanical study of Yercaud hills in the eastern Ghats of Tamil Nadu, India. *Anc. Sci Life.* **2011**, *30*, 104–109.
33. Sangeetha, M.K.; Raghavendran, H.R.B.; Gayathri, V.; Vasanthi, H.R. *Tinospora cordifolia* attenuates oxidative stress and distorted carbohydrate metabolism in experimentally induced type 2 diabetes in rats. *J. Nat. Med.* **2011**, *65*, 544–550. [CrossRef]
34. Gupta, R.; Sharma, V. Ameliorative effects of *Tinospora cordifolia* root extract on histopathological and biochemical changes induced by aflatoxin-b (1) in mice kidney. *Toxicol. Int.* **2011**, *18*, 94–98. [PubMed]
35. Ali, S.G.; Ansari, M.A.; Khan, H.M.; Jalal, M.; Mahdi, A.A.; Cameotra, S.S. *Crataeva nurvala* nanoparticles inhibit virulence factors and biofilm formation in clinical isolates of *Pseudomonas aeruginosa*. *J. Basic Microbiol.* **2017**, *57*, 193–203. [CrossRef] [PubMed]
36. Ali, S.G.; Ansari, M.A.; Khan, H.M.; Jalal, M.; Mahdi, A.A.; Cameotra, S.S. Antibacterial and antibiofilm potential of green synthesized silver nanoparticles against imipenem resistant clinical isolates of *P. aeruginosa*. *BionanoScience* **2018**, *8*, 544–553. [CrossRef]
37. Jalal, M.; Ansari, M.A.; Alzohairy, M.A.; Ali, S.G.; Khan, H.M.; Almatroudi, A.; Siddiqui, M.I. Anticandidal activity of biosynthesized silver nanoparticles: Effect on growth, cell morphology, and key virulence attributes of Candida species. *Int. J. Nanomed.* **2019**, *14*, 4667–4679. [CrossRef] [PubMed]
38. Ali, S.G.; Ansari, M.A.; Jamal, Q.M.S.; Almatroudi, A.; Alzohairy, M.A.; Alomary, M.N.; Al-Warthan, A. *Butea Monosperma* Seed Extract Mediated Biosynthesis of ZnO NPs and Their Antibacterial, Antibiofilm and AntiQuorum Sensing Potentialities. *Arab. J. Chem.* **2021**, *14*, 103044. [CrossRef]
39. Islam, R.; Sun, L.M.; Zhang, L.B. Biomedical Applications of Chinese Herb-Synthesized Silver Nanoparticles by Phytonanotechnology. *Nanomaterials* **2021**, *11*, 2757. [CrossRef]

40. Hu, F.; Song, B.; Wang, X.; Bao, S.; Shang, S.; Lv, S.; Fan, B.; Zhan, R.; Li, J. Green rapid synthesis of Cu2O/Ag heterojunctions exerting synergistic antibiosis. *Chin. Chem. Chem. Lett.* **2022**, *33*, 308–313. [CrossRef]
41. Ali, S.G.; Ansari, M.A.; Alzohairy, M.A.; Alomary, M.N.; AlYahya, S.; Jalal, M.; Khan, H.M.; Asiri, S.M.M.; Ahmad, W.; Mahdi, A.A.; et al. Biogenic gold nanoparticles as potent antibacterial and antibiofilm nano-antibiotics against *Pseudomonas aeruginosa*. *Antibiotics* **2020**, *9*, 100. [CrossRef]
42. El Shafey, A.M. Green synthesis of metal and metal oxide nanoparticles from plant leaf extracts and their applications: A review. *Green Process. Synth.* **2020**, *9*, 304–339. [CrossRef]
43. Ahmed, S.; Ahmad, M.; Swami, B.L.; Ikram, S. A review on plants extract mediated synthesis of silver nanoparticles for antimicrobial applications: A green expertise. *J. Adv. Res.* **2016**, *7*, 17–28. [CrossRef]
44. Jayaseelan, S.; Ramaswamy, D.; Dharmaraj, S. Pyocyanin: Production, applications, challenges and new insights. *World J. Microbiol. Biotechnol.* **2014**, *30*, 1159–1168. [CrossRef]
45. Elshaer, S.L.; Shaaban, M.I. Inhibition of quorum sensing and virulence factors of *Pseudomonas aeruginosa* by biologically synthesized gold and selenium nanoparticles. *Antibiotics* **2021**, *10*, 1461. [CrossRef] [PubMed]
46. Donlan, R.M. Biofilms: Microbial life on surfaces. *Emerg. Infect. Dis.* **2002**, *8*, 881–890. [CrossRef]
47. Harshey, R.M. Bacterial motility on a surface: Many ways to a common goal. *Annu. Rev. Microbiol.* **2003**, *57*, 249–273. [CrossRef] [PubMed]
48. Rather, P.N. Swarmer cell differentiation in *Proteus mirabilis*. *Environ. Microbiol.* **2005**, *7*, 1065–1073. [CrossRef] [PubMed]
49. Khan, F.; Manivasagan, P.; Lee, J.-W.; Pham, D.T.N.; Oh, J.; Kim, Y.-M. Fucoidan-stabilized gold nanoparticle-mediated biofilm inhibition, attenuation of virulence and motility properties in *Pseudomonas aeruginosa* PAO1. *Mar. Drugs* **2019**, *17*, 208. [CrossRef] [PubMed]
50. Allesen-Holm, M.; Barken, K.B.; Yang, L.; Klausen, M.; Webb, J.S.; Kjelleberg, S.; Molin, S.; Givskov, M.; Tolker-Nielsen, T. A characterization of DNA release in *Pseudomonas aeruginosa* cultures and biofilms. *Mol. Microbiol.* **2006**, *59*, 1114–1128. [CrossRef]
51. Ciofu, O.; Tolker-Nielsen, T. Tolerance and resistance of *Pseudomonas aeruginosa* biofilms to antimicrobial agents-How *P. aeruginosa* can escape antibiotics. *Front. Microbiol.* **2019**, *10*, 913. [CrossRef]
52. Hall, C.W.; Mah, T.F. Molecular mechanisms of biofilm-based antibiotic resistance and tolerance in pathogenic bacteria. *FEMS Microbiol. Rev.* **2017**, *41*, 276–301. [CrossRef]
53. Rajkumari, J.; Busi, S.; Vasu, A.C.; Reddy, P. Facile green synthesis of baicalein fabricated gold nanoparticles and their antibiofilm activity against *Pseudomonas aeruginosa* PAO1. *Microb. Pathog.* **2017**, *107*, 261–269. [CrossRef]
54. Qais, F.A.; Ahmad, I.; Altaf, M.; Alotaibi, S.H. Biofabrication of Gold Nanoparticles Using Capsicum annuum Extract and Its Antiquorum Sensing and Antibiofilm Activity against Bacterial Pathogens. *ACS Omega* **2021**, *6*, 16670–16682. [CrossRef] [PubMed]
55. Clinical Laboratory Standards Institute. *Performance Standards for Antimicrobial Susceptibility Testing*, 22nd ed.; Informational Supplement Document M100-S22; Clinical Laboratory Standards Institute: Wayne, PA, USA, 2012.
56. Ansari, M.A.; Khan, H.M.; Alzohairy, M.A.; Jalal, M.; Ali, S.G.; Pal, R.; Musarrat, J. Green synthesis of Al2O3 nanoparticles and their bactericidal potential against clinical isolates of multi-drug resistant *Pseudomonas aeruginosa*. *World J. Microbiol. Biotechnol.* **2015**, *31*, 153–164. [CrossRef] [PubMed]
57. Essar, D.W.; Eberly, L.; Hadero, A.; Crawford, I.P. Identification and characterization of genes for a second anthranilate synthase in *Pseudomonas aeruginosa*: Interchangeability of the two anthranilate synthases and evolutionary implications. *J. Bacteriol.* **1990**, *172*, 884–900. [CrossRef]
58. Chelvam, K.K.; Chai, L.C.; Thong, K.L. Variations in motility and biofilm formation of *Salmonella enterica* serovar Typhi. *Gut Pathog.* **2014**, *6*, 2. [CrossRef]
59. O'Toole, G.A.; Kolter, R. Initiation of biofilm formation in *Pseudomonas fluorescens* WCS365 proceedsvia multiple, convergent signalling pathways: A genetic analysis. *Mol. Microbiol.* **1998**, *28*, 449–461. [CrossRef]

Article

Investigations into the Antifungal, Photocatalytic, and Physicochemical Properties of Sol-Gel-Produced Tin Dioxide Nanoparticles

Sirajul Haq [1,*], Nadia Shahzad [2], Muhammad Imran Shahzad [3], Khaled Elmnasri [4], Manel Ben Ali [5], Alaa Baazeem [5], Amor Hedfi [5] and Rimsha Ehsan [1]

1. Department of Chemistry, University of Azad Jammu and Kashmir, Muazaffabad 13100, Pakistan
2. US-Pakistan Centre for Advance Studies in Energy, National University of Science and Technology (NUST), Islamabad 44000, Pakistan
3. Nanoscience and Nanotechnology Department, National Centre for Physics (NCP), Islamabad 44000, Pakistan
4. Higher institute of Biotechnology, University of Manouba, ISBST, BVBGR-LR11ES31, Biotechpole Sidi Thabet, Ariana 2010, Tunisia
5. Department of Biology, College of Science, Taif University, P.O. Box 11099, Taif 21944, Saudi Arabia
* Correspondence: cii_raj@yahoo.com

Abstract: Transmission electron microscopy (TEM), atomic force microscopy (AFM), X-ray diffraction (XRD), energy dispersive X-ray (EDX), scanning electron microscopy (SEM), diffuse reflectance spectroscopy (DRS), and Fourier transform infrared (FTIR) spectroscopy were applied to evaluate the tin dioxide nanoparticles (SnO_2 NPs) amalgamated by the sol-gel process. XRD was used to examine the tetragonal-shaped crystallite with an average size of 26.95 (±1) nm, whereas the average particle size estimated from the TEM micrograph is 20.59 (±2) nm. A dose-dependent antifun3al activity was performed against two fungal species, and the activity was observed to be increased with an increase in the concentration of SnO_2 NPs. The photocatalytic activity of SnO_2 NPs in aqueous media was tested using Rhodamine 6G (Rh-6G) under solar light illumination. The Rh-6G was degraded at a rate of 0.96×10^{-2} min for a total of 94.18 percent in 350 min.

Keywords: antifungal activity; tin dioxide; sol-gel; tetragonal; photocatalysis; solar-light

Citation: Haq, S.; Shahzad, N.; Shahzad, M.I.; Elmnasri, K.; Ali, M.B.; Baazeem, A.; Hedfi, A.; Ehsan, R. Investigations into the Antifungal, Photocatalytic, and Physicochemical Properties of Sol-Gel-Produced Tin Dioxide Nanoparticles. *Molecules* 2022, 27, 6750. https://doi.org/10.3390/molecules27196750

Academic Editors: Nagaraj Basavegowda and Kwang-Hyun Baek

Received: 17 September 2022
Accepted: 30 September 2022
Published: 10 October 2022

Publisher's Note: MDPI stays neutral with regard to jurisdictional claims in published maps and institutional affiliations.

Copyright: © 2022 by the authors. Licensee MDPI, Basel, Switzerland. This article is an open access article distributed under the terms and conditions of the Creative Commons Attribution (CC BY) license (https://creativecommons.org/licenses/by/4.0/).

1. Introduction

The fungi are the heterotrophic eukaryotes that are unable to make their own food. These multicellular eukaryotes are ubiquitous, thus fungal infections are common throughout the world. In humans, fungal infections are mostly caused when a fungus attacks over the low immunity area of the body that is adaptive to it. Fungi can live in plants, soil, air, water and in human body naturally [1]. Like other microbial organisms, some fungi are harmful, while some are useful. When a harmful fungus attacks the human body the victim complains of itching, swelling and redness depending on the attacked area of the body. Fungi cause both surface and systemic infections and can have lethal outcomes if diagnosed at the later stages [2].

The attack of pathogens, especially fungi, has put the food security at risk and, according to a rough estimate, almost one-third of annual crops are lost due to the attack and invasion of these harmful pathogens [3]. Economically valuable crops are harmed by pathogenic fungi at pre-harvest or post-harvest stages. The fungicides used for their control are imparting a damaging effect on both humans and the environment. Thus, silver nanoparticles have been reported for the control of phytopathogenic fungi as these NPs cause growth restriction of such fungi without disturbing the environment [4]. Solvothermally synthesized gold NPs have also been advertised for antifungal activity against the candida species [5]. The ZnO NPs synthesized by biological methods using extracts like *Allium cepa*, garlic, parsley, *Dolichos lablab* L. and *Sphingomonas paucimobilis* have also been reported in the literature showing affective fungal growth inhibition, mostly of candida species [6].

The reduction of organic dyes from water reservoirs are a major concern in this industrial world and the damages related to the presence of organic substances in aquatic environments is unmeasurable. Rhodamine 6G is a heterocyclic cationic polar dye belonging to the Xanthene family and has strong absorption in the visible region [7]. Rhodamine 6G in used in the field of hydraulics as fluorescent tracer to visualize flow patterns and also is commonly used also as a sensitizer. The discharge of rhodamine 6G into the aqueous medium is harmful for humans and longer term exposure results in multiple health issues such as vomiting, increase in heart rate, lung cancer, skin cancer and, in some cases, delay in physiological development. Therefore, it is highly necessary to degrade organic substances from waste water before they accumulate in the environment causing irreversible damage [8]. Photocatalysis involves the production of the hydroxyl radical and superoxide anion, which are generated by absorption of radiation by the catalyst. SnO_2 is widely used due to its nontoxic effect, stability and strong oxidizing properties [9].

SnO_2 is one of the best semiconductors, having shape dependent properties and a band gap of 3.6 eV [10]. The SnO_2 NPs have potential to degrade organic dye and help in the protection of the environment [11]. At the nano-scale, SnO_2 exhibits exceptional properties owing to its high surface area to volume ratio, which makes it a unique photocatalyst [12]. The nano-sized SnO_2 is an efficient catalyst for oxidation of organic compounds due to the presence of a high number of surface active groups [13]. The large surface area of SnO_2 NPs having a size below 10 nm has more reaction sites, which increases the photocatalytic efficacy, which might also be attributed to the large electron-hole pair separation [14]. The Sol-gel synthesis of SnO_2 NPs is preferred over other methods, because it is easy to handle, provides better control over the particle size and is economic [15].

The current research concerns the sol-gel synthesis of SnO_2 NPs for antifungal activity and photodegradation of rhodamine 6G. The as-manufactured SnO_2 NPs were characterized by manipulating SEM, XRD, EDX, AFM, TEM, DRS and FTIR spectroscopy. The antifungal activity was performed against the selected fungus species using the Agar well diffusion method. The selection of fungi is purely based on the availability of the fungal strain and its toxicity. The degradation of rhodamine 6G was brought under solar light irradiation and the reaction parameters were determined by a set of mathematical equations.

2. Results

2.1. XRD Analysis

The XRD pattern of SnO_2 NPs exhibited in Figure 1 shows the characteristic peaks along with corresponding hkl values for SnO_2 at 2θ 26.34 (110), 33.68 (101), 38.06 (200), 52.00 (211) and 65.21 (301), which harmonized with the diffraction bands listed in the JCPDS card no. 01-077-0449 assigned to the cubic geometry of crystals. The noisy XRD pattern with broad diffraction band suggests the presence of both amorphous and crystalline phase in the sample [16]. The sharpness of diffraction bands suggest that some portion of the synthesized material is highly crystalline while the varied width and intensity shows a wide range distribution of crystallite size. The average crystallite size for SnO_2 NPs enumerated by Debye-Scherrer equation is 26.95 nm with 0.39% imperfection, found in the crystal.

2.2. EDX Analysis

In the EDX spectrum of SnO_2 NPs (Figure 2), O is responsible for the peak at 0.3 keV, while Sn is responsible for a series of sharp bands in the 3.5–4 keV range, as well as a very tiny signal at 2.6 due to the presence of Cl. According to the EDX analysis, the synthesized SnO_2 NPs have a stoichiometric composition of Sn and O, with a trace of Cl as an impurity. According to EDX statistics, Sn, O, and Cl have weight percentages of 78.7, 20.2 and 1.1 percent, respectively.

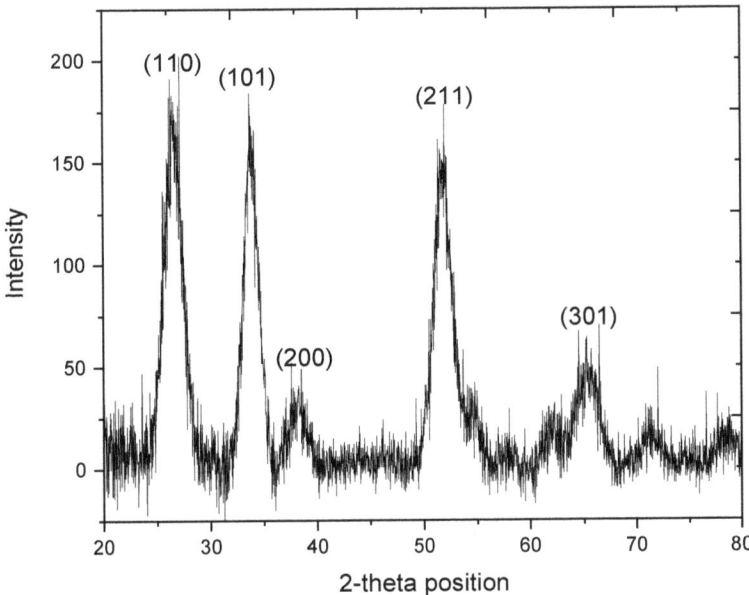

Figure 1. XRD pattern of SnO$_2$ NPs.

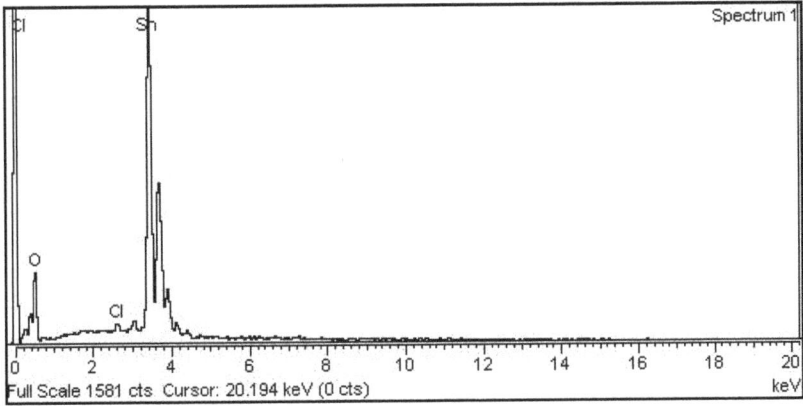

Figure 2. EDX pattern of SnO$_2$ NPs.

2.3. SEM Analysis

The structural analysis of SnO$_2$ NPs was carried out via SEM as shown in the low and high magnified micrographs (Figure 3a,b). The images reveal that flat shaped particles of different size are formed by the aggregation of small particles. Each flat shaped particle constitutes 2 to 9 small particles depending upon the size, and the cracks observed in the flat shaped particles are actually the boundaries of the aggregated particles. The size of the flat shaped particles predicted from SEM micrographs range from 78 to 114 nm with an average size of 98.56 nm.

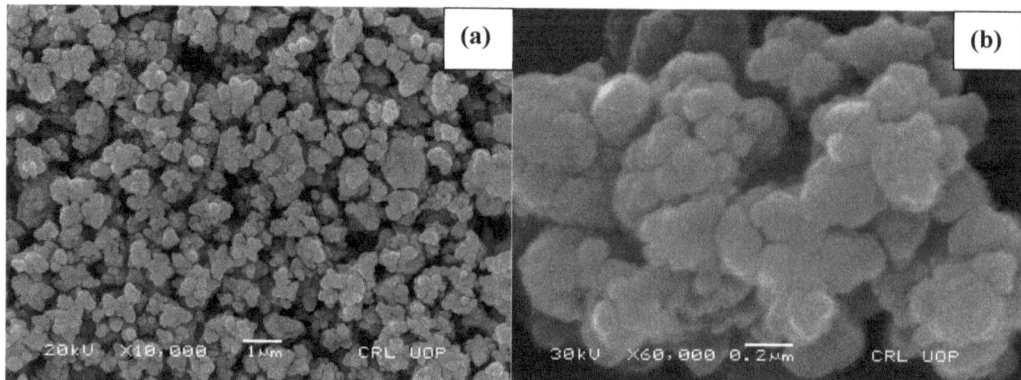

Figure 3. SEM micrographs of SnO$_2$ NPs. (**a**) ×10,000; (**b**) ×60,000.

2.4. TEM Analysis

The TEM micrograph SnO$_2$ NPs shown in Figure 4, exhibits two portions; one portion is formed due to the accumulation of the particles over one another forming a dark structure, whereas in the other portion the particles are somewhat evenly distributed and are closely connected with each other, leading to the formation of network structure. Although the shape of the particles are not uniform, many of the particles possess nearly spherical shape. It is also seen that the surface of the particles are smooth and have a wide range of size distribution. The particles' size measured by ImageJ software ranges from 13.24 nm to 30.88 nm with an average size of 20.59 nm.

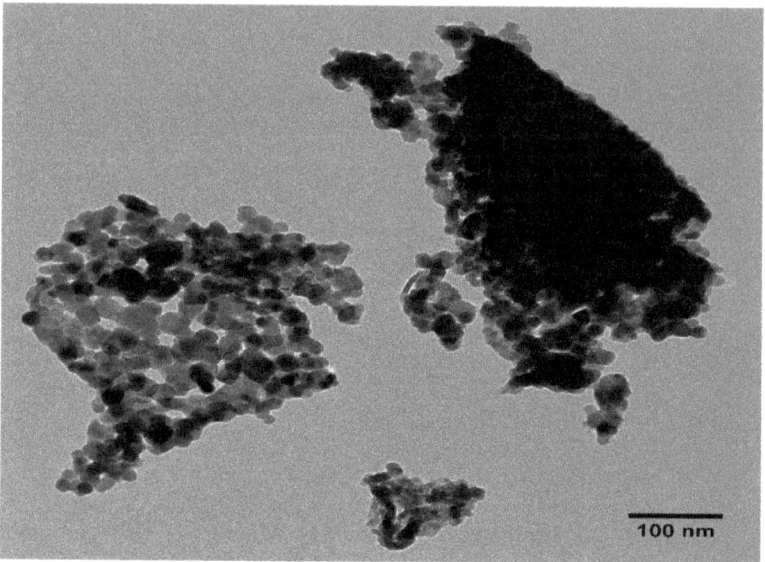

Figure 4. TEM micrograph of SnO$_2$ NPs.

2.5. AFM Analysis

The distribution of SnO$_2$ NPs of various sizes and shapes was analyzed via AFM in both 2-dimensions and 3-dimensions as shown in Figure 5. It is seen that the small particles are fused together, leading to the formation of bunch like structures. However, many tiny individual particles are also seen in the micrographs. The density of the particle is 0.920/μm² whereas the height of the particles ranges from 7.93 to 41.44 nm with average height of 23.95 nm. The particles' diameters, i.e., between 55.09 and 101.60 nm, with an average diameter of 72.73 nm.

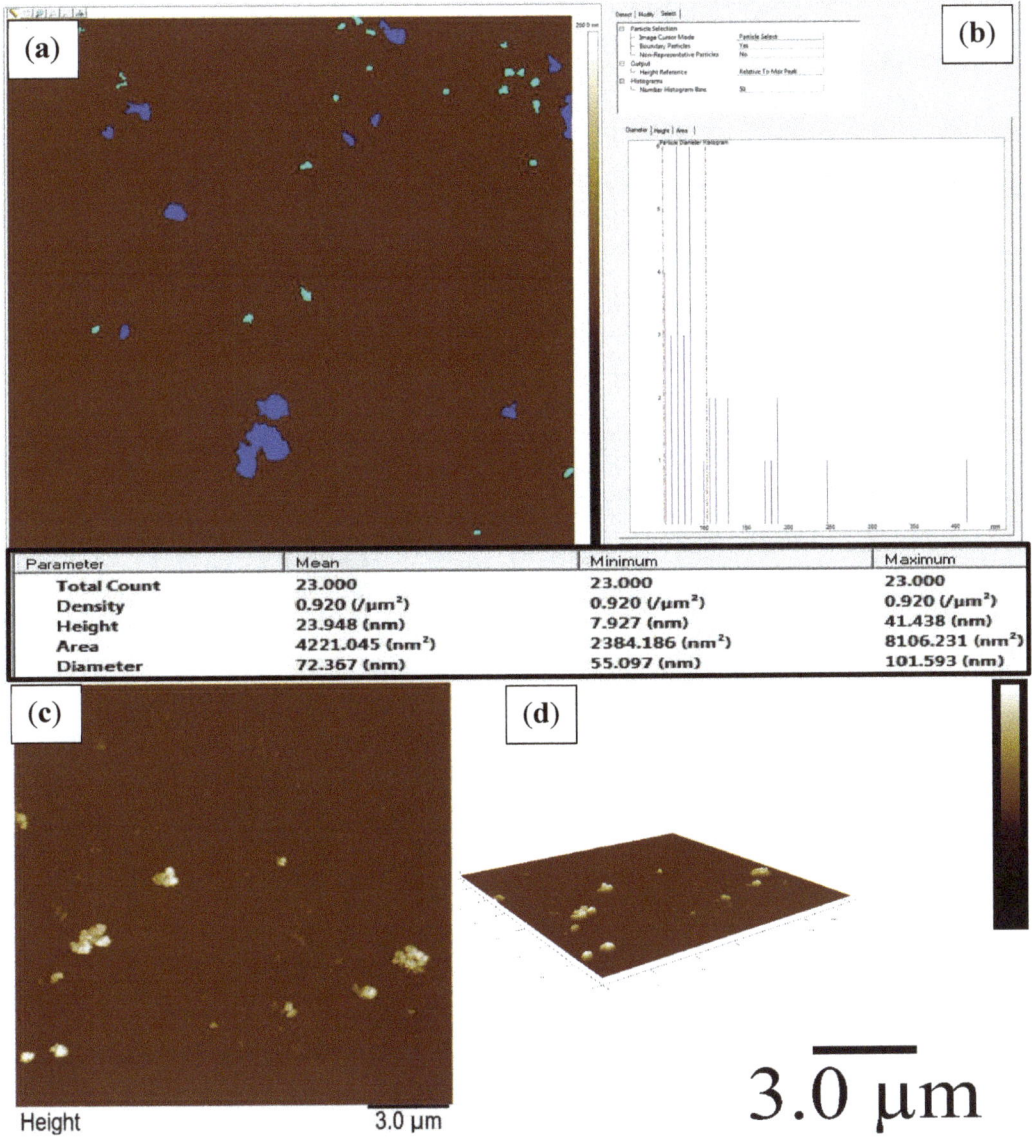

Figure 5. Selected area (**a**), histogram (**b**), 2-D (**c**), and 3-D (**d**) AFM micrographs of SnO₂ NPs.

2.6. DRS Analysis

The DRS spectrum of SnO$_2$ NPs (inset: Figure 6) shows greater absorbance in the UV range and a clear decrease was seen in absorbance with increasing wavelength, except for a depth occurring in the boundary line UV and visible region, which might be due to some structural defects. The Tauc plot (Figure 4) was drawn to calculate the band gap energy and was noted to be 3.65 eV, almost similar to that reported in the literature [17].

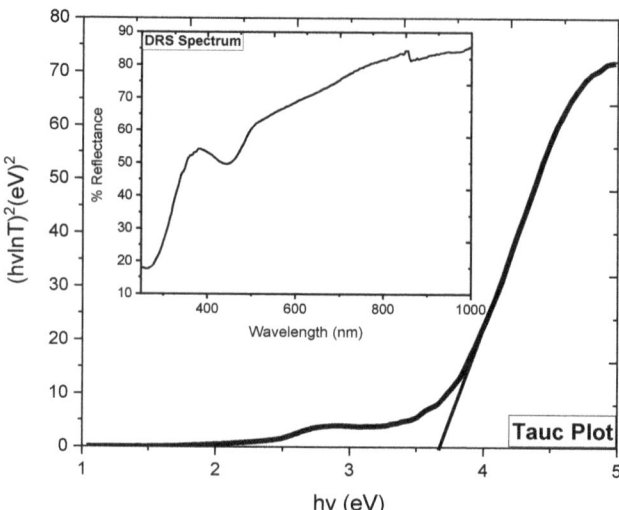

Figure 6. Tauc plot (inset: DRS spectrum) of SnO$_2$ NPs.

2.7. FTIR Analysis

The stretching and bending vibrations of the hydroxyl group are responsible for a broad band centered at 3248 cm^{-1} and another peak at 1627.90 cm^{-1} in the FTIR spectrum of SnO$_2$ NPs (Figure 7) [18]. The signal at 1383.31 cm^{-1} confirmed the existence of NO$_3$ in the sample, which might be attributable to the use of Sn(NO$_3$)$_2$ as a precursor in the synthesis. The peaks at 1140.44 and 1015.11 cm^{-1} are caused by Sn-OH crystal lattice vibrations [19]. The wide band in the range from 761–513 cm^{-1} is formed by the fusion of two bands at 692 and 601 cm^{-1}, which are ascribed to Sn-O-Sn and Sn-O vibrations, respectively [10].

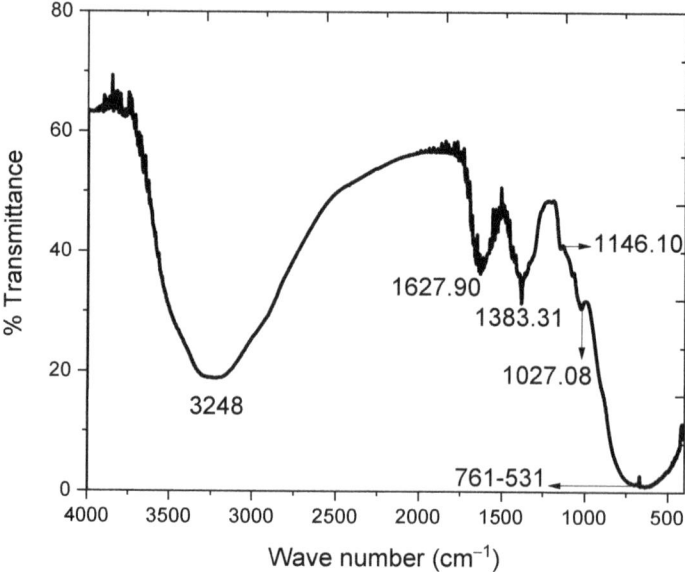

Figure 7. FTIR spectrum of SnO$_2$ NPs.

2.8. Antifungal Study

The antifungal activity of SnO$_2$ NPs was scrutinized against the selected fungi at different concentrations as shown in Figure 8 and the obtained data is tabulated in Table 1. The results shows that the activity of SnO$_2$ NPs increased along with increase in concentration in the well. At 40 µg/mL, no activity was shown against both fungi, but onward increase in concentration significantly inhibits fungus growth and the highest activity was found at 100 µg/mL. However, the activity of SnO$_2$ NPs was found to be less than the activity of the positive control, 6.1 mm and 6.3 mm for both species, respectively. The solvent was utilized as negative control and has no effect on the activity of SnO$_2$ NPs and positive control. The increase in the activity with increasing concentration is attributed to the larger number of particles present in the suspension, that provide more binding sites to interact with the fungi. It has been reported that most of the antifungal agents act in a non-specific way, either changing the permeability of the cell wall and cell membrane or disturbing the cytoplasmic composition/leakage of cytoplasmic fluid. They also act as enzyme inhibitors altering the biochemical nature, which leads to the death of organisms [20].

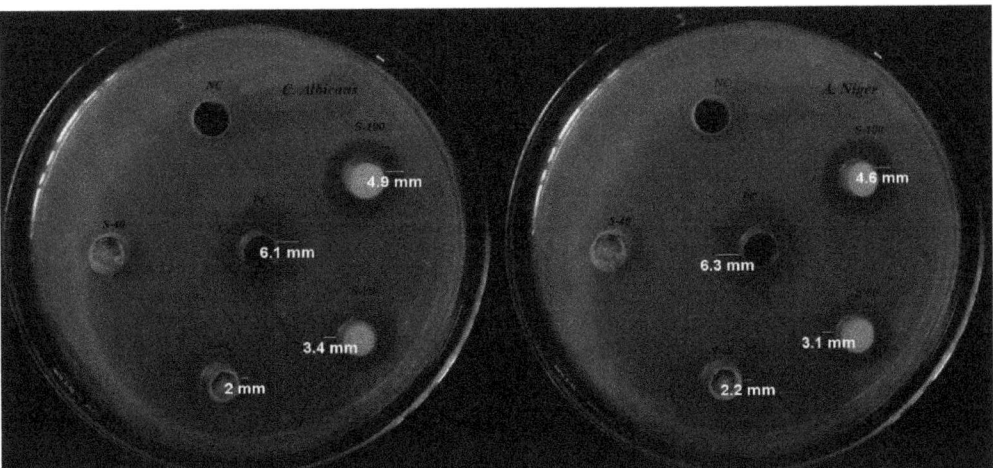

Figure 8. Experimental photographs of antifungal activity of SnO$_2$ NPs against selected fungi at different concentrations.

Table 1. Antifungal activity of SnO$_2$ NPs against the selected fungi and statistical analysis.

Species	Concentration (µg/mL)	Inhibition Zone (mm)	PS	NC	Variance (S2)	Standard Deviation (S)	Pearson Constant (<0.05)
C. Albicans	40	0	6.1	0	1.72	1.3	0.0054
	60	2					
	80	3.4					
	100	4.9					
A. Niger	40	0	6.3	0	1.5	1.2	0.0055
	60	2.2					
	80	3.1					
	100	4.6					

2.9. Photocatalytic Study

In the presence of SnO$_2$ NPs, the solar light induced degradation of Rh-6G was carried out in aqueous medium, and the visual deterioration was monitored by the fading hue of the dye solution over time. The degradation process was investigated experimentally using a double beam spectrophotometer, where a decrease in the absorbance maxima at 526 nm was noted as time passed, and the results are presented in Figure 9a [21]. The percentage degradation of Rh-6G was calculated using Equation (1), and the result was 94.18 percent in 330 min (Figure 9b, which was greater than previously reported [22]. The Langmuir-Hinshelwood kinetic model (Equation (2)) was manipulated to investigate the photocatalytic reaction kinetics, where the initial and end concentrations of Rh-6G are C_o and C_e, respectively, and k and t are the apparent constants [23]. The straight line produced by plotting $\ln C_o/C_e$ versus time (Figure 9c) with r^2 values of 0.844 suggests that the photocatalytic process is pseudo-first order. The photo-degradation rate constants for Rh-6G via SnO_2 NPs are enumerated from the slope of linear plots and are 0.999×10^{-2} min^{-1}.

$$\% \text{ Degradation} = \frac{C_o - C_t}{C_o} \times 100 \quad (1)$$

$$\ln(C/C_o) = -kt \quad (2)$$

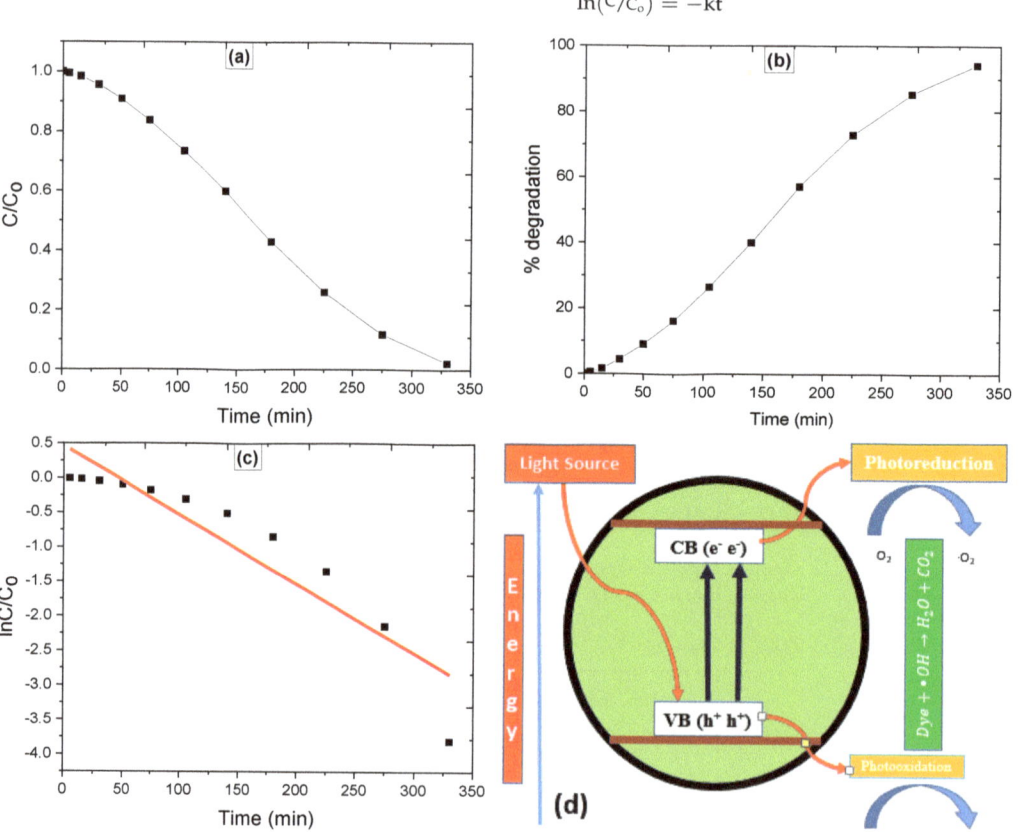

Figure 9. Photocatalytic parameters including, (**a**) = degradation profile, (**b**) = percentage degradation, (**c**) = kinetic plot and (**d**) = electron excitation and hole creation mechanism.

When light with an energy equal to or greater than the SnO_2 NPs band gap reaches the surface, the outermost electron is excited to the conduction band (CB), leaving a positive hole in the valence band (VB), as illustrated in Figure 9d. The positive holes interact with the water/hydroxyl group to produce hydroxyl radicals, which are powerful oxidizers that convert Rh-6G to H_2O and CO_2 [23]. The super oxide radicals, on the other hand, are produced by the interaction of an excited electron with absorbed oxygen, providing an additional source of hydroxyl radical and speeding up the oxidation of Rh-6G [24].

3. Materials and Methods

3.1. Materials

Sigma-Aldrich provided analytical grade chemicals such as Sn $(NO_3)_2$, hydrochloric acid and C_2H_5OH, which were utilized without any further purification. Deionized water was utilized to make all of the working solutions, and 15% nitric acid solution was manipulated to clean all the glassware before being bathed with the deionized water.

3.2. Synthesis of SnO_2 NPs

A 10 mM solution of $Sn(NO_3)_2$ was produced on dissolving 1.21 g in 500 mL deionized water, and 80 mL from this solution was combined with 20 mL of ethanol for the fabrication of SnO_2 NPs. The reaction mixture was stirred (250 rpm) and heated at 50 °C for 40 min at pH 2.5 by adding HCl solution. After forming a white gel and ageing it for 24 h, it was washed using deionized water and dried at 150 °C. For later usage, the white powder was kept in an airtight plastic bottle.

3.3. Characterization

The Panalytical X-Pert Pro X-ray diffraction model was used to investigate the crystal property, with XRD analysis in the 20°–80° 2-theta range and the Debye-Scherrer equation used to compute crystallite size. For morphological examination, a scanning electron microscope model JEOL 5910 (Japan) was utilized, and the particle size was determined using ImageJ software. At 20 keV, the energy dispersive X-ray model INCA 200 (UK) was utilized to assess the percentage composition and purity. The band gap energies were calculated for a reflectance spectrum collected using the diffuse reflectance spectroscopy model lambda 950 with a desegregating sphere in the wavelength range of 200–2500 nm. For the identification of surface functional groups, FTIR spectra in the region of 4000–400 cm^{-1} were acquired using a Nicolet 6700 (USA) spectrometer.

3.4. Antifungal Assay

The antifungal screening of SnO_2 NPs against *Aspergillus niger* (ATCC#16404) and *Candida albicans* (ATCC#10231) was carried out using the Agar well diffusion method. Four SnO_2 NPs suspensions was prepared by ultrasonic dispersion of 40, 60, 80 and 100 μg in 1 mL. The well was bored in the media using a sterile borer and each well individually was equipped with 100 μL of each suspension and was incubated at room temperature. The zone of inhibition was computed in millimeters (mm) after 7 days as the activity of SnO_2 NPs against the fungal species. The statistical analysis was carried out with 95% confidence interval using Microsoft Excel 2013 (Las Vegas, NV, USA).

3.5. Photocatalytic Assay

The experiment was carried out in a double-walled Pyrex reactor with a water input and exit under solar light. To avoid sun contact, 50 mL of Rh-6G solution (15 ppm) and 20 mg of catalyst (0.4 g/L) were added to the reactor for each reaction, and the reactor was enclosed in aluminum foil. After exposing the reaction to sunlight for a while, 3 mL of the sample was subjected to centrifugation for 4 min at 4000 rpm and examined with a double beam spectrophotometer (Thermo Spectronic UV 500). There was a decrease in absorbance maxima as the time passed.

4. Conclusions

A facile and single-step sol-gel process was operated for the fabrication of SnO_2 NPs, which was found to be more economical and time saving, with no use of toxic and expensive templates. Different methods were used to investigate the physicochemical characteristics, which revealed the development of a well-crystalline cubic-shaped crystallite in the nano-metric range. The different shapes and morphology of SnO_2 NPs were seen in microstructural analysis. The larger grain size might be due to the aggregation of various small particles. The EDX analysis confirmed the desired composition SnO_2 NPs and the presence of Cl as impurity in the sample might be due to the improper washing process. A significant antifungal activity was shown by the SnO_2 NPs against both the fungal species at higher concentrations. The 94.18 percent Rh-6G degraded in 330 min at a rate of 0.999×10^{-2} per min. The SnO_2 NPs' improved photocatalytic activity against Rh-6G was due to their tiny size and porous structure, as revealed by XRD, AFM and TEM analyses.

Author Contributions: Conceptualization, S.H. and M.I.S.; methodology, S.H., R.E. and N.S.; software, A.H. and M.I.S.; validation, M.B.A., M.I.S. and K.E.; formal analysis, S.H. and R.E.; investigation, N.S.; resources, M.I.S. and S.H.; data curation, K.E.; writing—original draft preparation, R.E. and S.H.; writing—review and editing, A.B., A.H., K.E. and M.B.A.; visualization, M.I.S.; supervision, S.H. and A.B.; project administration, N.S.; funding acquisition, A.B., M.B.A., A.H. and K.E. All authors have read and agreed to the published version of the manuscript.

Funding: This research was funded by Taif University Researchers Supporting Project number (TURSP-2020/295), Taif University, Taif, Saudi Arabia.

Institutional Review Board Statement: Not applicable.

Informed Consent Statement: Not applicable.

Data Availability Statement: All the data is enclosed in the manuscript.

Conflicts of Interest: The authors declare no conflict of interest.

Sample Availability: Samples of the compounds are available from the authors.

References

1. Alananbeh, K.M.; Al-Refaee, W.J.; Al-Qodah, Z. Antifungal Effect of Silver Nanoparticles on Selected Fungi Isolated from Raw and Waste Water. *Indian J. Pharm. Sci.* **2017**, *79*, 559–567. [CrossRef]
2. Mahdi, B.M. Review of Fungal Infection in Human Beings and Role of COVID-19 Pandemic. *Indian J. Forensic Med. Toxicol.* **2021**, *15*, 887–897.
3. Hashem, A.H.; Abdelaziz, A.M.; Askar, A.A.; Fouda, H.M.; Khalil, A.M.A.; Abd-Elsalam, K.A.; Khaleil, M.M. Bacillus megaterium-mediated synthesis of selenium nanoparticles and their antifungal activity against rhizoctonia solani in faba bean plants. *J. Fungi* **2021**, *7*, 195. [CrossRef]
4. Al-Otibi, F.; Perveen, K.; Al-Saif, N.A.; Alharbi, R.I.; Bokhari, N.A.; Albasher, G.; Al-Otaibi, R.M.; Al-Mosa, M.A. Biosynthesis of silver nanoparticles using Malva parviflora and their antifungal activity. *Saudi J. Biol. Sci.* **2021**, *28*, 2229–2235. [CrossRef] [PubMed]
5. Ahmad, T.; Wani, I.A.; Lone, I.H.; Ganguly, A.; Manzoor, N.; Ahmad, A.; Ahmed, J.; Al-Shihri, A.S. Antifungal activity of gold nanoparticles prepared by solvothermal method. *Mater. Res. Bull.* **2013**, *48*, 12–20. [CrossRef]
6. Abomuti, M.A.; Danish, E.Y.; Firoz, A.; Hasan, N.; Malik, M.A. Green synthesis of zinc oxide nanoparticles using salvia officinalis leaf extract and their photocatalytic and antifungal activities. *Biology* **2021**, *10*, 1075. [CrossRef]
7. Pino, E. Photocatalytic Degradation of Aqueous Rhodamine 6G Using Supported TiO_2 Catalysts. A Model for the Removal of Organic Contaminants from Aqueous samples. *Front. Chem.* **2020**, *8*, 365. [CrossRef]
8. Suwunwong, T.; Patho, P.; Choto, P.; Phoungthong, K. Enhancement the rhodamine 6G adsorption property on Fe_3O_4-composited biochar derived from rice husk. *Mater. Res. Express* **2020**, *7*, 025511. [CrossRef]
9. Silva, L.C.; Barrocas, B.; Jorge, M.E.M. Photocatalytic Degradation of Rhodamine 6G using TiO_2/WO_3 Bilayered Films Produced by Reactive Sputtering. In Proceedings of the 6th International Conference on Photonics, Optics and Laser Technology, Funchal, Portugal, 25–27 January 2018; pp. 334–340. [CrossRef]
10. Haq, S.; Rehman, W.; Waseem, M.; Shah, A.; Khan, A.R. Green synthesis and characterization of tin dioxide nanoparticles for photocatalytic and antimicrobial studies. *Mater. Res. Express* **2020**, *7*, 025012. [CrossRef]
11. Surendhiran, K.C.S.S.; Kumar, P.M.; Kumar, E.R.; Khadar, Y.A.S. Green synthesis of SnO_2 nanoparticles using Delonix elata leaf extract: Evaluation of its structural, optical, morphological and photocatalytic properties. *SN Appl. Sci.* **2020**, *2*, 1735. [CrossRef]
12. Garrafa-galvez, H.E.; Nava, O.; Soto-robles, C.A.; Vilchis-nestor, A.R. Green synthesis of SnO_2 nanoparticle using Lycopersicon esculentum peel extract. *J. Mol. Struct.* **2019**, *1197*, 354–360. [CrossRef]
13. Adnan, R.; Razana, N.A.; Rahman, I.A.; Farrukh, M.A. Synthesis and Characterization of High Surface Area Tin Oxide Nanoparticles via the Sol-Gel Method as a Catalyst for the Hydrogenation of Styrene Synthesis and Characterization of High Surface Area Tin Oxide Nanoparticles via the Sol-Gel Method as a Catal. *J. Chin. Chem. Soc.* **2010**, *57*, 222–229. [CrossRef]

14. Viet, P.V.; Thi, C.M.; Hieu, L.V. The High Photocatalytic Activity of SnO_2 Nanoparticles Synthesized by Hydrothermal Method. *J. Nanomater.* **2016**, *2016*, 4231046. [CrossRef]
15. Suvaitha, P.; Selvam, S.; Ganesan, D.; Rajangam, V.; Raji, A. Green Synthesis of SnO_2 Nanoparticles for Catalytic Degradation of Rhodamine B. *Iran. J. Sci. Technol. Trans. A Sci.* **2020**, *44*, 661–676. [CrossRef]
16. Haq, S.; Rehman, W.; Rehman, M. Modeling, Thermodynamic Study and Sorption Mechanism of Cadmium Ions onto Isopropyl Alcohol Mediated Tin Dioxide Nanoparticles. *J. Inorg. Organomet. Polym. Mater.* **2020**, *30*, 1197–1205. [CrossRef]
17. Gajendiran, J.; Rajendran, V. Synthesis and Characterization of Ultrafine SnO_2 Nanoparticles via Solvothermal Process. *Int. J. Phys. Appl.* **2010**, *2*, 45–50.
18. Yao, W.; Wu, S.; Zhan, L.; Wang, Y. Two-dimensional porous carbon-coated sandwich-like mesoporous SnO_2/graphene/mesoporous SnO_2 nanosheets towards high-rate and long cycle life lithium-ion batteries. *Chem. Eng. J.* **2019**, *361*, 329–341. [CrossRef]
19. Haq, S.; Rehman, W.; Waseem, M.; Rehman, M.U.; Khan, B. Adsorption of Cd^{2+} ions onto SnO_2 nanoparticles synthesized via sol-gel method: Physiochemical study. *Mater. Res. Express* **2019**, *6*, 105035. [CrossRef]
20. Anjaneyulu, Y.; Rao, R.P. Preparation, characterization and antimicrobial activity studies on some ternary complexes of Cu(II) with acetylacetone and various salicylic acids. *Synth. React. Inorg. Met.-Org. Chem.* **1986**, *16*, 257–272. [CrossRef]
21. Manjula, N.; Selvan, G.; Balu, A.R. Photocatalytic Performance of SnO_2:Mo Nanopowders Against the Degradation of Methyl Orange and Rhodamine B Dyes Under Visible Light Irradiation. *J. Electron. Mater.* **2019**, *48*, 401–408. [CrossRef]
22. Shah, A.; Haq, S.; Rehman, W.; Muhammad, W.; Shoukat, S.; Rehman, M. Photocatalytic and antibacterial activities of Paeonia emodi mediated silver oxide nanoparticles. *Mater. Res. Express* **2019**, *6*, 045045. [CrossRef]
23. Haq, S.; Shoukat, S.; Rehman, W.; Waseem, M.; Shah, A. Green fabrication and physicochemical investigations of zinc-cobalt oxide nanocomposite for wastewater treatment. *J. Mol. Liq.* **2020**, *318*, 114260. [CrossRef]
24. Ahmad, M.; Rehman, W.; Khan, M.M.; Qureshi, M.T.; Gul, A.; Haq, S.; Ullah, R.; Rab, A.; Menaa, F. Phytogenic fabrication of ZnO and gold decorated ZnO nanoparticles for photocatalytic degradation of Rhodamine B. *J. Environ. Chem. Eng.* **2021**, *9*, 104725. [CrossRef]

MDPI
St. Alban-Anlage 66
4052 Basel
Switzerland
Tel. +41 61 683 77 34
Fax +41 61 302 89 18
www.mdpi.com

Molecules Editorial Office
E-mail: molecules@mdpi.com
www.mdpi.com/journal/molecules

www.ingramcontent.com/pod-product-compliance
Lightning Source LLC
LaVergne TN
LVHW070728100526
838202LV00013B/1195